Recreational Hunting, Conservation and Rural Livelihoods

Conservation Science and Practice Series

Published in association with the Zoological Society of London

Wiley-Blackwell and the Zoological Society of London are proud to present our new *Conservation Science and Practice* volume series. Each book in the series reviews a key issue in conservation today. We are particularly keen to publish books that address the multidisciplinary aspects of conservation, looking at how biological scientists and ecologists are interacting with social scientists to effect long-term, sustainable conservation measures.

Books in the series can be single- or multi-authored and proposals should be sent to:

Ward Cooper, Senior Commissioning Editor, Wiley-Blackwell, John Wiley & Sons,
9600 Garsington Road, Oxford OX4 2DQ, UK
Email: ward.cooper@wiley.com

Each book proposal will be assessed by independent academic referees, as well as our Series Editorial Panel. Members of the Panel include:

Richard Cowling, Nelson Mandela Metropolitan, Port Elizabeth, South Africa
John Gittleman, Institute of Ecology, University of Georgia, USA
Andrew Knight, Nelson Mandela Metropolitan, Port Elizabeth, South Africa
Georgina Mace, Imperial College London, Silwood Park, UK
Daniel Pauly, University of British Columbia, Canada
Stuart Pimm, Duke University, USA
Hugh Possingham, University of Queensland, Australia
Peter Raven, Missouri Botanical Gardens, USA
Michael Samways, University of Stellenbosch, South Africa
Nigel Stork, University of Melbourne, Australia
Rosie Woodroffe, University of California, Davis, USA

Conservation Science and Practice Series

Recreational Hunting, Conservation and Rural Livelihoods: Science and Practice

Edited by

Barney Dickson, Jon Hutton and William M. Adams

A John Wiley & Sons, Ltd., Publication

This edition first published 2009, © 2009 by Blackwell Publishing Ltd

Blackwell Publishing was acquired by John Wiley & Sons in February 2007. Blackwell's publishing program has been merged with Wiley's global Scientific, Technical and Medical business to form Wiley-Blackwell.

Registered office
John Wiley & Sons Ltd, The Atrium, Southern Gate, Chichester, West Sussex, PO19 8SQ, UK

Editorial offices
9600 Garsington Road, Oxford, OX4 2DQ, UK
The Atrium, Southern Gate, Chichester, West Sussex, PO19 8SQ, UK
111 River Street, Hoboken, NJ 07030-5774, USA

For details of our global editorial offices, for customer services and for information about how to apply for permission to reuse the copyright material in this book please see our website at www.wiley.com/wiley-blackwell

The right of the author to be identified as the author of this work has been asserted in accordance with the UK Copyright, Designs and Patents Act 1988.

Wiley also publishes its books in a variety of electronic formats. Some content that appears in print may not be available in electronic books.

Designations used by companies to distinguish their products are often claimed as trademarks. All brand names and product names used in this book are trade names, service marks, trademarks or registered trademarks of their respective owners. The publisher is not associated with any product or vendor mentioned in this book. This publication is designed to provide accurate and authoritative information in regard to the subject matter covered. It is sold on the understanding that the publisher is not engaged in rendering professional services. If professional advice or other expert assistance is required, the services of a competent professional should be sought.

Library of Congress Cataloging-in-Publication Data
Recreational hunting, conservation, and rural livelihoods / edited by Barney Dickson, Jonathan Hutton, and Bill Adams.
    p. cm. – (Conservation science and practice series)
  Includes bibliographical references.
  ISBN: 978-1-4051-6785-7 (pbk: alk. paper)
  ISBN: 978-1-4051-9142-5 (hardback: alk. paper)
  1. Hunting. 2. Wildlife conservation. 3. Rural development. I. Dickson, Barney. II. Hutton, Jonathan. III. Adams, W. M. (William Mark), 1955-

  SK35.R3955 2009
  179'.3–dc22

                    2008029047

A catalogue record for this book is available from the British Library.

Set in Minion 10.5/12.5 by Newgen Imaging Systems (P) Ltd, Chennai, India
Printed and bound in Malaysia by Vivar Printing Sdn Bhd

1   2009

Contents

Preface and Acknowledgements

The contents represent the authors' own views and do not necessarily reflect the views and policies of UNEP or associated organizations.

This book includes a number of chapters derived from presentations made at a conference on 'Recreational Hunting, Conservation and Rural Livelihoods: Science and Practice' held in London in October 2006. This meeting was a Zoological Society of London (ZSL) Symposium in Conservation Biology, and was organised by the IUCN Species Survival Commission Sustainable Use Specialist Group (IUCN/SSC SUSG). The editors of the present volume worked with Kai-Uwe Wollscheid and Robin Sharp in preparing the conference. The book also reflects discussion at the two-day IUCN/SSC SUSG workshop discussing policy options for the regulation of hunting (and particularly the possible role of standards and certification) that followed that conference. This workshop was attended by conservation scientists and administrators, social scientists and a range of non-governmental organisations. It was organised by Jon Hutton, Kai-Uwe Wollscheid, Robin Sharp and Barney Dickson.

While the present book contains a number of chapters based on presentations to the 2006 conference, it has been conceived of and created as an independent volume. None of the organisations who supported the meetings have had a role in determining the book's content, or have sought to do so. As editors, we have been free to select authors and work with them to develop a book that addresses what we see as the key issues relating to recreational hunting, conservation and livelihoods. All editorial decisions have been made by the editors alone and any deficiencies or biases in the book are our responsibility. Doubtless others would have made different choices, but we hope that the balance of view that we have struck here, together with the critical and detailed consideration of hunting in this volume, will contribute to debate on this important issue, and move it forwards.

We would like to thank the ZSL for hosting the conference which led to this book, and to the IUCN/SSC SUSG for organising the event with ZSL. Joy Miller and Linda DaVolls, of ZSL, worked with skill and good grace.

ZSL is also thanked for hosting the IUCN/SSC SUSG workshop on policy options that followed the conference. We are grateful to the following additional organisations for supporting the conference and workshop: CIC (International Council for Game and Wildlife Conservation), Conservation Force, Dallas Safari Club, Safari Club International Foundation, Sand County Foundation. We thank Zsuzsanna Holtsuk, Robin Sharp and Kai Wollscheid for their tireless work in planning the two meetings. The following individuals provided invaluable advice and/or practical support: Bob Byrne, Rosie Cooney, Holly Dublin, Lee Foote, Milton Freeman, Brent Haglund, John J. Jackson III, Mike Jones, Georgina Mace, John Robinson, Chris Weaver and Mark Wright. We thank all those who made presentations at the conference and/or participated in the IUCN/SSC SUSG workshop.

We are grateful to Ward Cooper, Delia Sandford and Rosie Hayden at Wiley-Blackwell for their patience and attention to detail in the preparation of this book; to Ian Agnew (Department of Geography, University of Cambridge) for drawing the final maps and figures; and to ZSL for their continued support in this regard. Finally, we thank our own institutions for their support throughout this process: the Department of Geography at the University of Cambridge; Fauna & Flora International; Resource Africa; and the UNEP-World Conservation Monitoring Centre.

Barney Dickson, Jon Hutton and William M. Adams
August 2008

Contributors

William M. (Bill) Adams is Moran Professor of Conservation and Development at the University of Cambridge. He is based in the Department of Geography at the University, where he has taught since 1984. His research focuses on the social dimensions of conservation in Africa and the UK. He is a Trustee of Fauna & Flora International. Department of Geography, University of Cambridge CB2 3EN, UK. wa12@cam.ac.uk

Nicholas J. Aebischer is the Deputy Director of Research at the Game and Wildlife Conservation Trust. His research interests lie in understanding the processes driving declining farmland bird populations and in applying that understanding to their conservation. Game and Wildlife Conservation Trust, Fordingbridge, Hampshire SP6 1EF, UK. naebischer@gct.org.uk

Robert Arlinghaus is Junior Professor for Inland Fisheries Management at Humboldt University of Berlin and Leader of the Recreational Fisheries Research Laboratory at the Leibniz Institute of Freshwater Ecology and Inland Fisheries in Berlin. He studies the social, economic and ecological dimensions of recreational fisheries. Department of Biology and Ecology of Fishes, Müggelseedamm 310, 12587 Berlin, Germany. arlinghaus@igb-berlin.de

Rolf D. Baldus is an economist and German civil servant who worked in Tanzania as Director of the Selous Conservation Programme between 1987 and 1993 and as Government Advisor to the Wildlife Department from 1998 to 2005. He serves as President of the Tropical Game Commission in the International Council for Game and Wildlife Management (CIC). His publications are available from: www.wildlife-Baldus.com. Rolf.Baldus@bmz.bund.de

Vernon R. Booth is a Freelance Wildlife Management Consultant, Harare, Zimbabwe with 25 years experience in southern and eastern Africa. His expertise includes the economic and financial implications of wildlife conservation, management and utilisation (both consumptive and non-consumptive), and the management and administration of tourism and hunting concession areas, from both landowner and concessionaire perspectives, on private, communal and state land. verlyn5253@yahoo.co.uk

Nils Bunnefeld was a member of the Conservation Science Research Group in the Division of Biology, Silwood Park, Imperial College London where he did his PhD on red grouse. He is now a postdoctoral scientist working on moose at the Swedish University of Agricultural Sciences. Department of Wildlife, Fish and Environmental Studies, Swedish University of Agricultural Sciences, 901 83 Umeå, Sweden. nils.bunnefeld@vfm.slu.se

Brian Child is an Associate Professor in Geography at the University of Florida specialising in park management and community conservation. He has worked extensively in southern Africa, including supporting the CAMPFIRE program and game ranching in Zimbabwe, and community conservation and park management in Zambia. Center for African Studies & Department of Geography, University of Florida, PO Box 117315, Gainesville, FL 32611-7315, USA. BChild@africa.ufl.edu

Steven J. Cooke is Assistant Professor and Director of the Fish Ecology and Conservation Physiology Laboratory in the Department of Biology at Carleton University in Ottawa, Canada. His research focuses on under-standing the effects of recreational fishing on fish, fisheries, and the environment. Institute of Environmental Science and Department of Biology, Carleton University, 1125 Colonel By Drive, Ottawa, Canada ON K1S 5B6. scooke@connect.carleton.ca

David H.M. Cumming was previously Chief Ecologist and Deputy Director of Zimbabwe's Department of National Parks and Wild Life Management and later Conservation Director for WWF in Southern Africa. He is presently an Honorary Professor at the Percy FitzPatrick Institute, University of Cape Town, a Research Associate at the University of Zimbabwe, and an independent consultant in conservation and development. cummingdhm@gmail.com

Richard Davies is a self-employed wildlife business consultant. He started his conservation career in Zululand where he gained valuable field experience. He soon realised however that the challenges facing conservation were mostly socioeconomic. He has since focused on adding value to wildlife, improving management efficiency and the responsible, sustainable use of wildlife resources. PO Box 5693, Helderberg 7135, South Africa. rdavies@iafrica.com

Barney Dickson is Head of Climate Change and Biodiversity at UNEP World Conservation Monitoring Centre. In the past fifteen years he has worked on a range of conservation issues including conservation and poverty reduction and sustainable use. UNEP-WCMC, 219 Huntingdon Road, Cambridge CB3 0DL, UK. barney.dickson@unep-wcmc.org

Holly T. Dublin was until recently Chair of the IUCN Species Survival Commission. For over 20 years, she served as a Senior Conservation Adviser to WWF International. Her work focused on community-based natural resource management, sustainable use, protected area planning and biodiversity conservation across sub-Saharan Africa. C/o South African National Biodiversity Institute, Centre for Biodiversity Conservation, Private Bag X7, Claremont 7735, Cape Town, South Africa. holly.dublin@gmail.com

Adam Dutton is an environmental economist specialising in issues pertaining to the extractive use and trade of species. He is currently a D.Phil. candidate at the Wildlife Conservation Research Unit, Department of Zoology, University of Oxford, Tubney House, Tubney OX13 5QL, UK. adam.dutton@zoo.ox.ac.uk

Marco Festa-Bianchet is a Professor of Ecology and Conservation Biology at the University of Sherbrooke in Canada. He has been working on the population dynamics, evolutionary ecology and conservation of large herbivores, mainly mountain ungulates, for 27 years. Marco served for four years as Chair of the Committee on the Status of Endangered Wildlife in Canada and chairs the IUCN Caprinae Specialist Group. Département de biologie, Université de Sherbrooke, Sherbrooke, Québec, Canada J1K 2R1. m.festa@USherbrooke.ca

Michael R. Frisina is an Executive Director, August L. Hormay Wildlands Institute, Inc. and also an Adjunct Professor of Range Science at Montana State University and has held various positions with Montana Fish, Wildlife and Parks since 1969 where he is currently Range/Habitat Coordinator. He has M.S. and B.S. degrees in wildlife management and in 2004 was awarded an honorary doctorate in agriculture from Montana State University, Bozeman. Affiliation: Montana State University, Department of Animal and Range Sciences, Bozeman, MT, USA. Address: PO Box 4712, Butte, MT 59701, USA. habitat@bresnan.net

Kas Hamman has 36 years' experience in practical management and theoretical knowledge dissemination on wide-ranging aspects of biodiversity conservation. He has published a number of scientific and popular articles and has, over an extended period, focused his contribution on providing a scientific basis for sustainable biodiversity management to conservation managers in the Western Cape and South Africa. khamman@capenature.co.za

Jon Hutton is Director of the UNEP World Conservation Monitoring Centre based in Cambridge, UK. He is on the Steering Committee of IUCN's Species Survival Commission, Chair of its Sustainable Use Specialist Group and Honorary Professor of Sustainable Resource Use with the Durrell Institute of Conservation and Ecology. UNEP-WCMC, 219 Huntingdon Road, Cambridge CB3 0DL, UK. jon.hutton@unep-wcmc.org

Brian T.B. Jones is an environment and development consultant, working mainly in community-based conservation in Namibia and the SADC region. He worked for the Namibian Ministry of Environment and Tourism for ten years where he coordinated the Ministry's community-based conservation programme. Brian has an M.Phil. in Applied Social Sciences. PO Box 9455, Eros, Windhoek, Namibia. bjones@mweb.com.na

Robert E. Kenward is a Fellow of NERC Centre for Ecology and Hydrology. He served the International Association for Falconry and Conservation of Birds of Prey committee for 30 years, Raptor Research Foundation as a Director for 12 and now chairs Anatrack Ltd and IUCN-SSC European Sustainable Use Specialist Group, Wareham, UK. reke@ceh.ac.uk

Richard H. Lamprey is an ecologist, and has worked extensively in Kenya, Tanzania, Uganda, Ethiopia and Eritrea. His particular focus is in ecological monitoring and protected area planning. From 1995 to 2005 he worked with the Uganda Wildlife Authority in preparing a country-wide protected area system plan. He is now Fauna & Flora International's East Africa Technical Specialist. PO Box 20110-00200, Nairobi, Kenya. lamprey@infocom.co.ug

Nigel Leader-Williams has worked on issues related to hunting in both Africa and in UK. During the 1990s, he was Chief Technical Advisor to the Government of Tanzania, and made recommendations to reform the country's tourist hunting industry. He is currently Director of the Durrell Institute of Conservation and Ecology at the University of Kent, Canterbury, UK. N. Leader-Williams@kent.ac.uk

Raymond (Ray) Lee is the President of the Foundation for North American Wild Sheep. Prior to coming to the Foundation in 2000, Ray worked as the Big Game Management Supervisor for the Arizona Game and Fish Department. While in that position, he was instrumental in transplanting over 1000 bighorn sheep. The Foundation's International Headquarters are located at 720 Allen Avenue, Cody, Wyoming, USA. rlee@fnaws.org

Andrew J. Loveridge is a Postdoctoral Research Fellow at the Wildlife Conservation Research Unit, Department of Zoology, University of Oxford with research interests including the behavioural ecology and conservation of African carnivores and the value of sustainable utilisation of African wildlife to conservation. Wildlife Conservation Research Unit, Department of Zoology, University of Oxford, Tubney House, Tubney OX13 5QL, UK. andrew.loveridge@zoo.ox.ac.uk

Hector Magome has over 20 years experience in biodiversity conservation. He gained his PhD – which was awarded the Fiona Alexander Prize – from the Durrell Institute of Conservation and Ecology at the University of Kent. He is currently Executive Director of Conservation of South African National Parks. SANParks, PO Box 787, Pretoria 0001, South Africa. HectorM@ sanparks.org

Shane Patrick Mahoney is Executive Director of the Sustainable Development and Strategic Science Branch of the Department of Environment and Conservation, and founder and Executive Director of the Institute for Biodiversity, Ecosystem Science and Sustainability (IBES), Government of Newfoundland and Labrador, Canada. Conservation Visions, PO Box 37014, 720 Water Street West, St. John's, Newfoundland, Canada A1E 5Y2. conservationvisions@nf.sympatico.ca

E.J. Milner-Gulland is Professor of Conservation Science Research Group in Division of Biology at Imperial College London, based at the Silwood Park Campus. Her interests are in the conservation and sustainable use of wildlife, particularly the saiga antelope, and in optimising monitoring for community participation, detecting ecological trends and compliance. e.j.milner-gulland@imperial.ac.uk, www.iccs.org.uk

Arthur Mugisha has a PhD in protected areas management. He has worked in National Parks' Services East Africa for over 17 years and served as the chief executive of the Uganda Wildlife Authority. Currently, he is a technical specialist for Fauna & Flora International, affiliated to the International Gorilla Conservation Program. Ruth Towers, Ground Floor, 5A Clement Hill Road, PO Box 28217, Kampala, Uganda. Mugisha.arthur@gmail.com

Craig Packer is from the Department of Ecology, Evolution and Behavior, University of Minnesota, 1987 Upper Buford Circle, St. Paul, MN 55108, USA. packer@umn.edu

Dr Gil Proaktor is a member of the Conservation Science Research Group in the Division of Biology, Silwood Park, Imperial College London. Silwood Park, Buckhurst Road, Ascot, Berks SL5 7PY, UK. gil.proaktor03@imperial. ac.uk

Alison M. Rosser has more than 15 years experience of sustainable use and wildlife trade issues working with both IUCN and TRAFFIC East – Southern

Africa. She currently lectures on CITES and sustainable use issues at the Durrell Institute for Conservation and Ecology at the University of Kent, Canterbury, Kent CT2 7NR, UK. a.m.rosser@kent.ac.uk

Robin Sharp CB is Chair Emeritus of the IUCN/Species Survival Commission's European Sustainable Use Specialist Group, having been Chair from 1997 to 2007, and edits *Sustainable*, the global Group's newsletter. Before retiring in 1995 he was a senior civil servant in the UK Department of Environment. 30 Windermere Avenue, London NW6 6LN, UK. robisharp@googlemail.com.

R.J. (Bob) Smith has worked on conservation issues since 1996, based mainly in southern Africa and the UK. Most of his research has used the systematic conservation planning approach to develop conservation landscapes, but for the last 5 years he has also been interested in the impacts of corruption on conservation outcomes. He is currently a Research Fellow at the Durrell Institute of Conservation and Ecology at the University of Kent, Canterbury, UK. R.J.Smith@kent.ac.uk

Sardar Naseer A. Tareen is Chairman of the IUCN Central Asian Sustainable Use Group, serves as a member of WWF-Pakistan's governing body and is a member of the Balochistan Wildlife Board. He is a filmmaker and graduate in cinema graphics from the California Institute of Arts in Los Angeles. In 2002 he was honoured for his conservation achievements as one of Pakistan's 'Green Pioneers'. SUSG-CAsia, BRSP House, 5-A, Sariyab Road, PO Box 276, Quetta, Balochistan, Pakistan. naseertareen_susg@yahoo.com

William (Bill) Wall has extensive field experience in wildlife management including the development of conservation hunting programmes in Mexico, Russia and Central Asia, as well as private lands wildlife management strategies on corporate timberlands in the US. He is currently establishing an integrated sustainable approach to management of private native-corporation lands in Alaska. 198 Montana Dr., PO Box 988, Seeley Lake, MT 59868, USA. williamwall11@gmail.com

Kai-Uwe Wollscheid is Director General of the International Council for Game and Wildlife Conservation (CIC). He is Committee Member of the IUCN European Sustainable Use Specialist Group. He has an MSc in forestry and has worked on land use management, wildlife conservation and policy development in Asia, Africa and Europe. CIC Administrative Office, PO Box 82, H-2092 Budakeszi, Hungary. k.wollscheid@cic-wildlife.org

Introduction

Jon Hutton[1], William M. Adams[2] and Barney Dickson[1]

[1]UNEP-World Conservation Monitoring Centre, Cambridge, UK
[2]Department of Geography, University of Cambridge, UK

Hunting has long been a controversial issue in conservation. Human hunters have been responsible for the extinction of many species, from the depredations of the first human occupiers of new lands in the Pleistocene to the more recent actions of sailors, farmers and settlers involved in European imperial expansion. From the 16th century, sailors whose global reach exceeded their ability to store food stripped tropical islands not only of forests for timber, but of easily caught and stored meat such as tortoises, the dodo *Raphus cucullatus*, and later the great auk *Pinguinus impennis*. From the 18th century, the insatiable hunger for pelts and oil drove the extinction of many populations of whales, fur seals, sea otters, almost incidentally wiping out Steller's seacow *Hydrodamalis gigas*.

Historically, recreational hunting has played an important role in local extinctions, particularly of species already made rare (and hence perversely attractive) by other forces. Classic examples would be the hunting of game in the South African Cape and the near-extinction of the American bison. Fear of such pressures was a major factor in the establishment of conservation organisations at the start of the 20th century. The theme of hunter-turned-conservationist persisted through the 20th century, and with it debates about the extent to which conservation and hunting are compatible.

These debates have often focused on the biology of hunted populations and the sustainability of hunting: how many animals can be removed from a population without decline? The discussions have extended to the institutional questions of how hunting can be kept within scientifically defined limits, but have tended to be independent of the growing concerns about animal welfare

Recreational Hunting, Conservation and Rural Livelihoods: Science and Practice, 1st edition.
Edited by B. Dickson, J. Hutton and W.M. Adams. © 2009 Blackwell Publishing,
ISBN 978-1-4051-6785-7 (pb) and 978-1-4051-9142-5 (hb).

and rights, and questions of the ethics and appropriateness of recreational hunting.

In recent decades, however, as conservation has begun to address its social context, with concerns about the welfare of people in and around protected areas, and the needs of poor people in biodiversity-rich countries of the developing world, the questions asked about conservation and hunting have expanded:

- Can recreational hunting be done in such a way that it contributes to conservation rather than threatening animal populations?
- Can recreational hunting contribute to the reduction of poverty, especially the livelihoods of rural people?
- Can recreational hunting be justified ethically?

These three issues are far more complex than this bald formulation suggests; moreover, they are intricately linked together.

In the 21st century, therefore, it is not surprising that hunting for recreational purposes continues to be controversial. Hunters insist that their activity is an important conservation tool and, increasingly, it is suggested that recreational hunting can provide significant livelihood benefits in remote rural areas where opportunities are few and far between. However, these conservation and livelihood benefits are disputed. Meanwhile strong ethical concerns are raised about the morality of hunting for 'pleasure'.

For the conservationist, recreational hunting poses special challenges. From a biological perspective there are the effects of removing individual animals on population dynamics, population genetics, reproductive rates, life histories and the ecology of threatened species and ecosystems. In theory, the intensity of hunting can be adjusted so that it is made sustainable, and it is possible to conceive of arrangements (especially the sharing of income from professional hunting tourism businesses) such that recreational hunting could provide income for poor rural communities, and for cash-strapped conservation organisations. In practice, however, it can be difficult to control hunting, and even more difficult to establish the arrangements necessary to make sure that financial benefits go where they are most needed. The role of biology in the search for sustainability is more limited than generally appreciated, and the institutional, economic and political issues are complex and challenging.

This book explores these debates about recreational hunting, conservation and rural development. It discusses recreational hunting in both developing and developed world contexts, and addresses issues of science, ethics and livelihoods, as well as governance, policy and regulation. The different authors manifest a range of views on these issues; some are sympathetic to the value of recreational hunting while others are more neutral or sceptical.

In the first section of the book, Leader-Williams sets out some of the central themes of the book and discusses how the many forms of recreational hunting are to be defined. He outlines important controversies in debates about hunting in developed and developing countries, and considers the contribution of recreational hunting to rural livelihoods and to conservation. He suggests that those who wish to defend of recreational hunting should explain the benefits of the practice more clearly. Sharp and Wollscheid examine recreational hunting in North America, Europe and Australia. They outline the different forms it takes and note its significant economic scale. They also comment on the difficulty of finding systematic data about hunting and the need to address this information gap. Arlinghaus and Cooke extend the discussion of terrestrial hunting to recreational fishing. They argue that although recreational fishing was once assumed to be a benign phenomenon, this is not necessarily the case and they identify a variety of challenges that need to be addressed by new forms of adaptive management that integrate the findings of both biological and social sciences. Dickson surveys some of the main ethical arguments for and against recreational hunting. He notes that these arguments differ in their fundamental ethical starting points, but also suggests that each of them, considered on its own, is too simplistic. What is needed in order to make a proper assessment of the ethics of recreational hunting is a richer and more differentiated account of what we value in the natural world and why.

In the second section of the book – on the science of recreational hunting – Milner-Gulland et al. review some of the key scientific aspects of hunting that have implications for understanding its sustainability. These include the selective nature of much recreational hunting, the risk of over-exploitation and the uncertainty often associated with both monitoring and the socioeconomic context in which hunting takes place. Festa-Bianchet and Lee analyse the selective pressure of trophy hunting of bighorn sheep *Ovis canadensis*. They identify one case where such hunting has clearly had negative conservation impacts and analyse what is necessary to avoid such an outcome. They cite evidence that the appropriate management of recreational hunting can deliver benefits for conservation. Loveridge et al. examine the trophy hunting of the African lion *Panthera leo*. They describe situations where such hunting is not advisable but also argue that well regulated and well monitored hunting has only small impacts on lion populations and has the potential to contribute to the area of land available for the species. They discuss how to address some of the challenges of effective monitoring.

The third section of the book considers the question of the contribution of hunting to the livelihoods of rural people. Adams reviews the history of the relations between recreational hunting and local people. He notes that in contemporary Africa, while hunting can provide a valuable source of revenue for private landowners and for state-owned safari areas, the creation of such opportunities has often involved the dispossession of rural Africans. The challenge is to find ways of reversing this model. Frisina and Tareen describe a case study from Pakistan where the recreational hunting of dwindling populations of Suleiman markhor *Capra falconeri jerdoni* and Afghan urial *Ovis orientalis cycloceros* has recently played a central role in reversing the decline and has made a significant contribution to rural development. Jones analyses how community-based natural resource management in southern Africa has relied on recreational hunting as a source of both material and non-material benefits. He notes that the level of benefit can vary considerably and can be difficult to assess. There are, however, questions about how benefits are distributed within communities and national policy and legislation also has a big impact on the scale and nature of benefits.

The fourth section of the book comprises case studies of policy and practice in different contexts. Kenward discusses falconry. He outlines the history of the practice and considers its contribution to conservation and the challenges it faces. Aebischer's concern is with gamebird shooting in lowland UK. He argues that agricultural policies after 1945 were the main cause of the decline of key species, but contends that more recent policies offer a promising opportunity to reverse the decline and restore biodiversity. Lamprey and Mugisha describe the reintroduction of recreational hunting in Uganda. Their view is that it may be a successful tool for conservation, but only in open ecosystems and where the practice provides substantial benefits to local landowners and communities. Davies *et al.* analyse the attempt to combine both recreational hunting and photo-tourism in two South African protected areas. Contrary to accepted wisdom, the authors found that it has been possible to pursue both activities successfully in one of the areas.

The book then turns to an exploration of the way hunting is organised and governed. There is a particular emphasis on the different roles that the market and the State does and could play in shaping the nature and consequences of recreational hunting. Wall and Child discuss the institutional conditions for successful 'conservation hunting'. They argue that in many developing countries these conditions include the appropriate sharing of benefits, power and responsibility between landholders and simple monitoring systems that track a few key indicators of success such as data on trophy quality. Mahoney identifies what the key principles of what he sees as the successful North American model of recreational hunting. These include the maintenance of wildlife as a public trust resource and prohibiting commerce in dead wild-life and its products. Booth and Cumming examine the contrasting case of recreational hunting in southern Africa. There the key element has been the devolution of rights over wildlife resources to landholders, whether those are private landowners or local communities. Leader-Williams *et al.* consider how corruption may influence recreational hunting and assess what avenues are open to reform the governance of recreational hunting. The final section of the book considers two different policy approaches to recreational hunting. Rosser reviews the familiar regulatory approach at both national and inter-national levels. Child and Wall consider the potential of harnessing consumer choice, as exercised by hunters, to ensure the sustainability of hunting. Such choices would be informed by a system of certification. The authors argue that a simple and robust form of certification will best serve this goal. In the con-cluding chapter Adams *et al.* summarise the key challenges facing recreational hunting and consider how the debates about hunting, conservation and livelihoods may evolve in the future.

Conservation and Hunting

Conservation and Hunting: Friends or Foes?

Nigel Leader-Williams

Durrell Institute of Conservation and Ecology,
University of Kent, Canterbury, UK

Introduction

This chapter provides a brief introduction to the topic of recreational hunting. First, the chapter seeks a general understanding of the term 'recreational hunting'. Second, it raises some of the controversial debates that surround recreational hunting. Third, it outlines some of the scientific approaches that can underpin the biological sustainability of recreational hunting. Fourth, it discusses some of the debates over the contributions that recreational hunting can make to rural livelihoods. Fifth, it considers some contributions that recreational hunting makes to conservation. Finally, the chapter discusses some challenges that face those who wish to continue hunting for their recreation, and some of the opportunities now presented.

What is recreational hunting?

Hunting as sport or recreational pursuit has been described from the ear-liest histories and literatures (Adams, this volume, Chapter 8). However,

Recreational Hunting, Conservation and Rural Livelihoods: Science and Practice, 1st edition.
Edited by B. Dickson, J. Hutton and W.M. Adams. © 2009 Blackwell Publishing,
ISBN 978-1-4051-6785-7 (pb) and 978-1-4051-9142-5 (hb).

hunting has developed very differently across different periods of history and in different cultures. The aboriginal hunter, the ancient Assyrian king, the medieval poacher, the Victorian trophy hunter, and the modern sports hunter have all killed animals. However, they have not performed the same act. The weapons used, the game pursued, the reasons and justifications offered, the symbolic functions that their hunting has fulfilled, the legal restrictions that have applied to hunting, and the impacts on the ecological systems in which hunting has taken place, have all been different (King, 1991).

Much hunting that is practised for sport or recreation today targets large mammals and uses guns. Many terms are currently used to describe contemporary forms of hunting, including 'sport hunting', 'trophy hunting', 'tourist hunting', 'hunting tourism', 'field sports', and others. Equally, the over-arching term of 'recreational hunting' has entered the lexicon more recently, so I now seek some understanding of the term.

The *Oxford English Dictionary* defines the verb to hunt as 'to go in pursuit of wild animals or game' or 'to engage in the chase', and hunting as the action of the verb to hunt. Furthermore, the noun recreation is defined as 'the action of recreating oneself or another … by some pleasant occupation, pastime or amusement'. Taken in combination, recreational hunting might be construed as the 'pleasant occupation of going in pursuit of wild animals or game, or of engaging in the chase'.

This dictionary definition of hunting does not include killing the quarry. Indeed, anthropologists stress the importance of the 'sporting chance' to the true recreational hunter (Marvin, 2006). The quarry should present a challenge to the hunter, such that any killing occurs only at the end of a contest that is far from certain. The hope and intention of the true recreational hunter is to kill the quarry, but the skills used to find the quarry, and how the quarry is killed, are far more important than the fact the quarry is killed. Put another way 'one does not hunt in order to kill, but rather the reverse, one kills in order to have hunted' (Ortega & Gassett, 1968, in Marvin, 2006).

Anthropologists also stress how important are the specific cultural contexts in which different forms of recreational hunting are practised (Marvin, 2006). These include both premeditated actions performed before the hunt takes place, such as obtaining permission to hunt, travelling to the hunting area and seeking suitable quarry to kill legally, and the cultural norms associated with the hunt. After any kill, culturally determined actions may include collecting trophies from the hunted quarry, as well as gutting, tagging, butchering, freezing, and eating it. Put another way, the cultural context can shape decisions on when to hunt, what to hunt, how to hunt, what technology to use and how to dress (Moriarty & Woods, 1997). The latter is probably best epitomised by the traditional hunting pink of those who hunt foxes *Vulpes vulpes* on horseback with dogs. The khaki of the classic African hunting safari, and the tweed and waxed jackets of a European gamebird shoot, are somewhat less regimented, but nevertheless culturally distinctive, as well as practically suited to prevailing landscapes and weather conditions.

On this basis, I suggest that 'recreational hunting' refers to hunting where the hunter or hunters pursue their quarry for recreation or pleasure. The enjoyment of recreational hunters arises from the social and cultural norms associated with the hunt and from the sporting contest that occurs between hunter and quarry, which need not necessarily include killing the quarry.

Recreational hunters may seek a range of quarry species, from large mammals to small birds, and use a range of technologies in pursuit of their quarry, from more traditional bows and arrows, traps, dogs, falcons and ferrets, to more modern guns and rifles, and most recently to tranquillising dart guns. Besides enjoying the hunt, recreational hunters may also be seeking a trophy, and/or meat for consumption, and/or to make a direct or indirect contribution to population management and habitat conservation goals. Thus, recreational hunting is a multi-faceted activity that is variously motivated and takes place in many ecological and sociopolitical landscapes (see also Loveridge *et al.*, 2006).

Controversial debates?

Recreational hunting is the subject of considerable debate. Proponents and practitioners cite the conservation and socioeconomic benefits of recreational hunting. In contrast, opponents are concerned with issues of sustainability, and of ethics, animal welfare and animal rights.

Proponents claim that much recreational hunting has minimal biological impact, as it is generally selective of appropriate sex and age classes that do not significantly impact on the hunted population (Jackson, 1996). Thus, scientists have noted that yields from recreational hunting mostly tend to be very conservative and well below maximum sustained yields; that is, they involve off takes that should allow ongoing harvests in perpetuity (Caughley & Gunn, 1995). Indeed, practitioners have noted the recovery of game species such as white-tailed deer *Odocoileus virginianus*, pronghorn antelope

Antilocapra americana, wild turkey *Meleagris gallopavo*, Canada goose *Branta canadensis* and wood duck *Aix sponsa* in North America, through judiciously managed recreational hunting (Jackson, 1996).

Furthermore, proponents assert that recreational hunting allows the use of areas that game-viewing tourists would not visit. In turn, these areas remain under conservation management because of the economic incentives that hunting provides. Such areas include 'low' categories of protected areas (PAs) that allow forms of sustainable use. These are areas generally classified internationally by IUCN in their Categories IV to VI (IUCN, 1994), while commonly used national designations include game reserves and controlled hunting areas. Recreational hunting may also be an important use of private and communal lands that again remain under conservation management. In southern African countries like Zimbabwe and Namibia, the use of private and communal land areas for recreational hunting has doubled the areas under conservation management without the burden of the costs of this extra management falling on to already stretched State conservation agencies (Child, 1995). In turn, the use of such lands for recreational hunting can provide community benefits in remote rural areas (Jones, this volume).

Proponents also claim that the high financial returns derived from recreational hunting can provide important benefits to national exchequers and to local communities (Leader-Williams, 2000). The daily rates charged to recreational hunters who travel as tourists to developing countries are much higher per capita than are those generally charged for game-viewing tourism. Furthermore, hunting and trophy fees are set at hundreds and thousands of dollars per trophy, depending on the species killed, while park entrance fees are set in fives, tens and twenties of dollars. Therefore, game-viewing tourists need to be accommodated in much larger numbers than do hunters to achieve the same returns (Leader-Williams, 2000; Loveridge *et al.*, 2006). In turn, recreational hunters have much lower infrastructural requirements than game-viewing tourists, who may have considerable direct environmental impacts, for example through their need for lodges and roads, and for water extraction and waste disposal (Roe *et al.*, 1997).

Opponents of recreational hunting claim that much hunting has been biologically unsustainable (Loveridge *et al.*, 2006). Concerns have long been raised about its demographic impacts (Milner-Gulland *et al.*, this volume, Chapter 5) and more recently about its genetic impacts (Festa-Bianchet & Lee, this volume, Chapter 6). Indeed, some traditional forms of hunting practised for recreation are worryingly unsustainable, for example the millions of migratory birds shot and trapped annually by Mediterranean hunters (McCulloch *et al.*, 1992; www.birdlife.org/action/change/sustainable_hunting/index.html).

The indiscriminate slaughter of these migratory birds also flouts many of the ethical values associated with the notion of the sporting chance in recreational hunting. Likewise, forms of 'canned' hunting, where the quarry animal is also not afforded a sporting chance, are particularly open to attack by opponents. Other grounds for opposition arise from concerns over animal welfare and animal rights (Dickson, this volume, Chapter 4). Hunting by long chases may prove stressful to the hunted animal, as was suggested for red deer *Cervus elaphus* hunted by dogs in Exmoor (Bateson & Bradshaw, 1997), a finding that was subsequently contested (Harris *et al.*, 1999).

Taken overall, the position of the opponents of recreational hunting remains fixed, and they consider it to be anachronistic, unnecessary and morally unacceptable, often irrespective of scientific facts or evidence of conservation benefits. They may pursue their opposition at different political levels, ranging from the local to the national and international. Local opposition may result in protests and attempts to disrupt hunting activities. National opposition may result in attempts to achieve national bans of particular types of hunting through lobbying national parliaments, as recently occurred for hunting with dogs in Britain. Opposition at the international level may play out at biennial Conferences of the Parties to the Convention in International Trade in Endangered Species of Wild Fauna and Flora (CITES), during acrimonious debates over proposals to promote the conservation benefits of trophy hunting of species listed in Appendix I. At whatever level this opposition is played out, the debate can become highly politicised. For example, the government department that managed the parliamentary debates over hunting with dogs in England and Wales recently noted that an 'inordinate' amount of parliamentary time had been spent on this issue over last few years, over 240 hours since 1997 (www.defra.gov.uk/rural/hunting, posted on 28 September 2004).

Such controversy raises the question of what issues should be paramount in discussions of recreational hunting. For many in developed countries, remote from direct experience of living with wild animals, animal welfare and animal rights issues are of greatest importance (Dickson, this volume). Their positions remain firmly fixed, unswayed by any possible conservation gains or social benefits that recreational hunting may offer. Equally, some national policies in developing countries do not allow hunting either. India is opposed to hunting and sustainable use of animals based on its religious practices that revere the sanctity of animal life (Misra, 2002). Kenya has banned hunting and many forms of sustainable use, both for ethical reasons and because of the difficulty of effectively controlling the management of its once thriving safari-hunting industry (Price Waterhouse, 1996). Such nationally agreed policies clearly require respect from the proponents of recreational hunting. However, the different perspectives that proponents and opponents bring to debates on recreational hunting also present considerable challenges to constructive discussions of its future.

Which are the appropriate scientific approaches?

Over-hunting is one of the original evil quartet of factors that have been responsible for most documented recent extinctions (Caughley & Gunn 1995), and is a factor that continues to threaten species globally (IUCN, 2007). However, it not clear whether any species has been driven to extinction by what is now understood as recreational hunting. The passenger pigeon *Ectopistes migratorius*, once numbered in the millions in North America, became seriously depleted by a combination of habitat loss and unselective commercial shooting. The extent to which those hunting for recreation were responsible for its final demise in the wild is open to question (Schorger, 1955).

Nevertheless, because of concerns about over-hunting, much recent theory has been brought to bear on how to achieve biological sustainability (Loveridge *et al.,* 2006), including source–sink models, sustainable yields and quota setting, and reducing genetic losses while harvesting trophy males (Reynolds *et al.,* 2001; Loveridge *et al.,* 2006). Therefore, the science is now available on which to set conservative yield quotas for well-regulated recreational hunting, even for threatened species. For example, the southern white rhino *Ceratotherium simum simum* has been restored by strategies that included generally well-regulated, recreational hunting in South Africa (Adcock & Emslie, 1994; Leader-Williams, 2002). After their reduction to very low numbers in the early 1900s, white rhinos initially recovered within PAs. Soon after CITES came into force, white rhinos were listed in Appendix I in 1977. Because PAs were exceeding their carrying capacities for rhinos, white rhinos were then increasingly moved to private land. To provide incentives for landowners to keep rhinos, it was made possible for limited numbers of surplus white rhinos to be hunted on private land. Despite some initial regulatory problems

(Adcock & Emslie, 1994), the white rhino continued to recover so well that it was downlisted to CITES Appendix II in 1994 (Leader-Williams, 2002). Based on the successful approach followed for white rhinos, conservative quotas were agreed for the export of black rhino *Diceros bicornis* trophies from Namibia and South Africa in 2004 (Leader-Williams *et al.*, 2005).

So what science is needed? There can be little disagreement that improved local monitoring of many hunted populations is required. While the tagging of carcasses, the use of weigh stations, and the ageing and sexing of carcasses is a common practice within many North American and European hunts, the same cannot be said for the majority of hunted populations elsewhere, where quota setting is based on little more than guesswork (Severre, 1996). Nevertheless, there are areas of tension. Some hold that hunting quotas need to be based on detailed prior knowledge of the population biology of the hunted population, to avoid any uncertainty (Milner-Gulland & Ackakaya, 2001), or that, at the very least, quotas should be set using a precautionary approach. Others contend that quotas can be set through adaptive management that itself is practised in a precautionary manner (Rosser *et al.*, 2005).

The 'precautionary principle' notes that, when faced with scientific uncertainty, regulators should act in anticipation of harm to ensure that harm does not happen (Cooney, 2004). Nevertheless, it has remained unclear how this principle can be applied to recreational hunting, particularly when it is used as a tool to mitigate other possibly more serious impacts than limited trophy off take, for example through poaching, competition with livestock and habitat loss (Rosser *et al.*, 2005). Science cannot demonstrate that recreational hunting is safe before it has taken place, so hunting must happen before data can be gathered, and assessments of sustainability can only be retrospective. In turn, this provides a pivotal role for adaptive management, known less formally as the process of trial and error, whereby decisions and procedures are reviewed and the lessons learned are used to adjust the management system (Caughley & Gunn, 1995).

Adaptive management of hunting is crucial, given uncertainties over its ecological consequences, the likelihood of stochastic events, and social and economic changes. Nevertheless, to be effective, adaptive management requires resilient management institutions that are adaptable to changing biological and environmental conditions (Frisina & Tareen, this volume, Chapter 9).

Contributions to development?

From a socioeconomic and development perspective, there are two broad, but not exclusive, types of recreational hunting (Sharp & Wollscheid, this volume, Chapter 2):

- *local hunting,* where the hunter usually lives close to the hunting arena, and organises and pays appropriate fees for the hunting experience locally.
- *hunting tourism,* where the hunter travels some distance from home, often abroad, and is prepared to pay considerable sums of money, including to an intermediary supplier, to organise aspects of the hunt.

Many recreational hunters participate in the first type of hunting, while many fewer, but usually richer, recreational hunters take part in the second.

Both types of hunting have the potential to make considerable contributions to local development (Loveridge *et al.,* 2006), although only hunting tourism, where it takes place in developing countries, is directly relevant to international development. In terms of its socioeconomic benefits, hunting tourism attracts low volumes of international tourists, but can provide significant sums for conservation and foreign exchange for the national exchequer. Hunting tourism is less fickle than game viewing in times of trouble and unrest, and remains the least volatile option for wildlife use in unstable countries, such as those in the Congo Basin (Wilkie & Carpenter, 1999). In addition, recreational hunting can produce benefits for local communities, as with CAMPFIRE (Communal Areas Management Programme for Indigenous Resources) in Zimbabwe, ADMADE (Administration Management Design, which is a community-based approach to wildlife management) in Zambia, and Conservancies in Namibia (Hulme & Murphree, 2001). In the case of CAMPFIRE, sport hunting has been the primary (>90 per cent) source of revenue and, of this, 60 per cent has derived from the sale of elephant *Loxodonta africana* hunts. The balance derives from the sale of tourism lease rights, of hides and ivory, and of crocodile *Crocodylus niloticus* and ostrich *Struthio camelus* eggs (Bond, 1994, 2001). However, it is critical to determine if such benefits, whether distributed as household dividends or as community infrastructure, make people more tolerant of the costs of living with wildlife, and thereby create an incentive for conservation among local communities (Leader-Williams & Hutton 2005).

Contributions to conservation?

Conservation comprises actions that directly enhance the chances of habitats and species persisting in the wild. While conservationists agree over the need to conserve biodiversity, polarised debates often arise over whether protection or use is the best way of achieving this objective. However, some have asked whether a combination of both is possible, for additional conservation gains (Hutton & Leader-Williams, 2003).

Recent research on private land in a developed country has indeed shown that a combination of both protection and use through two forms of recreational hunting appears to provide additional conservation gains. Gamebird shooting is, and fox hunting was until recently, widely practised across lowland England. For their survival, the quarry species of gamebird shooters and fox-hunters require both woodland and hedgerow habitats and these are also required habitats for Britain's generally declining populations of woodland and farmland birds (RSPB, JNCC, WWT & BTO, 2005). But does participation by landowners in gamebird shooting and fox-hunting provide incentives to conserve key habitats on lowland farms? Measures of the extent of woodland cover on farms where the landowner was not involved in either activity, or was involved in hunting alone, in gamebird shooting alone, or in both activities, showed stepwise increases in the extent of woodland cover (Oldfield *et al.,* 2003). Indeed, on farms where the owners participated in both gamebird shooting and fox-hunting, woodland covered 10 per cent of their area, the same IUCN target for national coverage of PAs that conservationists have long aspired to achieve. It is important that similar research is carried out in other land ownership systems, and in countries of different development status, to determine if recreational hunting also provides an incentive to conserve in such situations. Equally, it is important to recognise that this research simply indicates a correlation and so does not demonstrate a cause-and-effect relationship. Therefore, further research based on the seminatural experiment created by the ban on hunting with dogs in Britain would be desirable.

Future challenges?

Recreational hunters around the world will increasingly need to face critical debates in defence of their sport, if other forms of recreational hunting are not to face the same fate as hunting with dogs in Britain. While scientists may benefit from the semi-natural experiments that changes in management could provide, hunters need to address issues that might affect the future of their sport in a timely manner. I now outline some of these challenges, noting some of the pressure points and possible ways forward.

Recreational hunting represents a considerable interdisciplinary challenge. For example, mathematically inclined biologists may focus on issues of biological sustainability relative to harvesting theory, while animal welfarists raise concerns about the fate of prey species during hunting, proponents of animal rights question whether the rights of prey species are violated and anthropologists debate changes in local cultural traditions. Furthermore, rural economists and sociologists may raise concerns about contemporary impacts on rural people and rural areas, while conservation biologists may be interested in examining the contributions of recreational hunting to wider species and habitat conservation. However, the wide range of specialist disciplines required for a holistic debate about the future of recreational hunting are rarely brought together in a collective dialogue. This can result in fractured, single-issue debates that are dominated by the best-organised position.

Likewise, recreational hunting has many and varied forms, ranging from the classic African foot safari to hunting on horseback with dogs in England. Each form is often highly context-specific, each with different participants and each with different constituencies and platforms of support, who generally speak without any common voice. Organised groups opposed to recreational hunting, whether to particular forms of hunting or to all recreational hunting, can pick off their targets one by one, without any likelihood of collective action or support by the various types of recreational hunter in defence of their different forms of sports. Again, this can lead to dominance of any such discussions by the best-organised, single position.

A key and emerging tactic of those opposed to recreational hunting is their use of increasingly political tools to voice their objections. As already noted, the use of national parliamentary time is one such tactic. Another is use of stricter domestic measures to prevent the import of trophies from otherwise internationally sanctioned trophy hunts abroad. CITES has mechanisms to agree quotas for species that are listed in Appendix I where those species are trophy hunted because of the conservation benefits that trophy hunting is believed to bring. Thus, hunting quotas have been agreed for species like African elephants, leopards *Panthera pardus*, white rhinos and markhor *Capra falconeri* at successive Conferences of the Parties to CITES, to which importing countries are party and vote upon. Yet, in response to domestic anti-hunting pressures, countries such as the United States can prohibit the import of trophies taken by hunters, therefore making it much less likely that US hunters will participate in such hunts in future. Such highly politicised actions continue, despite the evidence of the conservation benefits of recreational hunting, whether these comprise the biodiversity benefits of field sports in Britain (Oldfield *et al.*, 2003) or the recovery of markhor in Pakistan (Frisina & Tareen, this volume).

At the same time, recreational hunters require open minds to address some of the obvious pressure points that face the different forms of their various sports. Adequate controls are often lacking in many forms of recreational hunting. This lack of control can particularly affect high-value hunting tourism and may manifest itself through exceeding quotas, through lack of transparency in allocating hunting opportunities, and through flouting of foreign exchange regulations (Price Waterhouse, 1996). In turn, this may lead to concerns about the biological and socioeconomic sustainability of recreational hunting, and raise charges of corrupt practices (Leader-Williams *et al.*, this volume, Chapter 18). Thus, high-value recreational hunting has been the subject of long-term or temporary bans in several countries, for example Kenya and Uganda, respectively (Price Waterhouse, 1996; Lamprey & Mugisha, this volume, Chapter 13).

High value recreational hunting may experience similarly bad practices to those that affect game-viewing tourism operations, where transparency is no more likely in allocating opportunities for a lodge site, or where foreign exchange controls may be equally flouted. Nevertheless, the general public finds it much more difficult to understand that well-managed recreational hunting can contribute to conservation, given that individuals of species that conservationists seek to conserve are killed. Therefore, when control is lacking, recreational hunting is more likely to get a bad press than game-viewing tourism, which the public is more conditioned to believe is relatively benign. In addition, the general public is likely to believe that hunting is incompatible with game-viewing (Davies *et al.*, this volume, Chapter 14), in which more people take part, resulting in further bad publicity for hunting.

These pressure points throw up certain challenges for proponents of recreational hunting. First, can those who offer and regulate hunting tourism opportunities start to push their product closer towards the ideals of ecotourism? While many definitions exist for ecotourism, few encompass all the ideals for which true ecotourism should strive. They include: minimal environmental impacts on the natural resource base; financial contributions to conservation and protected area management; low social impact on, and financial contributions to, local communities; and raising awareness and contributing to conservation education (Quebec Declaration, 2002). All these ideals are within reach of well-run recreational hunting operations, but are rarely explicitly stated as goals. On the biological front, achieving sustainable quotas is key. On the socioeconomic front, reducing leakage and improving governance is key, which in turn could help ensure that appropriate financial contributions are available to support both conservation and local communities.

But how can these ideals be achieved? Two routes suggest themselves, certification and self-regulation. However, the many and varied forms of recreational hunting suggest that both routes will be complicated to implement, because it will be difficult to design any system(s) where one size fits all. Instead, it is most likely that context-specific, and locally- or nationally based approaches, will be best suited.

Certification has been proposed as an approach that could help move recreational hunters towards sustainable management (Lindsey *et al.*, 2007). Nevertheless, while certification schemes such as the Forest Stewardship Council (FSC) may work for certain natural products, it remains unclear how appropriate schemes could be drawn up for recreational hunting. It would be necessary to determine what aspect(s) of a recreational hunting operation would best be certified. For example, would the criterion be that a trophy has been taken as part of a sustainable harvest, mirroring the objectives of the FSC scheme, or that the outfitter runs his operation close to the ideals of the Sustainable Tourism Stewardship Council (Font *et al.*, 2003)? Furthermore, at what level should the chosen certification system be audited, and indeed, at what level(s) should (an) independent certification body(ies) be situated, and who should offer any resulting ecolabels? Finally and perhaps critically, who should pay for certification: the user or the regulator? While such difficult matters are resolved, it may be helpful to move towards local and context-specific efforts to self-regulate particular forms of recreational hunting.

In terms of self-regulation, a good example is the Independent Supervisory Authority for Hunting (ISAH) that was founded in 2000 to regulate hunting with dogs in England and Wales (www.isah.org.uk). ISAH was financed by subscriptions from its members but appointed an independent panel, in turn, to appoint an independent Chairman and Commissioners whose task was to set up and administer protocols for the regulation of hunting with dogs. The protocols that they established incorporated clear principles that were to guide their decisions, which included:

Humanity: the avoidance of unnecessary suffering;
Utility: the effective management of quarry species; and
Stewardship: sensitive management of living environment.

Despite its transparent approach, the establishment of the ISAH came too late to influence the political train of events set in motion by promises made during the British general election of 1997. Nevertheless, ISAH could provide a model of a nationally based, self-regulatory approach that could be adapted to local conditions elsewhere. Equally, future self-regulatory bodies should be established well in advance of any political processes that seek to debate the future of different forms of recreational hunting.

Conclusions

This chapter has offered a brief overview of the different and context-specific forms of recreational hunting, and discussed some of the views of its proponents and opponents. To its many opponents, the killing and hunting of quarry species is anachronistic and morally indefensible. To proponents, the traditional forms of recreational hunting can have considerable benefits for conservation and for rural economies. However, there are many and varied forms of recreational hunting, and opponents are very well organised and funded. Therefore, practitioners need to guard against complacency, to use self-regulation to address those aspects of recreational hunting that are most likely to cause public concern, and to explain the conservation benefits of recreational hunting clearly and unequivocally.

References

Adcock, K. & Emslie, R.H. (1994) The role of trophy hunting in white rhino conservation, with special reference to Bop parks. In *Rhinos as Game Ranch Animals*, eds. B.L. Penzhorn & N.P.J. Kriek, pp. 35–41. South African Veterinary Association, Onderstepoort.

Bateson, P. & Bradshaw, E.L. (1997) Physiological effects of hunting red deer (*Cervus elaphus*). *Proceedings of the Royal Society of London B*, 264, 1707–1714.

Bond, I. (1994) Importance of elephant hunting to CAMPFIRE revenue in Zimbabwe. *Traffic Bulletin*, 14, 117–119.

Bond, I. (2001) CAMPFIRE and the incentives for institutional change. In *African Wildlife and Livelihoods: the Promise and Performance of Community Conservation*, eds. D. Hulme & M. Murphree, pp. 227–243. James Currey, Oxford.

Caughley, G. & Gunn, A. (1995) *Conservation Biology in Theory and Practice.* Blackwells, Oxford.

Child, G. (1995) *Wildlife and People: the Zimbabwean Success.* Wisdom, Harare and New York.

Cooney, R. (2004) *The Precautionary Principle in Biodiversity Conservation and Natural Resource Management: An Issues Paper for Policy-Makers, Researchers and Practitioners.* IUCN – The World Conservation Union, Gland and Cambridge.

Font, X., Sanabria, R. & Skinner, K. (2003) Sustainable tourism and ecotourism certification: raising standards and benefits. *Journal of Ecotourism*, 2, 213–218.

Harris, R.C., Helliwell, T.R., Shingleton, W., Stickland, N. & Naylor, J.R.J. (1999) *The Physiological Responses of Red Deer (Cervus elaphus) to Prolonged Exercise Undertaken during Hunting: Joint Universities Study on Deer Hunting.* R&W Publications, Newmarket.

Hulme, D. & Murphree, M. (2001) (ed.) *African Wildlife and Livelihoods: The Promise and Performance of Community Conservation.* James Currey, Oxford.

Hutton, J.M. & Leader-Williams, N. (2003) Sustainable use and incentive-driven conservation: realigning human and conservation interests. *Oryx*, 37, 215–226.

IUCN – The World Conservation Union (1994) *Guidelines for Protected Areas Management Categories.* IUCN, Cambridge and Gland.

IUCN – The World Conservation Union (2007) *2007 IUCN Red List of Threatened Species.* IUCN, Cambridge and Gland.

Jackson, J.J. (1996) An international perspective on trophy hunting. In *Tourist Hunting in Tanzania,* eds. N. Leader-Williams, J.A. Kayera & G.L. Overton, pp. 7–11. IUCN, Gland and Cambridge.

King, R. (1991) Environmental ethics and the case for hunting. *Environmental Ethics*, 13, 59–85.

Leader-Williams, N. (2000) The effects of a century of policy and legal change upon wildlife conservation and utilisation in Tanzania. In *Conservation of Wildlife by Sustainable Use*, eds. H.H.T. Prins, J.G. Grootenhuis & T.T. Dolan, pp. 219–245. Kluwer Academic Publishers, Boston.

Leader-Williams, N. (2002) Regulation and protection: successes and failures in rhinoceros conservation. In *The Trade in Wildlife: Regulation for Conservation*, ed. S. Oldfield, pp. 89–99. Earthscan, London.

Leader-Williams, N. & Hutton, J.M. (2005) Does extractive use provide opportunities to offset conflicts between people and wildlife? In *People and Wildlife: Conflict or Co-existence?* eds. R. Woodroffe, S.J. Thirgood & A. Rabinowitz, pp. 140–161. Cambridge University Press, Cambridge.

Leader-Williams, N., Milledge, S., Adcock, K. *et al.* (2005) Trophy hunting of black rhino *Diceros bicornis*: proposals to ensure its future sustainability. *Journal of International Wildlife Law and Policy*, 8, 1–11.

Lindsey, P., Frank, L.G., Alexander, R., Mathiesen, A. & Romanach, S.S. (2007) Trophy hunting and conservation in Africa: problems and one potential solution. *Conservation Biology*, 21, 880–883.

Loveridge, A.J., Reynolds, J.E. & Milner-Gulland, E.J. 2006. Does sport hunting benefit conservation? In *Key Topics in Conservation Biology*, eds. D. Macdonald & K. Service, pp. 224–240. Oxford: Blackwell.

Marvin, G. (2006) Wild killing: contesting the animal in hunting. In *Killing Animals*, ed. Animal Studies Group, pp. 10–29. University of Illinois Press, Urbana and Chicago.

McCulloch, M.N., Tucker, G.M. & Baillie, S.R. (1992) The hunting of migratory birds in Europe: a ringing recovery analysis. *Ibis*, 134(Suppl 1), 55–65.

Milner-Gulland, E.J. & Akcakaya, H.R. (2001) Sustainability indices for exploited populations. *Trends in Ecology and Evolution*, 16, 586–692.

Misra, M. (2002) Evolution, impact and effectiveness of domestic wildlife bans in India. In *The Trade in Wildlife: Regulation for Conservation*, ed. S. Oldfield, pp. 78–85. Earthscan, London.

Moriarty, P.V. & Woods, M. (1997) Hunting ≠ predation. *Environmental Ethics*, 18, 391–404.

Oldfield, T.E.E., Smith, R.J., Harrop, S.R. & Leader-Williams, N. (2003) Field sports and conservation in the United Kingdom. *Nature*, 423, 531–533.

Ortega & Gassett (1968) *La Caza y Los Toros*. Revista de Occidente, Madrid. (Cited in Marvin, 2006.)

Price Waterhouse (1996) The trophy hunting industry: an African perspective. In *Tourist Hunting in Tanzania*, eds. N. Leader-Williams, J.A. Kayera & G.L. Overton, pp. 12–13. IUCN – The World Conservation Union, Gland and Cambridge.

Quebec Declaration (2002) www.world-tourism.org/sustainable/IYE/quebec/anglais/declaration.html

Reynolds, J.D., Mace, G.M., Redford, K.H. & Robinson, J.G. (2001) *Conservation of Exploited Species*. Cambridge University Press, Cambridge.

Roe, D., Leader-Williams, N. & Dalal-Clayton, D.B. (1997) *Take only Photographs, Leave only Footprints: The Environmental Impacts of Wildlife Tourism*. International Institute for Environment and Development, London.

Rosser, A.M., Tareen, N. & Leader-Williams, N. (2005) Trophy hunting and the precautionary principle: a case study of the Torghar Hills population of straight-horned markhor. In *Biodiversity and the Precautionary Principle: Risk and Uncertainty in Conservation and Sustainable Use,* eds. R. Cooney & B. Dickson, pp. 55–72. Earthscan, London.

RSPB, JNCC, WWT & BTO (2005) *The State of the UK's Birds 2005.* RSPB, Sandy.

Schorger, A.W. (1955) *The Passenger Pigeon: Its Natural History and Extinction.* University of Oklahoma Press, Norman, Oklahoma.

Severre, E.L.M. (1996) Setting quotas for tourist hunting in Tanzania. In *Tourist Hunting in Tanzania,* eds. N. Leader-Williams, J.A. Kayera & G.L. Overton, pp. 57–58. IUCN – The World Conservation Union, Gland and Cambridge.

Wilkie, D.S. & Carpenter, J.F. (1999) The potential role of safari hunting as a source of revenue for protected areas in the Congo Basin. *Oryx,* 33, 339–345.

An Overview of Recreational Hunting in North America, Europe and Australia

Robin Sharp[1] and Kai-Uwe Wollscheid[2]

[1]IUCN/European Sustainable Use Specialist Group
[2]International Council for Game and Wildlife Conservation,
Budakeszi, Hungary

Introduction

Recreational hunting is a significant social and economic phenomenon in most countries where leisure activities are fully developed, principally the richer countries of the world. Within these countries recreational hunting takes diverse forms with differing sets of regulations and norms governing where, when and how hunting takes place. This chapter begins by noting some of the distinctive features of recreational hunting in North America, Europe and Australia. It then moves on to summarise, for these areas, what is known about the number of people hunting, which animals they hunt, how much the hunters spend and where that money goes. In spite of the importance of recreational hunting as a leisure activity, there is very little consolidated information at the international level about recreational hunting and only limited information at national level. Until a full picture is available one can do no more than present an assortment of data about hunters and their activities, assembled at different times over the last few years and using statistical methods of hugely

Recreational Hunting, Conservation and Rural Livelihoods: Science and Practice, 1st edition.
Edited by B. Dickson, J. Hutton and W.M. Adams. © 2009 Blackwell Publishing,
ISBN 978-1-4051-6785-7 (pb) and 978-1-4051-9142-5 (hb).

variable quality. Nevertheless, even this limited and uneven information indicates the scale of recreational hunting in many developed countries.

Most of the hunting considered in this chapter is done from home, but some hunters travel within their own countries to hunt. In addition, a relatively small and often wealthy elite travels abroad for specially organised shoots or safaris. This latter type of 'tourist hunting' is outside the scope of this chapter, but is discussed in other chapters in this volume (Frisina & Tareen, Chapter 9; Jones, Chapter 10; Lamprey & Mugisha, Chapter 13; Booth & Cumming, Chapter 17).

Diversity of hunting

Within Europe, North America and Australia, recreational hunting takes many forms. In Europe, Pinet (1995) has distinguished four types of hunting tradition and assigns them to different 'regions' (see Table 2.1).

The North American and the Australian traditions draw on all these elements to a greater or lesser extent. Distinctive features of hunting in North America are that wildlife, including game, belongs to no-one and is not to be sold, while hunting is open to all, subject to access being granted where land

Table 2.1 **European hunting traditions.**

Hunting tradition	Main characteristics
Scandinavian	High ratio of hunters to general population; spontaneous leisure pursuit across all classes and from rural & urban areas; 'close to nature' philosophy.
Latin and Ireland	Forms the main pool of European hunters; mainly middle to lower income people from rural areas; pursues small game and migratory and non-migratory birds.
Anglo-Saxon	Low ratio of hunters to general population; 'sporting' approach; linked to land ownership; targets pheasants & partridges.
Continental 'north'	Germany, Netherlands etc.; low ratio of hunters to general population; high income, aristocratic tradition; complex codes of conduct; game management aspect.

Based on Pinet (1995).

is privately owned. In Australia, introduction of species desirable for hunting according to European norms links with the Anglo-Saxon tradition, but hunting for pest control is also a major activity.

Some recreational hunting takes the form of trophy hunting where the hunter seeks a physical token of the animal he (or, more rarely, she) has hunted, such as antlers, horns or skins. Trophy hunting is an important element in much tourist hunting. The idea of recording and exhibiting hunting trophies took hold in the first International Hunting Exhibition in Vienna in 1910 (Kujawski, 2005). Since then, elaborate systems of trophy measurement and recording have been developed. Lechner (1995) identifies five international trophy measurement systems or standards that are used today: International Council for Game and Wildlife Conservation (CIC); Rowland Ward; Boone & Crocket Club; Safari Club International (SCI); and Norman Douglas. Each of these systems differs in its geographical scope, its applicability to different sorts of animal, and the precise criteria that are employed.

Hunting ethics are often important to the self-understanding of hunters. The traditional view is that hunters should adhere to the principles of 'fair chase' which impose on the hunter certain constraints on the way in which game is pursued. Many national or regional hunters' organisations (e.g. the French Federation of Hunters, the Hunters' Central Organization in Finland and the Nordic Hunter Association, as well as the Rowland Ward Guild of Field Sportsmen and the Boone and Crocket Club) have adopted codes or rules of ethical conduct. These rules of behaviour intend to foster hunting ethics as a form of self-regulation above and beyond codified legislation.

Hunting in the United States

The most detailed information on recreational hunting comes from North America, where there is the largest number of hunters of any region. The US and Canadian authorities carry out regular, census-based surveys of wildlife use. The US Fish & Wildlife Service is the federal conservation agency responsible for fishing, hunting and wildlife watching. Since 1955 they have undertaken, through the US Census Bureau and at approximately five-year intervals, a survey to gather information on the number of anglers, hunters and wildlife-watching participants, and on how often they participate and how much they spend on their activities.

The most recent fully reported survey was carried out in 2001 (US Fish and Wildlife Service *et al.*, 2002). Detailed interviews were carried out with 25,070 'sportspersons' and 15,303 wildlife watchers. The survey uses carefully framed definitions, relates only to recreational activities and takes account of those who engage in more than one of the activities. The survey found that 82 million people took part in wildlife activities in 2001. Of these, 34.1 million fished, 13 million hunted and 66.1 million watched wildlife (these numbers do not sum because many people carried out more than one type of activity).

These 82 million people spent a total of US$108.4 billion, of which hunters were responsible for US$20.6 billion, those engaged in fishing spent US$35.6 billion, people doing both fishing and hunting spent US$13.8 billion and wildlife watchers were responsible for US$38.4 billion.

Of the US$20.6 billion spent by hunters, about 50 per cent went on equipment (e.g. guns, tents, camper vans, and bikes), 25 per cent on food, lodging and transport, and the remaining 25 per cent on other things such as books, membership, land purchase or rental, and licences. Over US$700 million was spent on equipment licences, tags, stamps and permits. Much of this money supports conservation directly, either because it goes to state wildlife agencies who spend it on conservation projects and hunting safety and education programmes, or because it is channelled through Ducks Unlimited – an 800,000 strong association of hunters and conservationists which conserves and manages wetlands for North America's wildfowl (Herring, 2006). One of the key instruments in this respect is the Federal Aid in Wildlife Restoration Program which was promoted early in the 20th century by Theodore Roosevelt and his colleagues as a response to uncontrolled hunting. This imposes a tax on gun purchases and requires the revenue raised to be devoted to conservation. The use of taxes paid by hunters to fund public conservation underpins the claim that the spectacular recovery of major species in the US during the 20th century is largely attributable to hunting (Jackson, 1996).

The US Survey demonstrates that what is there referred to as 'big game' dominates the hunting scene, attracting by far the most participants and the bulk of the expenditure. The broad categories of target species, with the activity and expenditure they attracted in 2001, are shown in Table 2.2.

While hunters are well represented in all age groups between 16 and 64, with only seven per cent aged 65 and over, the distribution between male and female is heavily skewed at 91 per cent to 9 per cent respectively. Hunters also come from all income groups but while overall they make up six per cent of the over 16 population, they are less well represented in groups with

Table 2.2 **Breakdown of game hunted in US.**

Quarry	Hunters (millions)	Days (millions)	Expenditure (US$bn)
Big game			10.1
Deer	10.3	133	
Wild Turkey	2.5	23	
Elk	0.9	6	
Bear	0.3	3	
Small game			1.8
Squirrel	2.1	22	
Rabbit & hare	2.1	23	
Pheasant	1.7	13	
Grouse/prairie chick	1.0	9	
Quail	0.9	8	
Migratory birds			
Ducks	1.6	18	1.4
Doves	1.5	9	
Geese	1.0	11	
Others (e.g. Groundhogs)	1.0	19	0.2
Unspecified			7.1
Total	13.0	228	20.6

Note: Columns showing numbers of hunters and days do not sum because many hunters hunt more than one category.

household incomes below US$25,000 per annum and more strongly represented at income levels over US$40,000, making up nine per cent of those with incomes between this figure and US$100,000. In terms of educational record hunters showed slightly above average numbers among those who had 12 years of school and one to three years of college. While only 1 per cent of black people and 2 per cent of other ethnic groups hunted, seven per cent of white people did.

Some indications about the trends in recreational hunting can be gained by comparisons made in the 2001 survey report with the 1996 survey findings and the initial results from the 2006 survey (US Fish & Wildlife Service *et al.*, 2007). There has been a decline in the number of hunters from about 14 million in 1996 to about 12.5 million in 2006. This decline has taken place mostly amongst the hunters of small game and, more recently, migratory birds. The number of those hunting 'big game' has remained fairly constant. This decline takes place against the background of an overall increase in the numbers engaging in wildlife-related activities. Between 2001 and 2006 these numbers went up from 82 million to 87 million.

Sixteen per cent of hunters travelled to another US state to hunt but there is no information in the 2001 Survey report on how many travelled outside the US. However joint work with Statistics Canada indicates that in 1996 Americans visiting Canada for wildlife related activities, including hunting, spent CA$705 million (Du Wors *et al.*, 1999).

Hunting in Canada

From 1981 to 1996 Statistics Canada has carried out major surveys for Environment Canada and the Canadian Wildlife Service at similar intervals to the US surveys, under the title 'National Surveys on the Importance of Nature to Canadians'. The Canadian Wildlife Service has a broad remit at federal level, similar to that of its US counterpart. The latest reported survey relates to activity in 1996 (Du Wors *et al.*, 1999). It was based on some 60,000 returned questionnaires. In its overall scope it includes 'outdoor activities in natural areas' and 'residential wildlife-related activities', which the US Survey does not. The introduction explains that the survey is conducted in order to guide policy development by demonstrating the values which Canadians attach to the range of activities described.

The survey shows that in 1996, 4.2 million Canadians took part in recreational fishing (18 per cent of the adult population) and that 1.2 million hunted (five per cent). These hunters spent a total of CA$824 million representing CA$692 per head. This compares to the CA$462 per head spent by recreational fishers. The expenditure per hunter is only around one third of the US figure. It is not clear why there is this difference and it may be due to the use of different methodologies in the two surveys. Canadian hunters were responsible for 14,200 jobs and for government tax revenues of CA$384 million (Leigh *et al.*, 2000). This revenue is not earmarked for conservation.

The proportion of hunters hunting different types of animal is roughly similar to that found in the US. Thus 72 per cent of hunters hunt large mammals, 23 per cent small mammals, 24 per cent waterfowl and 38 per cent other birds. Of the 1.2 million hunters, 15 per cent were women, a somewhat higher percentage than in the United States. As in the US, hunters had slightly above average wealth and education compared with the population as a whole.

Similarly to the US there has been a decline in the number of hunters. In 1981 it was estimated that there were 1.8 million hunters in Canada but this had fallen to 1.2 million in 1996.

Information about the numbers of animals killed and licences issued annually appears to be meticulously recorded but is only consolidated at the state level. Just to give one example from the latest British Columbia statistics, in 2002 the state issued 440,000 hunting licences and raised revenue of CA$8.3 million from them (Anon., 2002).

To sum up the position of the two North American countries, hunting, along with fishing and wildlife watching, is treated by government authorities as a wildlife-related activity, being managed and regulated under the same umbrella. Although there seems to be a downward trend, large numbers of people still hunt and contribute billions of dollars to the economy, some of which is, in the US, applied via earmarked taxation directly to conservation. In contrast to some regions or countries where hunting is seen as controversial, the Canadian government report suggests that there is substantial potential for increased participation in hunting, noting that twice as many people expressed interest as were actively participating.

Hunting in Europe

In Europe, the other major region where recreational hunting flourishes as an organised activity, the information available is much more patchy. The only attempt to put the available information about individual countries together in comparable form has been made by FACE (Federation of Associations for Hunting and Conservation of the EU), and is displayed on its website (FACE, 2007). The FACE data for 2007 provide information about 25 EU member states and 10 other European countries where FACE has members. This indicates that there are just under seven million hunters in 35 countries, not including Russia. Some of the data for individual countries appears to be well supported by evidence (such as the number of hunting licenses issued), while for other countries the evidence is much less substantial.

Table 2.3 **Hunting numbers in selected European countries.**

Country	Number of hunters	% of total population
Ireland	350,000	8.9
Cyprus	45,000	6.4
Finland	290,000	5.8
Norway	190,000	4.75
Spain	980,000	2.3
France	1,313,000	2.1
UK	800,000	1.3
Italy	750,000	1.2
Germany	340,000	0.4
Poland	100,000	0.3
Netherlands	30,000	0.2

From FACE (2007).

A selection of figures from FACE (2007) is sufficient to illustrate the range of numbers of hunters and participation rates among European countries (see Table 2.3).

By way of comparison it is worth noting that the US and Canadian surveys cited earlier also show large variations in participation rates between regions in the US (from 12 per cent in the West North Central region to two per cent in the Pacific region) and between individual provinces in Canada (from 15.1 per cent in Newfoundland to 3.2 per cent in British Columbia). The reasons for these variations are not discussed in the North American survey reports, but urban/rural population balance and lifestyles are likely to play a part. However in the case of European countries distinctive hunting traditions may provide part of the explanation.

A comparison of five national economic studies suggested to Pinet that in the early 1990s the average European hunter was spending €1500 on hunting and related activities. In fact the studies showed an enormous variation in hunters' expenditure, ranging from €5800 in Belgium, through €1200 in France to €350 in Ireland, and did not employ a standard methodology. Nevertheless, Pinet (1995) estimated that European hunters were spending a total of €9.88 billion annually. In the view of the present authors this figure should be treated with considerable caution, but there is no current alternative estimate. Pinet further estimated from aggregated national statistics that on average 65 hunters supported one full-time equivalent job. This amounted to 101,300 jobs for 6,585,000 hunters in Europe in 1995.

A recent consultants' study of sport shooting in the UK in 2004 (PACEC, 2006) provides what is probably the most detailed and thorough survey of its kind in Europe to date. In looking at the findings it should also be recalled that while elsewhere in the world 'hunting' generally means pursuing a live wild quarry with a rifle or other weapon, in Britain country sports have always been divided into 'hunting, shooting and fishing'. In this context 'hunting' means hunting with hounds, while 'shooting' means shooting at game birds or mammals and pest birds.

The study indicated that there were 480,000 sport shooters in the UK, spending a total of £2 billion annually. This amounts to £4166 per shooter. It was calculated to generate the equivalent of 70,000 full-time jobs and to lead to £250 million being spent on habitat management.

These are remarkable figures when set against those for North America and Europe already cited. In particular the average gross spend per shooter is around five times as high as that of the average US hunter. The difference is to be explained by the specific features of sport shooting in the UK. The majority of game birds shot are artificially reared and driven to the guns by beaters, thus not being truly wild and, in the view of some, not truly hunted. A sports shooter in the UK might pay £1500 for one day's pheasant (*Phasianus colchicus)* and partridge (*Alectoris rufa* and *Perdix perdix)* shooting of this kind and since these are the major game species shot and around 80 per cent of the shooting of these species is 'driven', this explains why the expenditure figures are so different from those for North American hunters. At the same time the expenditure on habitat management is significant and it may be the first time that this contribution has been properly quantified.

In order to remove the artificial element (i.e. driven grouse (*Lagopus lagopus)*, pheasant and partridge) from this analysis and to enable more realistic international comparisons a member of the PACEC team kindly supplied the authors with a special estimate for sport shooting of fully wild species in the UK. The quarry species consisted of lowland game, deer, wildfowl and avian and mammalian pests. It was found that 410,000 shooters participate in this activity, spending a total of £420 million (equivalent to £1024 per person). This was calculated to generate 12,000 full-time equivalent jobs (5300 direct and 6700 in the supply chain) (Nic Boynes, personal communication). This expenditure figure compares well with similar data from the rest of Europe and North America.

Summing up the situation in Europe, we can say that, as in North America, there are millions of hunters contributing billions of euros to the economy, much of it in rural areas, and supporting the maintenance of wildlife-friendly habitats. However, there are enormously different ratios of hunters to the general population in different regions of Europe and very strong contrasts between the type and amount of regulation imposed. It is rare for hunting, fishing and wildlife-watching to be regulated by the same agency.

Hunting in Australia

In their study of recreational hunting from an international perspective for the Australian Sustainable Tourism initiative, Bauer & Giles (2002) summarise information about the number of hunters in Australia, their expenditure and the type of hunting they engage in. Whereas the Aboriginal inhabitants depended wholly on hunting for their livelihoods, the European settlers drove them from much of their land and enacted laws that seriously restricted their rights. They then introduced a range of deer species to make their own recreational hunting more interesting. These deer species, kangaroos and rabbits that are hunted as pests, introduced species that have escaped into the wild, such as pigs and buffaloes, and certain game birds comprise most of the target species for recreational hunters. There is little formal infrastructure for hunting and no clear 'philosophy' as in the case of North America and many European countries.

Nevertheless, in 1997 a National Hunting Policy Working Group, made up of the major hunting bodies, estimated over one million people or around five per cent of Australians to be hunters (Bauer & Giles, 2002). The Working Group also estimated that in 1997 hunters were generating expenditure of A\$1 billion (US\$780 million in 2006) through access fees and licences, purchase of vehicles and equipment, and downstream related employment. Of this, AU\$325 million was reckoned to go to regional communities, presumably those where hunting took place. Most hunting is on private land, remote from the main urban settlements, and access to it depends on hunters enjoying good relations with landowners and managers.

These rural properties provide mostly unguided access for hunters, though there is a development of guided services, which also cater for the small

Table 2.4   **Expenditure by deer hunters in Australia in 1995.**

Item	Expenditure (AU$ million)
Trip costs & supplies	36.05
Overseas trips	8.41
Equipment	58.44
Total	102.90

From Bauer & Giles (2002).

number of hunters from overseas visiting Australia. Access and service costs are reported to range from AU$50 per day for duck shooting to AU$500 for partridge and pheasant hunts. Organised buffalo and banteng hunts in the Northern Territory and Queensland may cost up to AU$10,000. Cause reported in 1995 on a survey which concluded that there were some 17,500 serious deer hunters in Australia spending over AU$100 million annually – see Table 2.4 (Bauer & Giles, 2002).

The survey also found that annual expenditure per head was AU$5880, while 8.9 per cent of these deer hunters took trips abroad to hunt, more than two-thirds going to New Zealand. Deer harvests overall are estimated at 15,000–20,000 annually.

Summary of hunting in major developed regions

The information given in this chapter about the number of hunters hunting in their own countries in the major developed regions presents only a very rough overview. It has been collected on different bases at various times, ranging from rigorous surveys conducted by census bureaux through licence or membership numbers to broad guesstimates based on small samples. It is full of gaps and uncertainties, especially in relation to economic and conservation impacts. And we have not presented any information on the Russian Federation, China, South-East Asia, Central and South America or New Zealand. Nevertheless the data from North America, Europe and Australia are sufficient to show that recreational hunting in these regions is a significant land use that may be favourable to wild species and their habitats, involving

Table 2.5 **Summary of hunters in major developed regions.**

Country/region	Number of hunters (millions)	Expenditure (US$ bn)
United States	13.0	20.6
Canada	1.2	0.72
Europe (of 35)*	7.0	10.66
Australia	1.0	0.78
Total	22.2	32.76

*25 EU member states and 10 countries where FACE has members.

millions of people and generating billions of dollars, euros and pounds to support livelihoods (see Table 2.5).

By contrast the hunting tourists taking organised trips to South Africa, Namibia, Zimbabwe, Botswana and Tanzania to hunt high-value trophies are numbered in hundreds or low thousands (Barnett & Patterson, 2005). While these tourist hunters spend far more per head than stay-at-home hunters and present many of the specific conservation issues discussed in other chapters, their global total spend is much less than that of their non-tourist colleagues. Even in South Africa, which attracts by far the largest number of incoming hunters of any African country, Damm (2005) estimates that while over 7000 foreign hunters contributed US$130 million to the economy in 2003/4, around 200,000 resident hunters were spending US$452 million. This is not to belittle the importance of tourist hunting, since where it does take place it is highly significant in relation to direct expenditure on conservation and in its potential for boosting rural incomes and providing incentives for local people to collaborate in wildlife management.

Conclusion

This chapter has attempted to draw together in a very brief compass an account of the main types of recreational hunting as practised by hunters from some major developed countries and to provide an outline socioeconomic overview of their activities. The scale of the phenomenon and the large expenditure by recreational hunters are noteworthy. An inference from the difficulty of finding systematic data on hunting is that hunters seem scarcely conscious of the need in today's world to provide an objective and balanced description of what they are doing and its environmental and social impact. This is surprising when some forms of hunting are under political attack in contexts where the wider public have highly partial images of what most hunting is all about. There is an urgent need for standardised and regularly updated as well as commonly accessible data across all hunting countries.

References

Anon. (2002) *Big Game Statistics for the 2002/03 Season.* Ministry of Water, Land and Air Protection, Fish and Wildlife Recreation and Allocation Branch, British Columbia http://www.env.gov.bc.ca/fw/wild/documents/hunter_stats_2002.pdf

Barnett, R. & Patterson, C. (2005) Sport Hunting in the Southern African Development Community (SADC) Region: an Overview. TRAFFIC East/Southern Africa, Johannesburg.

Bauer, J. & Giles, J. (2002) *Recreational Hunting: an International Perspective.* CRC for Sustainable Tourism Pty Ltd Griffith University, Gold Coast, Queensland.

Damm, G. (2005) Hunting in South Africa: facts, risks and opportunities. *African Indaba*, 3, 4–5.

Du Wors, E.M., Villeneuve, F.L., Filion, R. *et al.* (1999) *The Importance of Nature to Canadians: Survey Highlights.* Environment Canada, Ottawa.

Federation of Associations for Hunting and Conservation of the EU (FACE) (2007) *Hunting in Europe.* http://www.face-europe.org/fs-hunting.htm

Herring, H. (2006) Hunter, angler, conservationist. *The Nature Conservancy Magazine*, Autumn 2006. http://www.nature.org/magazine/autumn2006/features/art18601.html

Jackson, J. (1996) An international perspective on trophy hunting. In *Tourist Hunting in Tanzania*, eds. N. Leader-Williams, J.A. Kayera & G.L. Overton, pp. 7–11. Occasional Paper of the IUCN Species Survival Commission No. 14. IUCN, Gland & Cambridge.

Kujawski, O.E.J. (2005) *Jagdtrophäen – Gewinnung, Behandlung, Bewertung.* BLV Jagdpraxis. BLV Buchverlag, München.

Lechner, E. (ed.) (1995) *Jagdreisen weltweit.* Fink-Kümmerly + Frey, München.

Leigh, L., Du Wors, E.M., Villeneuve, R. *et al.* (2000) *The Importance of Nature to Canadians: The Economic Significance of Nature-Related Activities.* Environment Canada, Ottawa.

PACEC (Public and Corporate Economic Consultants) (2006) *The Economic and Environmental Impact of Sport Shooting in the UK.* A report prepared on behalf of the British Association for Shooting & Conservation, the Countryside Alliance and the Country Landowners' Association in association with the Game Conservancy Trust. PACEC, Cambridge.

Pinet, J.-M. (1995) *The Hunter in Europe*. http://www.face-europe.org/huntingineurope/ Pinet%20Study/Pinet_study_EN.pdf

US Fish & Wildlife Service *et al.* (2002) *2001 National Survey of Fishing, Hunting and Wildlife-Associated Recreation*. US Department of the Interior, Fish & Wildlife Service and US Department of Commerce, US Census Bureau, Washington, DC.

US Fish & Wildlife Service *et al.* (2007) 2006 National Survey of Fishing, Hunting and Wildlife-Associated Recreation: preliminary results. http://federalaid.fws.gov/ surveys/surveys.html

(3)

Recreational Fisheries: Socioeconomic Importance, Conservation Issues and Management Challenges

Robert Arlinghaus[1] and Steven J. Cooke[2]

[1]Department of Biology and Ecology of Fishes, Leibniz-Institute of Freshwater Ecology and Inland Fisheries, Berlin, Germany & Inland Fisheries Management Laboratory, Faculty of Agriculture and Horticulture, Humboldt-University at Berlin, Germany
[2]Institute of Environmental Science and Department of Biology, Carleton University, Ottawa, Canada

Introduction

Human exploitation of fish and other aquatic animals such as crustaceans and mammals is virtually ubiquitous on Earth and since ancient times has provided humanity with food, income and other social goods such as recreation. Fishing where the primary objective is not to produce food or generate income through the sale or trade of fishing products is commonly termed 'recreational fishing' or 'angling' (Arlinghaus *et al.*, 2007). In most developed or industrialised societies of the temperate regions, recreational fisheries have long represented the major use of aquatic wildlife, thus constituting the dominant fishing activity in limnetic surface waters (Arlinghaus *et al.*, 2002). Furthermore, the importance of recreational fishing in many coastal areas (Coleman *et al.*, 2004) and less developed nations (Cowx, 2002) is increasing rapidly.

Recreational Hunting, Conservation and Rural Livelihoods: Science and Practice, 1st edition.
Edited by B. Dickson, J. Hutton and W.M. Adams. © 2009 Blackwell Publishing,
ISBN 978-1-4051-6785-7 (pb) and 978-1-4051-9142-5 (hb).

This paper treats recreational fishing as the aquatic complement to the recreational hunting of terrestrial organisms (primarily birds and mammals). Its objectives are to: (i) present a universal definition of recreational fishing and common capture methods; (ii) report the global magnitude of recreational fishing participation; (iii) identify conservation issues in recreational fishing; and (iv) discuss the management challenges in reconciling fish resource use and conservation in recreational fishing. This chapter discusses both freshwater and marine environments but, due to space limitations and the diversity of conditions in different parts of the world, only selected examples are provided.

Definitions

Definitions of recreational fishing vary depending on the origin and the cultural perception of the activity (Aas, 2002). Many of the definitions proposed by researchers and international fisheries bodies (e.g., Food and Agricultural Organization of the United Nations (FAO), 1997; Aas, 2002; Pitcher & Hollingworth, 2002) are not suitable to describe all forms of recreational fishing. For example, most of the definitions of recreational fishing make reference to the motives of recreational fishers such as fishing for 'enjoyment' (Policansky, 2002), 'sport' (FAO, 1997) or 'fun' (Pitcher & Hollingworth, 2002). However, the motives of recreational fishers are diverse and differ from one person to another (Fedler & Ditton, 1994). Hence, specific motives such as 'sport' are unsuitable for defining recreational fishing as they do not embrace the attitudes and value systems of *all* recreational fishers.

We propose that recreational fishing can be defined generically by focusing on the most basic and essential level of human needs, i.e. physiological needs essential for survival, such as nutrition. If fishing contributes substantially to meeting an essential physiological need and if that physiological need cannot be easily met by a substitute activity to fishing, it is not recreational. The following definition for recreational fishing emerges:

> Recreational fishing is fishing for aquatic animals that do not constitute the individual's primary resource to meet essential physiological needs.

This definition is sufficiently broad to include other animals beyond fish, such as lobsters and crabs, avoids referring to individual motives, does not discriminate against particular methods of fish capture (e.g. angling, gill netting), does not preclude the catch being taken for personal consumption (as long as the catch is not the primary resource to meet essential physiological needs), and does not discriminate against non-Western cultures (cf. Aas, 2002), but does distinguish commercial and purely subsistence fishing from recreational fishing. It does this because the purpose of commercial and subsistence fishing is to generate products for sale, trade or immediate consumption in order to meet primary physiological needs. Further, an employee of a company that offers guidance to tourist anglers would not be a recreational fisher, even if that person uses the same tackle, at the same spot and at the same time of the day as the tourist.

In the definition, the term 'primary resource' is important because there are recreational anglers who sell part of the catch to offset the costs of fishing (Cowx, 2002) or who trade fish informally with friends in return for other services. There are also many recreational anglers in Western societies who complement household diets with fish (Arlinghaus, 2004), and thus contribute to meeting basic physiological needs. But they do not depend primarily on the catch to meet essential physiological needs. Those needs can be met by activities other than fishing.

In Western culture where there is a clearer demarcation between working time and free time, and fishing products are usually exchanged on domestic or export markets, recreational fishing can be more succinctly defined in the following way:

> Recreational fishing is fishing for aquatic animals that are not traded on domestic or export markets.

From the perspective of the individual angler, recreational fishing is therefore non-commercial fishing, but there are of course many commercial activities, such as the gear industry, that result from and are dependent on recreational fishing.

Capture methods

Recreational fishing is today predominantly conducted by angling methods, i.e. hook and line fishing. Recreational angling methods include hand line fishing, pole fishing (without a reel) or the standard fishing by rod, line and reel. Hence, in Western societies of the temperate regions 'recreational fishing' is typically used synonymously for 'angling' (Arlinghaus *et al.*, 2002, 2007), simply because non-angling fishing methods, such as gill nets, are used only locally, e.g. in the Nordic countries.

The principle of capturing fish by angling is simple. The aquatic animal to be caught finds itself unable to get rid of a hooking device (e.g. a baited hook or a bait combined with a twine or a spider web) such that the animal can be lifted out of its element by means of a line. Angling typically aims at the mouth region of fishes. However, there is also a less common (and sometimes illegal) angling that purposely hits the fish along the body axis with a hook (typically called 'snagging'). Other capture methods employed in recreational fishing include spear fishing, bow fishing, rifle fishing, hand fishing, i.e. doodling, netting (fyke, gill), other forms of trapping fish etc. (Figure 3.1), but they are overall less common than angling techniques and only of local importance.

Figure 3.1 **Recreational fishing techniques. (a) Spear fishing for snapper around patch reefs in the Caribbean (Credit: Andy Danylchuk). (b) Bow fishing for common carp in the US (Credit: Ohio Division of Wildlife). (c) Fly fishing for bonefish in the Bahamas with fishing guide in foreground (Credit: Steven Cooke). (d) Hand-lining for cichlids in Malawi (Credit: Setsuko Nakayama). (e) Seventy-five teams of anglers in Ontario about to begin a two day competitive angling event (i.e. tournament) for smallmouth bass (Credit: Steven Cooke). (f) Mako shark *Isurus oxyrinchus* captured aboard a marine recreational fishing charter off Massachusetts, USA (Credit: Greg Skomal).**

Figure 3.1   **Continued.**

Participation and socioeconomic benefits

Providing global estimates of participation in recreational fishing is difficult due to the lack of representative data from many regions of the world. Moreover, each of the available studies uses different sampling frames or defines recreational fishing differently. The best available data compiled here nevertheless suggest some remarkable trends in participation. For example, hotspots of recreational fishing participation are found in relatively sparsely populated but freshwater- or coastline-rich regions, such as the Scandinavian countries, Australia and North America (Figure 3.2). Recreational fishing participation in each of these 'participation hotspots' comprises more than 10 per cent of the adult population, with record values of close to 50 per cent recreational fishing participation found in the adult population of

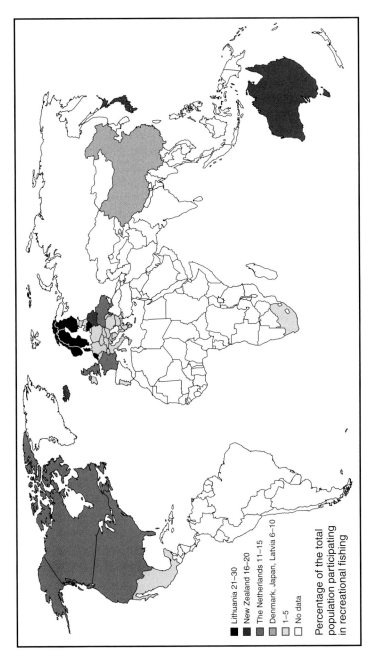

Figure 3.2   Global overview of percentage of the total population in different countries participating in recreational fishing. White fields indicate lack of data. The best available information was used, but there are problems with coverage and sampling validity in the different studies used to extract participation figures.

Edited by B. Dickson, J. Hutton and W.M. Adams. ISBN 978-1-4051-6785-7 (pb) and 978-1-4051-9142-5 (hb).

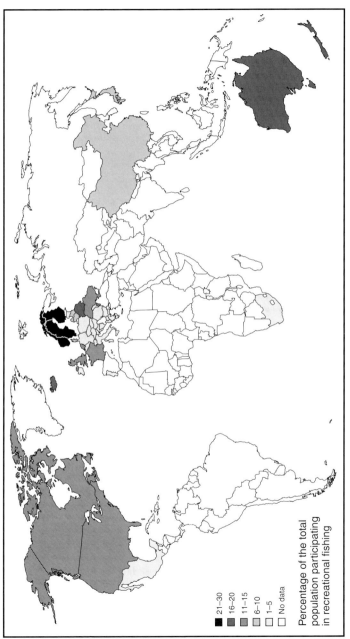

Figure 3.2 Global overview of percentage of the total population in different countries participating in recreational fishing. White fields indicate lack of data. The best available information was used, but there are problems with coverage and sampling validity in the different studies used to extract participation figures.

Norway. If available data are related to the total population, values of reported participation in recreational fishing range up to 32 per cent in Norway with a mean value of 10.6 per cent of the adult population. Cooke & Cowx (2004) extrapolated from Canadian values that 0.73 billion people can be expected to fish recreationally worldwide (11.5 per cent of the global population). The low levels of participation in certain countries that are reported in the literature are likely to be the result of weak data rather than low participation overall. For example, recreational fishing participation is often reported to be low in southern Europe and many Eastern European countries, although it is well known that substantial number of people participate in this type of fishing, maybe more than in Central European countries.

The socioeconomic and ecological benefits generated by recreational fishing were comprehensively reviewed by Weithman (1999) and Arlinghaus *et al.* (2002), and are therefore only briefly mentioned here. These benefits include not only an increased quality of life for the recreational angler but also income accrued at local, regional and national levels in fishing-expenditure-dependent commercial activities (e.g. tackle shops). In North America, recreational anglers directly support fisheries management, conservation efforts and outdoor recreation opportunities through excise taxes and the purchase of licences, stamps, and equipment registration fees. Moreover, activity by recreational anglers can provide resources for the rural development of coastal areas of less developed nations. Kearney (1999) suggested that the conservation-conscious recreational fishing community represents one of the greatest potential forces for the conservation of aquatic biodiversity. Indeed, in some countries, fishery stakeholders such as angling associations have pushed governments to formulate environmental legislation and were the driving forces for several subsequent legal revisions (Kirchhofer, 2002). A Water Protection Act initiated by fishery associations in Switzerland in 1955 served as an important milestone in freshwater conservation. There are also examples of anglers´ conservation associations (e.g. the Anglers´ Conservation Association formed in 1948 in the UK) that have fought legal actions against environmental pollution (Bate, 2001). However, not all measures adopted by traditional recreational fisheries managers are beneficial to the conservation of fish populations, as will be shown below. Nevertheless, a relatively high proportion of society keeps in contact with nature through linkages with recreational fisheries and consequently tends to be more sensitive to environmental issues than the majority of an increasingly urban population. This awareness is central for an ecosystem approach to fisheries management and sustainability (Arlinghaus, 2006a).

Conservation issues

Recreational fishing is facing a number of conservation issues, many of which were reviewed recently in Cooke & Cowx (2006) and Lewin *et al.* (2006). A brief summary of these global syntheses follows.

High exploitation

Excessive exploitation of animal populations though elevated anthropogenic mortality represents one of the most prominent conservation issues shared by hunting and fishing (Mace & Reynolds, 2001). Recreational fishing has only recently been recognised as a significant contributor to global fish declines, particularly on inland waters (Cooke & Cowx, 2004, 2006), and some clear examples have now been presented (Post *et al.*, 2002; Coleman *et al.*, 2004). In recreational fishing, annual exploitation rates up to 80 per cent of the average standing stock have been reported, but there is huge variation from fishery to fishery, ranging from close to zero to very high values (Lewin *et al.*, 2006). As long as the level of fishing is within the compensatory potential of the exploited species (for example, enhancement of growth and fecundity in response to human-induced population declines), extinction is very unlikely. There are no obvious examples where recreational fishing alone has led to the extinction of a species (Arlinghaus *et al.*, 2002). However, this last statement does not preclude the potential for population declines or changes in structure of the exploited species (Lewin *et al.*, 2006). The most important factors determining whether fishing causes a decline of fish populations are the life history of the species and the amount of angling effort. An impact is more likely if the exploited species has low fecundity and high age at maturation and is easily catchable due to the species' association with easily identifiable habitat structures (e.g. submerged macrophytes) (Post *et al.*, 2002).

Selective harvest and trophy issues

Recreational fishing is usually selective with respect to species, size, ages, sex or behavioural and physiological traits (Lewin *et al.*, 2006; Cooke *et al.*, 2007). Selective exploitation results from factors such as the species preferences of anglers, fishing regulations that allow harvest of particular sizes or species, and the differential vulnerability of specific phenotypes (for example, faster growing fish, larger fish or bolder individuals) to the angling gear. Within some recreational fishing communities, there is also much emphasis placed on 'trophies', which are selectively targeted and may be removed from the system. Trophies are typically individuals that are exceptionally large, or have other phenotypic characteristics that make them attractive to the angler. Concern exists that harvest (or indirect mortality) of trophy, or more generally large, individuals (Birkeland & Dayton, 2005) can have demographic or evolutionary effects on fish and wildlife populations (Jørgensen *et al.*, 2007). In recreational fishing it is difficult to target a trophy, although there are many tourist operators that focus on providing anglers with the opportunity to fish for trophies. Populations may be managed to produce trophies or may do so intrinsically (for example, due to local climate and productivity). Unfortunately, many of the fish species that are currently rare and considered trophies also have life-history characteristics such as late age at maturity and low fecundity (Reynolds *et al.*, 2001) that make them vulnerable to exploitation.

In addition to ecological changes such as truncation in age and size structure (Lewin *et al.*, 2006), high and selective exploitation may result in evolutionary responses in the exploited population if some phenotypes experience higher mortality than others and thus are less able to reproduce (Jørgensen *et al.*, 2007). Although noted for many years, the potential for fishing- or hunting-induced evolutionary changes has not been universally appreciated by fisheries and wildlife managers. This is in part due to the difficulty of providing conclusive evidence of this phenomenon in wild stocks because of phenotypic plasticity in life history or other adaptive traits and the associated difficulty in ascribing observed phenotypic changes to fishing-induced evolution. However, the prerequisites for evolutionary changes in fish populations in response to recreational fishing exist. These prerequisites include local adaptation, heritable population variation and a high and selective fishing mortality (Lewin *et al.*, 2006), and the evidence that fishing might result in evolutionary changes is mounting (Jørgensen *et al.*, 2007). Cooke *et al.* (2007) found that largemouth bass *Micropterus salmoides* differed in vulnerability to angling; individuals of high angling vulnerability differed from individuals with low vulnerability in a number of physiological and behavioural traits indicating the selective effect of angling on these traits. The notion that fishing can be selective should be more fully embraced by fish and wildlife managers in the future.

Harvest or disturbance during the reproductive period

Many recreational fisheries intentionally or accidentally target fish during the reproductive period. This can have negative consequences on individual fitness, reproductive success and recruitment (Cooke *et al.*, 2002). For example, Atlantic salmon *Salmo salar* are targeted during their upriver spawning migrations. Although many fish are released, warm water temperatures can lead to high levels of catch-and-release mortality (Wilkie *et al.*, 1996). Smallmouth bass *Micropterus dolomieu* and largemouth bass provide sole paternal care during which time the parental male must fan the developing offspring in their nest and defend the offspring from potential brood predators. Removal of parental males leads to nest predation and increased rates of nest abandonment (Suski *et al.*, 2003). Catching and releasing fish shortly before reproduction can also lead to reduced reproductive output and larval size-at-hatch (Ostrand *et al.*, 2004), which is why many jurisdictions have implemented close seasons to protect fish during the reproductive period.

Sub-lethal effects, wounding, and the one that 'got away'

In recreational fishing, there are many fish that are landed and released with the assumption that they will survive. This regulatory or voluntary catch-and-release can involve billions of fish world-wide (Cooke & Cowx, 2004; Arlinghaus *et al.*, 2007). However, given that post-release mortality rates range from about zero to near 90 per cent (Muoneke & Childress, 1994; Bartholomew & Bohnsack, 2005), the assumption that the released fish survive is not always valid. Nevertheless, anglers can make decisions on gear type (such as hook type, bait/lure or line strength, sometimes with direction in the form of regulation), which can influence the level of fishing-induced stress and injury. For example, organic or live baits tend to result in fish being hooked more deeply than artificial lures (Arlinghaus *et al.*, 2008). In addition, the use of barbless hooks can reduce the time needed to remove the fish from the hook and thus reduce harmful air exposure, and the use of circle hooks can minimise deep hooking (Cooke & Suski, 2005).

In fishing, there are always some fish that 'get away'. These are called drop-offs or break-offs. In drop-offs, the fish comes off the hook(s) prior to being landed. Break-off implies that the line was broken and that the hook(s) and bait or lure are left in the fish. It is sometimes assumed that breaking off will lead to reductions in growth (and potentially mortality) if foraging is retarded. However, there are also studies that showed that hooks can be evacuated if left inside the fish (Tsuboi *et al.*, 2006). Overall, there is considerable evidence that mortality of caught-and-released fish and the associated sub-lethal impacts can be low or greatly reduced if the fish is appropriately handled and not released under unfavourable conditions (e.g., at very high water temperatures, or after being captured from deep water; Arlinghaus *et al.*, 2007).

Pollution and environmental disturbance

Recent syntheses have revealed that recreational fishing can lead to environmental pollution and disturbance in a number of ways (Cooke & Cowx, 2006; Lewin *et al.*, 2006). For example, recreational anglers use boats to access inland and coastal habitats. Powerboats can cause the degradation of localised habitats, particularly in near-shore and inland environments, through wave-induced erosion of shorelines and increased suspension of sediment; they may also disturb both fish and wildlife (Wolter & Arlinghaus, 2003). In addition, angling can disturb wildlife (especially birds) if access to waters or shoreline is uncontrolled (Cryer *et al.*, 1987). Trampling of vegetation can cause alterations of habitat and negatively impact on invertebrates (Mueller *et al.*, 2003).

Fishing activities generate litter, leading to localised habitat alteration. Litter such as fishing line and hooks can become entangled in various wildlife species, which can result in injury or mortality (e.g., Nemoz *et al.*, 2004). However, the most contentious issue is the deposition of lead from fishing sinkers, which can create localised pollution and, if ingested (particularly by waterfowl), may result in lead poisoning and death (Scheuhammer *et al.*, 2003).

In addition to inadvertent pollution, there is also 'intentional' pollution arising from using bait to attract animals. In fishing, ground-baiting (with cereals, maggots or other bait) or chumming, the process of distributing bait in water to attract fish, is common in both freshwater and marine environments. The excess nutrients from this activity (Arlinghaus & Mehner, 2003; Niesar *et al.*, 2004) can lead to a deterioration in water quality and a reduction in benthic fauna (Cryer & Edwards, 1987).

Supplementation, stocking and introductions

To support recreational fishing, many natural resource agencies supplement or stock endemic populations with additional animals that are either transferred from other locales or have been raised in captivity. Although an important management tool, stocking itself may cause conservation problems. For example, largemouth bass are widely stocked to enhance fishing opportunities in the United States. However, there is a growing body of evidence to suggest that fish stocking can cause problems for the recipient population (Molony *et al.*, 2003) and the broader ecosystem (Holmlund & Hammer, 2004), particularly if the stocked fish are not locally adapted to their new environment (Philipp *et al.*, 2002). Outbreeding depression can arise in the recipient population leading to reduced growth rates, problems with immunocompetancy, and mortality (Philipp *et al.*, 2002; Goldberg *et al.*, 2005). In addition, captive-bred animals may be inferior to their wild conspecifics leading to changes in the recipient population (Molony *et al.*, 2003).

Many species of fish have been introduced outside their endemic range in an effort to create new recreational angling opportunities. Rainbow trout *Oncorhynchus mykiss*, carp *Cyprinus carpio* and largemouth bass are three of the most widely introduced species. In some cases they have caused notable changes in community structure (reviewed in Cambray, 2003). Stocking and particularly introductions are among the most abused management tools in contemporary wildlife and fisheries management. Thorough risk analyses should be carried out before these practices are implemented (Arlinghaus *et al.*, 2002).

Management challenges

The conservation issues identified above, coupled with the great popularity of recreational fishing, call for improved management to address the most contentious issues. Below, we present an abbreviated list of the most important management challenges faced by contemporary recreational fisheries stakeholders.

Declining participation: A common problem for recreational fishing in some countries is declining license sales (and presumably participation rate), particularly in North America (Fedler & Ditton, 2001). These declines are the result of demographic change and an increasing urban population in which rural lifestyles and activities such as hunting and fishing are becoming less popular (Arlinghaus, 2006b). Considering that many of the proponents of angling and hunting are now from urban rather than rural environments (Franklin, 1998), there is interest in enhancing opportunities for urban fishing (Arlinghaus & Mehner, 2004) in an attempt to reverse the decline.

Stakeholder conflicts: Fishing requires space and interacts with wild living organisms. At times, anglers and others engaged in recreation occupy the same space, generating conflict intrasectorally (Arlinghaus, 2005). However, one of the greatest sources of conflict in the future is likely to be fish welfare and the more fundamental and ideologically driven animal rights movement (Arlinghaus, 2005; Arlinghaus *et al.*, 2007). For these stakeholders, hunting a wild animal for the sake of recreation is typically perceived as a cruel activity that should stop. This perspective conflicts with the value and belief system of many recreational anglers (Arlinghaus, 2005). There are jurisdictions such as Germany where the only accepted reason for fishing for recreation is the appropriation of food, and practices such as the use of live bait, competitive fishing or voluntary catch-and-release are not tolerated (Arlinghaus, 2007). Because of the similarities between hunting and fishing, we suggest that the way forward is to enhance the interaction between anglers, hunters and fish and wildlife managers in order to find ways to address animal welfare concerns in practice.

Controlling effort and harvest: In order to address the conservation issues identified in this paper, it is necessary to control or limit fish mortality. Although both fish and wildlife managers make extensive use of harvest regulations such as bag limits, the actual effect of these traditional harvest regulations is debatable (Radomski *et al.*, 2001). Most regulations focus on controlling actual harvest or other sources of fish mortality through the use of size-based harvest limits and daily bag limits. This does not necessarily reduce total annual fish mortality as overall effort is not controlled. Nevertheless, there is some potential for self-regulation through dynamic fish–angler interactions. That is, some anglers will leave a fishery when fish abundance declines, and resume fishing if fish populations rebound. However, the potential for depensatory processes (i.e. a positive relation between the per capita population growth rate and population size at low population sizes rendering the population vulnerable to extinction) limits the self-regulatory capacity of some fish (and angler) populations (Post *et al.*, 2002). There are instances where harvest controls alone have been ineffective in sustaining recreational fishing, resulting in calls to limit entry (and thus directly control effort) (Cox *et al.*, 2002). To control angling effort and harvest recently the use of licence allocation lotteries and implementation of aquatic protected areas have been proposed. In the latter case, all fishing activities are typically prohibited, thus effectively reducing effort (and harvest) to zero. We predict a greater use of both harvest regulations and effort control means in the management of fish and wildlife and that a combination of these tactics will be locally effective depending on environmental conditions and the biology of the species targeted (Carpenter & Brock, 2004).

Compliance: Fundamental to the function of management regulations for fish and wildlife is the need for high levels of compliance. Unfortunately, fishing suffers from less than perfect compliance (Muth & Bowe, 1998; Sullivan, 2002), at times making management strategies falter or fail (Sullivan, 2002). Recent research efforts have focused on identifying a typology of the motivations for poaching which are common to both the hunting and angling sectors. Muth & Bowe (1998) suggest that understanding the motivations behind noncompliance provides opportunities for developing strategies such as education to counter this activity.

Outlook

It is somewhat surprising that fish and wildlife populations continue to suffer overexploitation given that the fundamental principles underlying the regulation of exploitation have been known for many years (Sutherland, 2001). Although the principle of sustainability is entrenched in fishing philosophy (Arlinghaus *et al.*, 2002), difficulties in developing scientifically sound and effective management strategies remain. One major constraint in recreational fishing is the diffuse nature of the activity, which makes it impossible to monitor the status of the exploited population accurately in space and time. However, for many years it was assumed that recreational fishing was a benign phenomenon. This assumption is beginning to be challenged with increasing research conducted on the topic.

Arlinghaus & Cooke (2005) stressed that recreational fishing should be studied in the same way as commercial fishing. However, despite the many parallels that provide the grounds for unified efforts to conserve exploited wildlife (Cooke & Cowx, 2006), it is also important to realise the differences between commercial fishing and recreational fishing and between recreational fishing and recreational hunting. For example, it is virtually impossible to develop monitoring systems for all recreational fisheries worldwide. This suggests the need for a 'rule of thumb' management approach that is framed in terms of active adaptive management and based on 'learning by doing' (Pereira & Hansen, 2003; Arlinghaus, 2006a). However, it is also necessary to realise the social and economic benefits generated by both recreational fishing and hunting and the important capacity of these stakeholders to contribute resources and human capital to successful management and to fish and wildlife conservation. In many cases, recreational fishing can take place with minimal impact on the resources; however, a certain degree of biological impact needs to be accepted as long as this does not affect entire populations. However, this statement should not be misinterpreted to mean that laissez-faire management is the way of the future. In fact, some changes in approaches and management direction are sorely needed. For example, fisheries stakeholders might conclude that due to the largely irreversible alteration of habitat structure and function in most river fisheries in central Europe, enhancement of fish and wildlife populations through stocking, introductions, and translocations is the most appropriate strategy for increasing the sustainability of these activities. Unfortunately, this approach fails to recognise and address the causal agents that are preventing the development of sustainable recreational fish and wildlife populations, and only serves as a band-aid approach. There is growing interest in developing more integrative fish and wildlife management strategies that are adaptive to changing conditions in the long term (Arlinghaus, 2006a). To facilitate such approaches, there is an urgent need to integrate biological and social sciences with the aim of providing insights into the dynamics of the entire social–ecological system of recreational fisheries. A greater incorporation of lessons from social science is necessary in order to understand the human constraints on reconciling resource use and conservation (Arlinghaus, 2004, 2005, 2006a). It remains to be seen whether increased funds and human capital are made available to pursue this route.

Acknowledgements

RA thanks Thomas Klefoth for help with Figure 3.2. SJC acknowledges the financial support of the Canadian Foundation for Innovation, the Ontario Research Foundation, and the Natural Sciences and Engineering Research

Council of Canada. RA acknowledges support through the Adaptfish-project funded by the Leibniz-Community in Germany.

References

Aas, Ø. (2002) The next chapter: multicultural and cross-disciplinary progress in evaluating recreational fisheries. In *Recreational Fisheries: Ecological, Economic and Social Evaluation,* eds. T.J. Pitcher & C.E. Hollingworth, pp. 252–263. Blackwell Science, Oxford.

Arlinghaus, R. (2004) *A Human Dimensions Approach towards Sustainable Recreational Fisheries Management.* Turnshare Ltd., London.

Arlinghaus, R. (2005) A conceptual framework to identify and understand conflicts in recreational fisheries systems, with implications for sustainable management. *Aquatic Resources, Culture and Development,* 1, 145–174.

Arlinghaus, R. (2006a) Overcoming human obstacles to conservation of recreational fishery resources, with emphasis on Europe. *Environmental Conservation,* 33, 46–59.

Arlinghaus, R. (2006b) Understanding recreational angling participation in Germany: preparing for demographic change. *Human Dimensions of Wildlife,* 11, 229–240.

Arlinghaus, R. (2007) Voluntary catch-and-release can generate conflict within the recreational angling community: a qualitative case study of specialised carp, *Cyprinus carpio,* angling in Germany. *Fisheries Management and Ecology,* 14, 161–171.

Arlinghaus, R. & Cooke, S.J. (2005) Global impact of recreational fisheries. *Science,* 307, 1561–1562.

Arlinghaus, R. & Mehner, T. (2003) Socio-economic characterisation of specialised common carp (*Cyprinus carpio* L.) anglers in Germany, and implications for inland fisheries management and eutrophication control. *Fisheries Research,* 61, 19–33.

Arlinghaus, R. & Mehner, T. (2004) A management-oriented comparative analysis of urban and rural anglers living in a metropolis (Berlin, Germany). *Environmental Management,* 33, 331–344.

Arlinghaus, R., Mehner, T. & Cowx, I.G. (2002) Reconciling traditional inland fisheries management with sustainability in industrialized countries, with emphasis on Europe. *Fish and Fisheries,* 3, 261–316.

Arlinghaus, R., Cooke, S.J., Lyman, J. *et al.* (2007) Understanding the complexity of catch-and-release in recreational fishing: an integrative synthesis of global knowledge from historical, philosophical, social, and biological perspectives. *Reviews in Fisheries Science,* 15, 75–167.

Arlinghaus, R., Klefoth, T., Kobler, A. & Cooke, S.J. (2008) Size-selectivity, capture efficiency, injury, handling time and determinants of initial hooking mortality of angled northern pike (*Esox lucius* L.): the influence of bait type and size. *North American Journal of Fisheries Management*, 28, 123–134.

Bartholomew, A. & Bohnsack, J.A. (2005) A review of catch-and-release angling mortality with implications for no-take reserves. *Reviews of Fish Biology and Fisheries*, 15, 129–154.

Bate, R. (2001) *Saving Our Streams: the Role of the Anglers' Conservation Association in Protecting English and Welsh Rivers.* The Institute of Economic Affairs and Profile Books, London.

Birkeland, C. & Dayton, P.K. (2005) The importance in fishery management of leaving the big ones. *Trends in Ecology and Evolution*, 20, 356–358.

Cambray, J.A. (2003) Impact on indigenous species biodiversity caused by the globalisation of alien recreational freshwater fisheries. *Hydrobiologia*, 500, 217–230.

Carpenter, S.R. & Brock, W.A. (2004) Spatial complexity, resilience, and policy diversity: fishing on lake-rich landscapes. *Ecology and Society*, 9, 8. [online journal http://www.ecologyandsociety.org/vol9/iss1/art8]

Coleman, F.C., Figueira, W.F., Ueland, J.S. & Crowder, L.B. (2004) The impact of United States recreational fisheries on marine fish populations. *Science*, 305, 1958–1960.

Cooke, S.J. & Cowx, I.G. (2004) The role of recreational fishing in global fish crises. *BioScience*, 54, 857–859.

Cooke, S.J. & Cowx, I.G. (2006) Contrasting recreational and commercial fishing: searching for common issues to promote unified conservation of fisheries resources and aquatic environments. *Biological Conservation*, 128, 93–108.

Cooke, S.J. & Suski, C.D. (2005) Do we need species-specific guidelines for catch-and-release recreational angling to conserve diverse fishery resources? *Biodiversity and Conservation*, 14, 1195–1209.

Cooke, S.J., Schreer, J.F., Dunmall, K.M. & Philipp, D.P. (2002) Strategies for quantifying sublethal effects of marine catch-and-release angling – insights from novel freshwater applications. *American Fisheries Society Symposium*, 30, 121–134.

Cooke, S.J., Suski, C.D., Ostrand, K.G., Philipp, D.P. & Wahl, D.H. (2007) Physiological and behavioral consequences of long-term artificial selection for vulnerability to recreational angling in a teleost fish. *Physiological and Biochemical Zoology*, 80, 480–490.

Cowx, I.G. (2002) Recreational fishing. In *Handbook of Fish Biology and Fisheries Vol. 2: Fisheries*, eds. P.J.B. Hart & J.D. Reynolds, pp. 367–390. Blackwell Science, Oxford.

Cox, S.P., Beard, T.D. & Walters, C. (2002) Harvest control in open-access sport fisheries: Hot rod or asleep at the reel? *Bulletin of Marine Science*, 70, 749–761.

Cryer, M. & Edwards, R.W. (1987) The impact of angler groundbait on benthic invertebrates and sediment respiration in a shallow eutrophic reservoir. *Environmental Pollution*, 46, 137–150.

Cryer, M., Whittle, G.N. & Williams, R. (1987) The impact of bait collection by anglers on marine intertidal invertebrates. *Biological Conservation*, 42, 83–93.

Food and Agricultural Organization of the United Nations (FAO) (1997) Inland fisheries. *FAO Technical Guidelines for Responsible Fisheries*, 6, 1–38.

Fedler, A.J. & Ditton, R.B. (1994) Understanding angler motivations in fisheries management. *Fisheries*, 19(4), 6–13.

Fedler, A.J. & Ditton, R.B. (2001) Dropping out and dropping in: a study of factors for changing recreational fishing participation. *North American Journal of Fisheries Management*, 21, 283–292.

Franklin, A. (1998) Naturalizing sports: hunting and angling in modern environments. *International Review for the Sociology of Sport*, 33, 355–366.

Goldberg, T.L., Grant, E.C., Inendino, K.R., Kassler, T.W., Claussen, J.E. & Philipp, D.P. (2005) Increased infectious diseases susceptibility resulting from outbreeding depression. *Conservation Biology*, 19, 455–462.

Holmlund, C.M. & Hammer, M. (2004) Effects of fish stocking on ecosystem services: an overview and case study using the Stockholm archipelago. *Environmental Management*, 33, 799–820.

Jørgensen, C., Enberg, K., Dunlop, E.S. *et al.* (2007) Managing evolving fish stocks. *Science*, 318, 1247–1248.

Kearney, B. (1999) Evaluating recreational fishing: managing perceptions and/or reality. In *Evaluating the Benefits of Recreational Fisheries*, ed. T.J. Pitcher, pp. 9–14. The Fisheries Centre, Vancouver.

Kirchhofer, A. (2002) The role of legislation, institutions and policy making in fish conservation in Switzerland: past, present and future challenges. In *Conservation of Freshwater Fish: Options for the Future*, eds. M.J. Collares-Pereira, I.G. Cowx & M.M. Coelho, pp. 389–401. Blackwell Science, Oxford.

Lewin, W.C., Arlinghaus, R. & Mehner, T. (2006) Documented and potential biological impacts of recreational angling: insights for conservation and management. *Reviews in Fisheries Science*, 14, 305–367.

Mace, G.M. & Reynolds, J.D. (2001) Exploitation as a conservation issue. In *Conservation of Exploited Species*, eds. J.D. Reynolds, G.M. Mace, K.H. Redford & J.G. Robinson, pp. 4–15. Cambridge University Press, Cambridge.

Molony, B.W., Lenanton, R., Jackson, G. & Norriss, J. (2003) Stock enhancement as a fisheries management tool. *Reviews in Fish Biology and Fisheries*, 13, 409–432.

Mueller, Z., Jakab, T., Toth, A. *et al.* (2003) Effects of sports fisherman activities on dragonfly assemblages on a Hungarian river floodplain. *Biodiversity and Conservation*, 12, 167–179.

Muoneke, M.I. & Childress, W.M. (1994) Hooking mortality: a review for recreational fisheries. *Reviews in Fisheries Science*, 2, 123–156.

Muth, R.M. & Bowe, J.F. (1998) Illegal harvest of renewable natural resources in North America: toward a typology of the motivations for poaching. *Society and Natural Resources*, 11, 9–24.

Nemoz, M., Cadi, A. & Thienpont, S. (2004) Effects of recreational fishing on survival in an *Emys orbicularis* population. *Biologia*, 59, 185–189.

Niesar, M., Arlinghaus, R., Rennert, B. & Mehner, T. (2004) Coupling insights from a carp (*Cyprinus carpio* L.) angler survey with feeding experiments to evaluate composition, quality, and phosphorus input of groundbait in coarse fishing. *Fisheries Management and Ecology*, 11, 225–235.

Ostrand, K.G., Cooke, S.J. & Wahl, D.H. (2004) Effects of stress on largemouth bass reproduction. *North American Journal of Fisheries Management*, 24, 1038–1045.

Pereira, D.L. & Hansen, M.J. (2003) A perspective on challenges to recreational fisheries management: summary of the symposium on active management of recreational fisheries. *North American Journal of Fisheries Management*, 23, 1276–1282.

Philipp, D.P., Claussen, J.E., Kassler, T.W. & Epifanio, J.M. (2002) Mixing stocks of largemouth bass reduces fitness through outbreeding depression. *American Fisheries Society Symposium*, 31, 349–363.

Pitcher, T.J. & Hollingworth, C.E. (eds.) (2002) *Recreational Fisheries: Ecological, Economic and Social Evaluation.* Blackwell Science, Oxford, pp. 1–16.

Policansky, D. (2002) Catch-and-release recreational fishing: a historical perspective. In *Recreational Fisheries: Ecological, Economic and Social Evaluation*, eds. T.J. Pitcher & C.E. Hollingworth, pp. 74–94. Blackwell Science, Oxford.

Post, J.R., Sullivan, M., Cox, S. *et al.* (2002) Canada's recreational fisheries: the invisible collapse? *Fisheries*, 27(1), 6–15.

Radomski, P.J., Grant, G.C., Jacobson, P.C. & Cook, M.F. (2001) Visions for recreational fishing regulations. *Fisheries*, 26(5), 7–18.

Reynolds, J.D., Jennings, S. & Dulvy, N.K. (2001) Life histories of fishes and population responses to exploitation. In *Conservation of Exploited Species*, eds. J.D. Reynolds, G.M. Mace, K.H. Redford, & J.G. Robinson, pp. 148–168. Cambridge University Press, Cambridge.

Scheuhammer, A.M., Money, S.L., Kirk, D.A. & Donaldson, G. (2003) Lead fishing sinkers and jigs in Canada: review of their use patterns and toxic impacts on wildlife. Technical Report No.: CW69–1/108E. Canadian Wildlife Service, Ottawa.

Sullivan, M.G. (2002) Illegal angling harvest of walleyes protected by length limits in Alberta. *North American Journal of Fisheries Management*, 22, 1053–1063.

Suski, C.D., Svec, J.H., Ludden, J.B., Phelan, F.J.S. & Philipp, D.P. (2003) The effect of catch-and-release angling on the parental care behaviour of male smallmouth bass. *Transactions of the American Fisheries Society*, 132, 210–218.

Sutherland, W.J. (2001) Sustainable exploitation: a review of principles and methods. *Wildlife Biology*, 7, 131–140.

Tsuboi, J., Morita, K. & Ikeda, H. (2006) Fate of deep-hooked white spotted charr after cutting the line in a catch-and-release fishery. *Fisheries Research*, 79, 226–230.

Weithman, A.S. (1999) Socioeconomic benefits of fisheries. In *Inland Fisheries Management in North America*, 2nd edn., eds. C.C. Kohler & W.A. Hubert, pp. 193–213. American Fisheries Society, Bethesda, Maryland.

Wilkie, M.P., Davidson, K., Brobbel, M.A. *et al.* (1996) Physiology and survival of wild Atlantic salmon following angling in warm summer waters. *Transactions of the American Fisheries Society*, 125, 572–580.

Wolter, C. & Arlinghaus, R. (2003) Navigation impacts on freshwater fish assemblages: the ecological relevance of swimming performance. *Reviews in Fish Biology and Fisheries*, 13, 63–89.

The Ethics of Recreational Hunting

Barney Dickson

UNEP-World Conservation Monitoring Centre, Cambridge, UK

Introduction

In many developed countries recreational hunting is a controversial and politically contested issue. Campaigning against such hunting is a central activity for some well established organisations. In Britain this includes the League Against Cruel Sports (http://www.league.org.uk) and in the United States, the Humane Society of the United States (http://www.hsus.org/). One reason for the fierceness of the debate is that the arguments about recreational hunting are embroiled in disputes about other types of hunting and in broader arguments about our moral relationship with animals and with nature. Indeed, some disputes about recreational hunting also touch on issues of race, class and gender. Recreational hunting serves as a lightning rod for a host of different concerns.

Some of the disputes about recreational hunting are broadly empirical. What *are* the attitudes of hunters? How much does a hunted animal suffer in the course of the hunt? Does recreational hunting generate benefits for conservation? But other disputes turn on moral questions, often of a quite fundamental nature. Conflicting stances on hunting may be anchored in different starting positions on such questions as which moral concepts are central and which entities have moral significance.

Recreational Hunting, Conservation and Rural Livelihoods: Science and Practice, 1st edition.
Edited by B. Dickson, J. Hutton and W.M. Adams. © 2009 Blackwell Publishing,
ISBN 978-1-4051-6785-7 (pb) and 978-1-4051-9142-5 (hb).

This chapter offers an overview of some the main arguments that have been advanced for or against recreational hunting. The emphasis is on setting out the underlying moral basis of each argument. This leaves little space to discuss some of the specific issues that may arise regarding particular aspects and types of recreational hunting. The attempt to describe the philosophical roots of different positions is speculative in some cases, but not in others. Some of the arguments put forward, especially by those opposed to recreational hunting, have been informed by the work of professional philosophers. The chapter begins by setting out the two most well known and systematic attempts to describe our moral obligations to animals – the animal welfare position of Peter Singer and the defence of animal rights by Tom Regan. While the ethics of recreational hunting are not the main concern of either theorist, they both maintain that their account implies that recreational hunting is wrong. Following that, a very different sort of objection that focuses on the attitudes and character of hunters is considered. This leads naturally into a consideration of views that hold that the attitudes of hunters provide the grounds for a defence of recreational hunting. The final position that is considered, and one that connects most closely to the concerns of many other chapters in the book, is the claim that recreational hunting practices can be justified by the conservation gains they deliver.

Since there is a lack of common ground between the antagonists, disputes about recreational hunting sometimes generate considerable heat, but little detailed argument. The concluding section offers some reflections on this. It is suggested that many of the positions outlined here are too simple. Each of them focuses on one morally relevant feature of hunting, but ignores a range of other considerations. What is needed is a way of thinking about recreational hunting that does justice to the moral complexity of the issue.

Animal welfare

Peter Singer is an academic philosopher well known for his attack, on animal welfare grounds, on practices such as factory farming and animal experimentation (Singer, 1993, 1995). He does so on the basis of a type of utilitarianism and he follows Jeremy Bentham, one of the founders of utilitarianism, in thinking that this moral theory requires us to take account of the welfare of animals as well as of humans. Singer holds that his utilitarianism entails that recreational hunting is wrong.

The most basic notion within Singer's theory is that of 'having an interest'. The boundaries of moral considerability are to be set by determining which creatures have interests. His view is that the capacity to feel pleasure and pain – what is often but misleadingly referred to as sentience – is both a necessary and a sufficient condition of having interests. If a creature cannot feel pleasure or pain then it has no interests, whereas if it is sentient then it has, at the very least, an interest in avoiding pain. Some creatures have other interests that, Singer thinks, can be understood in terms of having their preferences satisfied. Nevertheless, it is sentience that sets the boundaries of moral considerability. All and only sentient creatures count ethically. Since many non-human animals are sentient in this sense, they count, morally. Thus Singer expands the moral circle beyond its traditional boundary, to include animals other than humans.

As a utilitarian Singer thinks that the fundamental criterion of right action is the maximisation of good consequences. He holds that good consequences are to be understood in terms of satisfying peoples' interests and that therefore, at least at some level of moral reasoning, one should aim to maximise the satisfaction of interests. In doing this, we should be guided by the principle of equal consideration of interests. An equivalent interest counts equally, whoever's interest it is. In doing the calculations, the pain of a horse should count as much as the equivalent pain of a human baby.

It is a feature of Singer's criterion of moral considerability that entities such as species and ecosystems do not count, morally, for they are not sentient and so do not have interests. Species and ecosystems do not, themselves, give rise to moral obligations to conserve them. Nevertheless, Singer suggests that we may be required to do what, in fact, does conserve species and ecosystems because that will often be essential to avoid harming those entities – such as the individual members of some species – that do have interests. In his example, of a threatened valley with conservation value, 'when a proposed dam would flood a valley and kill thousands, perhaps millions of sentient creatures, these deaths should be given great importance in any assessment of the costs and benefits of building the dam' (Singer, 1993). So in order to protect those creatures we should prevent the proposed dam and that will have the consequence of conserving the ecosystem as well.

Singer condemns factory farming and much animal experimentation because of the pain suffered by the animals involved, pain that is not outweighed by benefits to others. The fact that many of the animals are also killed plays, at most, only a supplementary part in his argument against these practices. For him what makes these practices wrong is primarily the suffering they cause rather than the deaths that result. While he also condemns recreational hunting he does not himself develop the case against it any detail. The remarks that he does make on the topic suggest that while part of his objection to recreational hunting hinges on the suffering it causes, this argument from suffering may not be as strong as in the case of intensive farming. He asks,

> [w]hy, for instance, is the hunter who shoots a deer for venison subject to more criticism than the person who buys a ham at the supermarket? Overall, it is probably the intensively reared pig who has suffered more. (Singer, 1995)

While some types of hunting may cause more suffering than shooting deer, he seems to recognise that his argument against recreational hunting must turn on the wrongness of killing wild animals.

Singer admits that his utilitarian account of what is wrong with killing animals is more complicated than his account of why it is wrong to cause them to suffer. It makes use of a distinction between creatures that are persons and those that are not. This is not the same as the distinction between humans and non-humans. Rather, a person is defined as a rational and self-conscious being and some non-human animals are persons in this sense, while many others, although they are sentient, are not. He suggests that most mammals, perhaps even all of them, are persons, while birds are probably not. Killing a person will normally be worse than killing a non-person because a person will typically have many future-oriented preferences that will remain unsatisfied if they are a killed, while a non-person, that does not see itself as an entity with a future, will have far fewer such preferences. Nevertheless, killing a sentient creature is still likely to be wrong in so far as its life is pleasant because its death will lead to a loss of the pleasant experiences it would have had in the future. However, the particular version of utilitarianism that Singer espouses entails that if the animal is replaced by another animal leading an equally pleasant life, and certain other conditions are satisfied, killing will not be wrong, as it leaves the total amount of pleasant experiences in the world unchanged.

In the light of this account of the wrongness of killing animals, Singer's theory implies that the recreational hunting of mammals (i.e. animals that are persons) is likely to be wrong. The hunting of sentient animals that are not persons, such as birds, may also be wrong, if the life of the animal is pleasant and it is not replaced by another animal leading an equally pleasant life. This account of why hunting birds is wrong is vulnerable to the counter-argument that in cases where recreational hunting practices do contribute to conservation it can be contended that birds that are killed *are* replaced by other birds leading equally pleasant lives. Indeed, for Singer, in all cases, whether we are considering animals that are persons or sentient animals that are not persons, the judgements about the wrongness of hunting may be outweighed if the killing leads to more interests being met overall. As with other forms of utilitarianism, the judgements about particular actions (or types of action) are dependent on the overall consequences of those actions.

Animal rights

Tom Regan is another academic philosopher who has elaborated a substantive moral theory that implies that at least some forms of recreational hunting are wrong (Regan, 1984, 2001). His view is that there are certain types of animal that possess rights and that recreational hunting of those animals violates their rights.

There are some broad similarities between Regan's views and Singer's. Like Singer, Regan extends the boundaries of what is morally considerable beyond humans to include some non-human animals, with important implications for our treatment of animals in a range of areas. And like Singer his view is individualistic in the sense that it is our treatment of individual animals that matters morally, and he rejects the idea that holistic entities such as species and ecosystems give rise directly to obligations to protect them. Notoriously, he describes view that one should sacrifice individuals for the sake of these 'wholes' – a view that is, arguably, central to the justification of recreational hunting on the grounds of its contribution to conservation – as environmental fascism (Regan, 1984).

Alongside these similarities, there are some significant differences between Singer and Regan that emerge from the specific nature of their theories. In outline, Regan's argument that certain animals have rights is as follows. First, it is said that some types of animals are 'subjects of a life'. An animal is a subject of a life if it can perceive and remember, if it has desires and preferences, if it is able to act intentionally and if it has a sense of the future. He describes such animals as possessing a unified psychological presence (Regan, 2001). Regan thinks this is true of mammals more than one year old. Second, it is claimed that subjects of a life have equal inherent value. It is then contended that anything with inherent value has an equal right to treatment respectful of that value. Foremost amongst these rights is the right not to be killed. It is concluded that recreational hunting that involves killing such animals – mammals of more than one year old – violates this right and is therefore wrong.

Regan then goes on to argue that we should also treat other animals as if they were rights-holders and so concludes that the hunting of those animals should also be abandoned. These honorary rights-holders include birds, and mammals less than one year old. He gives two reasons for making this extension. First, since there is some doubt as to exactly which animals are subjects of a life, it therefore behoves us to give the benefit of the doubt to those animals that are on the border of subjecthood. Second, by treating these animals as if they are rights-holders we avoid fostering the belief that animals are just resources to be used for human purposes.

Regan is insistent that the alleged benefits of hunting, whether these are taken to be the satisfactions that humans gain from hunting (such as physical exercise or communing with nature) or the overall reduction in animal suffering that may result from the reduction in animal population sizes, do not constitute an adequate defence of the practice. In part this is because either those benefits can be achieved in other ways, or because hunting does not actually deliver them. But more fundamentally, Regan holds that even if hunting was the only way to achieve those benefits, it still would not justify hunters in violating the rights of individual animals.

This points to the fundamental difference with Singer's view. Singer, as a utilitarian, holds that what is right is determined by weighing up the consequences of different courses of action. Thus, if hunting does generate sizeable satisfactions for humans, or significant reductions in animal suffering, then these are facts that ought, in principle, to be taken into account. But for Regan this type of 'weighing up' is utterly misconceived. Generating these sorts of benefits cannot justify violating the rights of individuals. Rights are to be respected even if it leads to an overall reduction in preference satisfaction. In this sense, Regan's theory is more strongly individualist than Singer's. He maintains that rights of the individual cannot be sacrificed to advance the greater good. This is why his theory delivers more clear-cut answers to questions about hunting – at least for the central cases of rights-bearers. Of course, while clarity is a virtue, it may be of little value if it is bought at the expense of a mistaken theory. We will return to this in the final section.

Virtues and vices

There is a very different sort of objection to recreational hunting from those that have just been considered. This objection is quite common and often figures in informal contexts. It is the view that what is wrong with recreational hunting is that it is done for fun. Roy Hattersley expresses this view in connection with fox-hunting. This is a practice in which trained dogs pursue foxes, followed by human hunters usually on horseback. Hattersley writes:

> I have long supported whoever it was who said that the real objection to fox-hunting is the pleasure that hunters get out of it... If killing foxes is necessary for the safety and survival of other species, I – and several million others – will vote for it to continue. But the slaughter ought not to be for fun.   (Quoted in Scruton, 1996)

This objection has been subject to much less careful elaboration than the arguments based on animal welfare or animal rights. It can be understood in different ways. On one view, the objector already has reasons for thinking that recreational hunting is wrong (perhaps of the sort advanced by Singer or Regan) and is simply indicating that undertaking the activity for fun makes matters worse.

However, Hattersley's formulation suggests that doing it for fun turns what otherwise might be permissible into something that is wrong. On this reading the motive is central to what is wrong with the activity and, in treating the killing of wild animals as a suitable form of recreation, the hunter is responding to those animals in a morally reprehensible way. He or she is displaying a disposition to act and a pattern of emotional responses that should be deplored. The objector might go on to say that the attitudes of respect and reserve towards wild animals would be more appropriate, together with an inclination to avoid interfering in the lives of these animals if possible. If animals are to be killed – and Hattersley indicates that in some circumstances this may be justified – then the appropriate emotion would be one of regret rather than exultation.

To criticise recreational hunting in this way is to take the discussion into the field of virtue ethics. This is a rather different approach from either utilitarianism or rights theory. Instead of focusing on the different amounts of pleasure and pain (or preference satisfaction) caused by possible actions, or on what is required to respect the rights of affected creatures, the emphasis is on assessing how a person responds (both emotionally and practically) to the morally relevant features of the world around them.

However, understanding hunting within the framework of virtue ethics does not necessarily entail a negative assessment of hunting. Roger Scruton, for example, accepts that if those engaged in fox-hunting did indeed take pleasure specifically in the suffering caused to foxes, rather than in the activity of fox-hunting as a whole, that would indeed be vicious (Scruton, 1996). But he contests whether this is the case. In opposition to Hattersley, he presents fox-hunting as displaying the virtues of traditional social solidarity, respect for the hunted animal and concern for the countryside (Scruton, 1998).

Indeed, positive ethical evaluations of recreational hunting based on an evaluation of the attitudes and dispositions manifested in the practice are found in a wide variety of times and places. MacKenzie, in his study of hunting in the British Empire, has shown how, in the latter part of the 19th century, the rules and traditions of big game hunting were consciously elaborated to constitute this as a suitable recreation for the colonial elite (MacKenzie, 1988). From the perspective of this group, sport hunting, with its notion of 'fair chase', was seen as ethically superior to other, more instrumental forms such as commercial hunting and the subsistence hunting practised by many of the Empire's subjects. On this view, and in stark contrast to Hattersley's stance, the non-instrumental nature of big game hunting provides support for a positive rather than a negative judgement of its moral qualities.

The different sides in these various disputes agree that what matter, ethically, are the patterns of action and emotional response displayed by individuals in relation to hunting. That this is a rich soil – perhaps richer than the protagonists realise – is one lesson of Marks's study of recreational hunting in Scotland County, North Carolina (Marks, 1991). He examines a wide range of hunting practices, including the hunting of foxes, deer, quail, doves, rabbits and racoons, each with its own customs, rules and values. Marks is much less interested in passing judgement on these activities than in understanding them. Nevertheless, in the course of his study he shows how many hunters make evaluative judgments about what they are doing, about the way in which their peers conduct themselves and about other types of hunting.

The utilitarian and rights-based critics of recreational hunting are likely to say that this focus on the character of hunters is misplaced. They think that what matters is what happens to the animals themselves. Regan, for example, is quite explicit that in attacking recreational hunting he is not criticising the character of those who engage in the practice. His concern is with the fate of the hunted animal and whether its rights are violated:

> Though those who participate in it need not be cruel or evil people … what they do is wrong… To make a sport of hunting or trapping them [animals] is to do what is wrong because it is to fail to treat them with the respect they are due as a matter of strict justice.   (Regan, 1984)

Some defenders of recreational hunting may share this unease with the assumption that the central question concerns the moral character of the recreational hunters themselves. A common defence of the practice turns not on the dispositions of hunters, but on the positive consequences of recreational hunting for conservation.

Conservation

As with other positions for and against recreational hunting it is possible to make a distinction between the empirical and evaluative parts of the claim that recreational hunting can be justified through its contribution to conservation.

The empirical part of the claim – that recreational hunting does indeed provide benefits to conservation – is one that is considered in many other chapters of this book (for example, Chapters 2, 5 and 6). Typically this claim is made only of specified recreational hunting practices or is qualified by adding the phrase 'in the right circumstances'.

Hunting may benefit conservation either directly or indirectly. There may be a direct benefit in limiting the size the population being hunted. This may be because the increasing numbers are damaging the surrounding habitat; or because the hunted population is itself an alien species that the conservationist would like to see removed altogether. Indirect benefits may result from the income generated from recreational hunting, which can then be invested in conservation. This can happen through different routes. It may be that hunters have to purchase licences or permits from statutory agencies, who then invest the proceeds in conservation. Or it may be that a landowner charges hunters to hunt on his or her land and that this income, in the right circumstances, provides the landowner with an incentive to invest in the conservation of the hunted species.

Of course, these benefits for conservation may not always be realised. There are situations where the economic incentives are such as to favour mining the resource, rather than conserving it (Swanson, 1994), while in other cases there may be more subtle, detrimental effects on populations and evolutionary processes (Milner-Gulland *et al.*, this volume, Chapter 5). Nevertheless, some argue that with proper regulation and management, it is possible to address many of these potentially negative side-effects (Loveridge *et al.*, 2006; Festa-Bianchet & Lee, this volume, Chapter 6).

The empirical claim that recreational hunting benefits the conservation of species and ecosystems only provides a justification for hunting if conservation is itself valuable. This ethical premise is often implicit in the arguments put by defenders of recreational hunting, but it stands in need of justification.

The value of species and ecosystems may be understood instrumentally, in terms of their role in furthering a range of human goods, such as economic development and recreation. Alternatively, species and ecosystems may be understood as possessing intrinsic value that gives rise to ethical obligations to conserve those species and ecosystems. This second view will be briefly explored here.

The nature of intrinsic value has been much discussed within environmental philosophy. J. Baird Callicott endorsed the view, at one time, that only the biotic community as a whole possesses intrinsic value and that individuals only have value in so far as they contribute to that whole. Drawing, as he saw it, on the work of Aldo Leopold, he said that the 'the good of the biotic community is the ultimate measure of moral value, of the rightness or wrongness of actions' (Callicott, 1980). This allowed him to make a straightforward defence of those types of hunting that contribute to conservation by reducing the population size of over-abundant species. Indeed, he thought 'to hunt or kill a white tailed deer in certain districts may not only be a ethically permissible, it might actually be a moral requirement, necessary to protect the local environment, taken as a whole, from the disintegrating effects of cervid population explosion' (Callicott, 1980).

Others, while holding that species and ecosystems may have intrinsic value, deny that they are the sole possessors of such value, and offer various accounts of what that value consists in. Rolston holds that the value of a species is to be understood in terms of its character as a dynamic natural kind with its own more or less distinct integrity (Rolston, 1988). Sober suggests that the value of natural things is to be understood as similar to that that of aesthetic objects, although the exact nature of the parallel is not spelt out (Sober, 1986).

Callicott himself subsequently revised his position to allow that individuals, as well as the biotic community as a whole, had value and generated obligations (Callicott, 1988, 1995). However, none of these more complex positions say very much about how obligations to individuals, including individual wild animals, are to be balanced against obligations to conserve species and ecosystems. Yet an answer to this question is fundamental to the defence of recreational hunting – a practice that involves killing individual animals – on the grounds of its contribution to the conservation of the whole. Thus, the ethical foundation they offer for a defence of recreational hunting on conservationist grounds is incomplete.

Conclusions

Several different ethical arguments that entail conclusions about the recreational hunting have been outlined here. These positions differ in their fundamental ethical assumptions. It can therefore seem that our challenge is to choose between these different starting points. However, a point made by a number of critics – who do not agree amongst themselves – is that at least some of the positions described here (notable those advocated by Singer, Regan and the early Callicott) are too simplistic; starting from an impoverished base, each ignores relevant ethical considerations.

Sorabji describes the theories of Regan and Singer as 'one-dimensional' (Sorabji, 1993); Nussbaum says of Singer's utilitarianism that it is 'homogenising' (Nussbaum, 2004); Anderson argues that three approaches (animal rights, animal welfare and environmentalism) each raises relevant moral considerations, and none on its own recognises all the relevant sources of value (Anderson, 2004); Sober says that the narrow environmentalist view, that holds that only wholes are of value, is just as monolithic as some versions of individualism (Sober, 1986); and Callicott himself criticises his own earlier view as being too narrow (Callicott, 1988, 1994).

If these criticisms are well founded, then there is no necessity to choose one of the positions discussed here. Two respects in which we may need a richer, more differentiated account than offered by any of these positions can be highlighted. The first concerns the question of whether we have direct obligations both to individuals and to ecological wholes such as species and ecosystems. Callicott's original view, that only the biotic community as a whole has value, seems radically implausible and is certainly contrary to many of our ethical intuitions. The converse view, shared by both Singer and Regan, that only individuals have value and that any obligation we have to conserve ecological wholes can be reduced to our obligations to individuals, while having much stronger roots in our ethical traditions, nevertheless gives rise to counter-intuitive results. Thus, as Sorabji puts it, Regan must hold that the last two honey buzzards have no greater entitlement to our protection than a dog (Sorabji, 1993). The individualists do try to argue that we should be concerned with the fate of ecological wholes because their conservation – and this is an empirical claim – is required by our obligations to individual animals. But it is not clear that this empirical claim is well founded and, even if it is, it seems a roundabout and misconceived explanation of why we should conserve species and ecosystems. It is more plausible to suppose that we have obligations both to individuals and to the ecological wholes, even if we still lack a persuasive account of the strength and nature of the latter type of obligation.

The second, connected issue where a more differentiated account is needed concerns the distinction between domesticated and wild animals. The intuitions of many people appear to be that this distinction is an important one. In many societies we are inclined to think that we are obliged to supply domesticated animals with the necessaries of life, such as food and shelter, to protect them from predation, the elements and disease, and to care for them when injured or ill. Indeed, many of these obligations are enshrined in law. But we seem to think that we have a rather different set of obligations to wild animals. There our intuitions are increasingly that, in the light of the value we accord to natural biological, ecological and evolutionary processes, we have an obligation to sustain the conditions in which they can live out their lives, as free as possible from certain sorts of human intervention. Singer and Regan seem to accept the force of these intuitions, and argue that their own theories imply something similar. However it is not clear that this argument is warranted. At root, neither of their theories accord any place to the domesticated/wild distinction. They derive our obligations to animals from the nature of those animals considered in abstraction from their place within or without human society. Both theories place the well-being of individual animals at their centre, which leaves them with the problem of, firstly, how to respond to the undoubted harm that is suffered by many animals in the wild in the normal course of their lives, and secondly, how to account for the value we place on natural processes that include suffering and death for individual wild animals. Their theories seem to imply that we should be intervening to protect these animals and they have to offer *ad hoc* explanations of why this is not actually so.

These reflections, which suggest that we have direct obligations to conserve species and ecosystems as well as to individuals, and that the distinction between domesticated and wild animals is a morally important one, are very far from constituting a complete theory of our moral relations with animals and nature. But they do indicate one direction in which the argument about recreational hunting might go. A defender of recreational hunting might contend, in the light of our obligations to respect wild animals and the ecosystems in which they live out their lives, and in the absence of a strong duty of care to individual wild animals, that recreational hunting may be justified if it contributes to conserving species and broader ecological processes. Conversely, the opponent of recreational hunting may accept similar premises but still argue that recreational hunting itself represents an unwarranted form of intervention in the lives of individual animals and the ecosystems in which they live. In any case, one of the minor virtues of recreational hunting may be that debates about its merits bring into focus fundamental questions about what we value in the natural world and why.

References

Anderson, E. (2004) Animal rights and the values of nonhuman life. In *Animal Rights: Current Debates and New Directions*, eds. C. Sunstein & M. Nussbaum, pp. 277–298. Oxford University Press, Oxford.

Callicott, J.B. (1980) Animal liberation: a triangular affair. *Environmental Ethics*, 2, 311–338.

Callicott, J.B. (1988) Animal liberation and environmental ethics: back together again. *Between the Species*, 5, 163–169.

Callicott, J.B. (1995) Animal liberation: a triangular affair, with Preface (1994). In *Environmental Ethics*, ed. R. Elliott, pp. 29–59. Oxford University Press, Oxford.

Loveridge, A.J., Reynolds, J.C. & Milner-Gulland, E.J. (2006) Does sport hunting benefit conservation? In *Key Topics in Conservation Biology*, eds. D. MacDonald & K. Service. Blackwell, Oxford.

Marks, S.A. (1991) *Southern Hunting in Black and White*. Princeton University Press, Princeton, New Jersey.

MacKenzie, J.M. (1988) *The Empire of Nature: Hunting, Conservation and British Imperialism*. Manchester University Press, Manchester.

Nussbaum, M.C. (2004) Beyond "Compassion and Humanity": Justice for Nonhuman Animals. In *Animal Rights: Current Debates and New Directions*, eds. C. Sunstein & M. Nussbaum, pp. 299–320. Oxford University Press, Oxford.

Regan, T. (1984) *The Case for Animal Rights*. Routledge, London.

Regan, T. (2001) *Defending Animal Rights*. University of Illinois Press, Urbana & Chicago.

Rolston, H. (1988) *Environmental Ethics*. Temple University Press, Philadelphia.

Singer, P. (1993) *Practical Ethics*, 2nd edn. Cambridge University Press, Cambridge.

Singer, P. (1995) *Animal Liberation*, 2nd edn. Pimlico, London.

Sober, E. (1986) Philosophical problems for environmentalism. In *The Preservation of Species*, ed. B.G. Norton, pp. 173–194. Princeton University Press, Princeton, New Jersey.

Sorabji, R. (1993) *Animal Minds and Human Morals*. Duckworth, London.

Scruton, R. (1996) *Animal Rights and Wrongs*. Demos, London.

Scruton, R. (1998) *On Hunting*. Yellow Jersey Press, London.

Swanson, T. (1994) *The International Regulation of Extinction*. Macmillan, Basingstoke.

Science

$$\textbf{5}$$

The Science of Sustainable Hunting

E.J. Milner-Gulland[1], Nils Bunnefeld[1,2] and Gil Proaktor[1]

[1]Conservation Science Research Group, Silwood Park Campus, Imperial College London, UK
[2]Department of Wildlife, Fish and Environmental Studies, Swedish University of Agricultural Sciences, Umeå, Sweden

Introduction

As in any other branch of natural resource exploitation, the science of recreational hunting revolves around sustainability – how can we assess the effect of hunting on populations, and how can we use that assessment to improve management? The underlying science of sustainable hunting is well understood. Much of the development of the field has been carried out within fisheries science (Clark, 1990; Hilborn & Walters, 1991), where the focus has been on the use of sophisticated techniques to model the effects of fishing on stock sizes and hence to make management recommendations for optimising yields over time. The models focus on the interactions between the fished stock, fishing effort (inputs) and catches (outputs), and are necessarily bio-economic. Wildlife managers, by contrast, have fewer problems with estimation of stock sizes and may not be managing to maximise commercial yields; hence their science has historically been much more biologically focused (Sinclair *et al.*, 2005). In both cases, there is often the assumption that there

Recreational Hunting, Conservation and Rural Livelihoods: Science and Practice, 1st edition.
Edited by B. Dickson, J. Hutton and W.M. Adams. © 2009 Blackwell Publishing,
ISBN 978-1-4051-6785-7 (pb) and 978-1-4051-9142-5 (hb).

is a management authority and that it can influence hunting rates and choose sustainable levels of hunting.

For hunted species that are not subject to major commercial harvesting operations, there is often no manager (or at least, no effective manager), so hunting is *de facto* unregulated. Although the species may be of value to individuals for food or sale, there may be limited access to large markets, user rights may be weak or resource productivity may be low and diffuse. Hence there is no incentive for governments, communities or private individuals to control harvesting. An open access harvesting situation ensues that may result in the 'tragedy of the commons' (Hardin, 1968). Sustainability comes about only by default, if hunting happens to be at a low level due to a lack of efficient technology or of a significant market for the off take. As human demand increases, hunting is likely to become unsustainable, leading to conservation problems such as the bushmeat crisis (Robinson & Bennett, 2000; Davis & Brown, 2007). When the species become of conservation concern there may be an attempt to introduce management for sustainability into the system *post hoc*. This could involve the enforcement of laws prohibiting hunting or attempts to develop a legal framework for regulating off take.

Placing recreational hunting into this context, we focus in this chapter on three areas in which recreational hunting may differ from hunting for other uses. We bear in mind that 'recreational hunting' itself is a broad term that covers a range of experiences and motivations (Loveridge *et al.*, 2006; Leader-Williams, this volume, Chapter 1). But in general, recreational hunting:

(**a**) may often be selective for particular individuals, distinguished by some phenotypic trait like big horns or large body size;
(**b**) by definition does not simply have monetary value to the hunter, and hence is not driven solely by the profit motive; and
(**c**) is more likely to be subject to some form of management than non-recreational hunting, whether this be formal or based on hunters' codes of conduct.

The fundamentals of harvesting theory

Harvesting theory is built on the foundation of the logistic growth curve, which describes density-dependent growth of wildlife populations

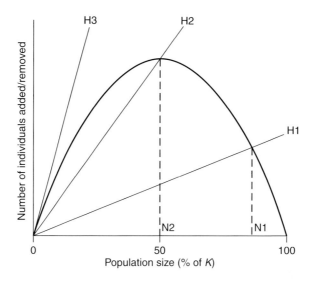

Figure 5.1 **Sustainability of hunting in the standard Schaeffer model (Clark, 1990).**
The parabola represents the number of new individuals added to the population
at a given population size, assuming logistic growth. The curves H1–H3 represent
different hunting rates, assuming that the number killed is a constant proportion
of the population size (e.g. H1 might represent 5% of the population). Any curve
that cuts the population growth parabola represents a sustainable equilibrium, at
which the number added to the population equals the number removed and the
population size is constant. H3, which does not cut the population growth curve, is
an unsustainable harvest rate. H2 represents the rate producing the maximum sus-
tainable yield (MSY). *Note:* **N1 is the population size associated with hunting rate**
H1, and N2 is the population size associated with hunting rate H2.

(Clark, 1990; Begon *et al.*, 1996; Milner-Gulland & Mace, 1998). Sustainable
hunting removes the same number of individuals as are added to the population through natural population growth. Populations grow when they are
below their carrying capacity, because the level of resources allows reproduction or immigration to exceed mortality or emigration rates (Figure 5.1). It
follows from this that, in theory, a population can sustain hunting into the
long term at any size above zero and below carrying capacity, with the sustainable off take peaking at intermediate population sizes. So there is not one
'sustainable' size at which populations can produce a particular sustainable

level of off take; rather, there is a trade-off to be made between population size and off take levels. Both can vary continuously, from zero to carrying capacity in the case of population size, and from zero to the maximum sustainable yield (at half carrying capacity) in the case of off take.

Of course real life is not so simple. Even considering a single species that grows according to the logistic equation, environmental and demographic stochasticity and observation errors mean that off takes and population sizes vary over time. If the hunted population is considered within a realistic ecological setting, things become much more complicated. Biodiversity encompasses the genetic, population and ecosystem levels. Each of these needs to be non-declining in order for sustainability to be achievable, and so it is not adequate to consider sustainability using a model that focuses on only one measure, the total population size. At the genetic level, we need to consider the effect of hunting on the adaptive landscape, and how it affects trait frequency in the population. At the population level, hunting affects different age, sex and stage classes, and there are interactions between hunting and behaviour. At the ecosystem level, we need to consider the effects of hunting on species interacting with the target species, and on the structure and function of the community as a whole (ecosystem services such as pollination, nutrient cycling and water regulation). For example, a study of the effect of bushmeat hunting on tree recruitment in a Bolivian forest showed a highly significant reduction in seedling mortality from trampling in heavily hunted forests that were practically devoid of large mammals (Roldan & Simonetti, 2001).

In practical terms we can only consider the sustainability of recreational hunting in terms of the main effects that it has on biodiversity – the question being to what extent hunting impacts primarily on population size, and to what extent other components of diversity are also significantly affected.

The effect of selectivity on sustainability

Selective hunting of individuals with desirable characteristics, such as large antlers, may have both ecological and evolutionary effects. As an example of ecological effects, in the saiga antelope *Saiga tatarica*, heavily male-biased hunting led to the failure of a substantial proportion of females to conceive, exacerbating an already steep population decline caused by the direct effects of uncontrolled hunting (Milner-Gulland *et al.*, 2003). If animals are targeted according to a desirable characteristic which has a heritable component then evolutionary change may also occur. Selectivity for particular alleles shifts their frequency in the population, and consequently may lead to a loss of genetic diversity. The imposition of additional mortality on one component of life history alters the adaptive landscape, and so may generate evolutionary change. For example, if additional mortality is imposed on adult animals that typically have high survival, we might expect this to bias individuals towards investing in early reproduction (Hirshfield & Tinkle, 1975; Law & Grey, 1989).

Evidence for the secondary effects of selective hunting on population ecology and evolution (over and above the prime effect of reducing population size) is still patchy, but growing (e.g. Jennings *et al.*, 1998; Law, 2000; Kokko *et al.*, 2001; Saether *et al.*, 2003). Evolutionary changes in body size at first reproduction have been documented in some commercial fish species, such that individuals from heavily fished populations begin reproduction at lighter weights than normal (Hutchings, 2005). There is evidence for selection for large horns by recreational hunters causing reduced horn and body size in bighorn sheep *Ovis canadensis* (Festa-Bianchet & Lee, this volume, Chapter 6). The extent to which these effects actually compromise sustainability, and their importance in recreational hunting systems, is as yet unclear. In cases where no effects have been found, it is hard to know whether this is because they are negligible, or because they are long-term and hard to detect, given the short timescales over which management operates.

We present two case studies of the biological effects of recreational hunting which take very different approaches to the problem; the first takes a modelling approach to the effect of hunting on the evolution of components of the life history strategy, the second is an empirical study of the effects of shooting on the population ecology of the target species and its parasite.

Life history trade-offs in red deer hinds

Modelling allows us to formulate hypotheses about the likely long-term evolutionary effects of hunting, which may then be testable using empirical data. A key feature of a female ungulate's reproductive strategy is the weight at which she first gives birth – too light and she risks mortality both for herself and her calf; too heavy and she has missed opportunities to reproduce and enhance her lifetime reproductive success. We developed an individual-based model of a red deer *Cervus elaphus* hind population, parameterised to represent the well studied population of red deer on the island of Rum (Clutton-Brock *et al.*, 1982; Milner-Gulland *et al.*, 2004). The model evolved weight-specific reproductive strategies in a stochastic density-dependent environment. We then imposed selective harvesting on these females, and allowed the population to evolve a new set of reproductive strategies.

Selective harvesting led to a substantial evolutionary response, which varied depending on the weight class of females targeted (Figure 5.2). Without harvesting, the optimal weight at first reproduction was 81 kg (corresponding to an age of three to four years, depending on food resources). Non-selective harvesting at 10 per cent of the population a year led to a lighter mean age at first reproduction (72 kg). This was due to the shift in the trade-off between

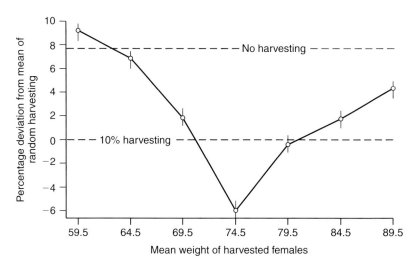

Figure 5.2   **The effect of selective harvesting on the mean weight at first reproduction of a simulated red deer population. The additive effect of selection for heavier or lighter animals is shown as a percentage change from the mean weight at first reproduction of a population harvested unselectively at a rate of 10% per annum. Thus negative values indicate a lighter mean weight at first reproduction, and positive values a heavier weight than is seen when harvest is unselective. We assume that selective harvesting focuses only on a particular weight class; the weight classes tested in the simulations are shown on the *x*-axis. Horizontal lines indicate the outcome for a 10% randomly harvested population and one subjected to no harvesting.**

mortality and reproduction – if the female has a risk of being harvested, it is beneficial to reproduce earlier despite the cost of doing this, to ensure that she has offspring. If harvesting targeted either very light or very heavy females, the evolutionary response of the population was to evolve back towards heavier mean weights at first reproduction; if animals were harvested before they had matured enough to breed or in weight classes that very few animals had reached, the evolutionary effect of harvesting was minimal. If animals were targeted at weights just above the non-selective mean weight, then the population evolved to reproduce at lighter weights, which gave them the opportunity to breed before experiencing heavy mortality. However if hunting was simply selective for heavier or lighter animals, rather than being explicitly focused on one weight class, then the additional evolutionary effect of selectivity was minor (see Proaktor *et al.*, 2007 for more details).

Unintentional selection in red grouse shooting

The red grouse *Lagopus lagopus scoticus* is an economically important recreationally hunted species in the UK. It is of particular ecological interest due to its fluctuating population dynamics (Hudson, 1992; Haydon *et al.*, 2002). The cause of these fluctuations, at least in part, is infestation by the nematode *Trichostrongylus tenuis* (Hudson *et al.*, 1998). Up to 50 per cent of the population is shot each year. However contrary to theoretical predictions, this heavy mortality does not appear to dampen population fluctuations (Hudson & Dobson, 2001).We carried out a detailed study of the age and sex selectivity of shooting in Northern England in order to understand the effect that this might have on demography. In this area, beaters drive grouse towards a line of shooting butts, and animals are shot as they fly over the butts. There is no intentional selection among the animals flying over, but selection might be introduced through those flying being a biased sample of the population as a whole, or through people targeting individuals flying alone (Hudson, 1986).

The age and sex composition of shot animals was compared to that of the population just before shooting started. There was indeed bias in the age–sex composition of the shot sample, but interestingly the direction of the bias depended on the number of grouse shot, and hence on the population density (Cattadori *et al.*, 2003). At low population densities, more older animals and more females were shot than expected, while the opposite was true at high

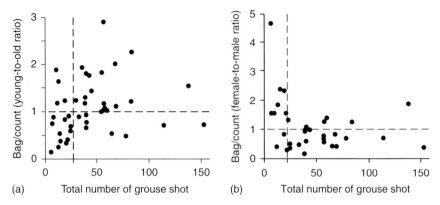

Figure 5.3   (a) The young-to-old ratio of the grouse bag for shooting drives in the North Pennines, UK, in August–September, divided by the young-to-old ratio estimated by population counts in the same areas in July, is shown for different bag sizes. A bag/count young-to-old ratio >1 means that a higher proportion of young animals were shot than occurred in the population before shooting. A tree model identified a node at 27.5 grouse, above which significantly more young grouse were shot than expected from the population count in July, before shooting begins , and below which significantly more old grouse were shot. (b) The female-to-male ratio of the bag divided by the count is shown for different bag sizes. A tree model identified a node at 21.5 grouse, above which old males were significantly more susceptible to shooting than old females, and below which old females were significantly more susceptible.

densities (Figure 5.3). Given that older animals have higher parasite burdens (Hudson, 1992), this means that at high densities, shooting is likely to increase the mean parasite burden in the population by removing the less burdened younger animals. Rather than dampening down fluctuations, this process may well exacerbate them.

This study suggests for one fairly simple system that recreational hunting can have effects on the dynamics of species other than the target species, in this case a parasite, and can thus contribute to the cycles that influence the ecology of the heather moorland ecosystem. Additionally, because hunters are not intentionally selective, this demonstrates that we cannot afford to ignore the indirect effects of hunting, even when hunting appears to be targeting all individuals equally.

The influence of non-monetary motives on supply and demand

All goods represent a range of values both to producers and consumers. The most obvious and easily measured, and often also the dominant, values are monetary. However there are a range of non-monetary values that contribute to a good's total economic value, which may include option values, bequest values, existence values, indirect use values such as watershed protection, and recreational values (Edwards-Jones *et al.*, 2000). All these values contribute to the utility (i.e. the satisfaction or happiness) derived from a good. Like many goods based on natural services, recreational hunting, by definition, has a high non-market value component to the utility derived from it. However, because the fundamental metric of economics is utility, not money, this fact does not make a fundamental difference to economic analyses. There are also other actors in the commodity chain (the owners of the wildlife, the sellers of the hunting trip package), who are likely to be working more straightforwardly to a profit motive.

The long-term sustainability of a hunting system depends on the dynamics of supply and demand. The quantity of a good (in this case a recreational hunting experience) demanded by hunters at a given price is influenced by the substitutability of the good and the hunters' incomes. If hunters place a high value on a particular hunting trophy and there are no direct substitutes, then they will pay whatever it takes to get that trophy and the quantity demanded will stay at much the same level regardless of its price (demand is *inelastic*). If the good is easily substitutable, and is not crucial to well-being, then changes in price will have a big effect on the quantity demanded (*elastic* demand). Examples of inelastic demand might be the demand for the 'big five' game species, and examples of elastic demand might be the demand for hunting trips to unusual places to hunt a range of less charismatic species. An interesting twist on demand for recreational hunting is that as a species becomes rarer, demand may become progressively more inelastic, as hunters seek the kudos of killing a particularly rare species (Hall *et al.*, 2008). Alternatively, if hunters have a code of practice that frowns on hunting threatened species, perhaps backed up with a sustainability labelling system (Child & Wall, this volume, Chapter 20), then the opposite effect might occur.

The supply curve for hunted species has the unusual property of being backward-bending in open access systems (Clark, 1990). This is because population growth is density dependent (Figure 5.1). As price increases for normal goods, suppliers have the incentive to put more input into production, and the amount produced rises correspondingly. For hunted species, this is true at first, but as the population size drops below the level that produces the maximum sustainable yield, increasing input (hunting effort) leads to decreasing output (catch). The high price stimulates a high effort level which leads to a depleted population, and so a low catch rate. This property of the supply curve for hunted species has a fundamental impact on the sustainability of hunted systems (Ling & Milner-Gulland, 2006). The degree to which it bends backwards depends on the ownership structure of the resource – if anyone can hunt (an open access system) then it bends back as in Figure 5.4(b). If the resource is privately owned, then the degree of backwards bending depends on the owner's discount rate (Clark, 1990). At the extreme of a zero discount rate, the curve is like a normal supply curve, as the owner will never wish to exploit at a level that depletes the population (Figure 5.4(a)).

Putting supply and demand together, we can see that very different outcomes are plausible even within the narrow category of recreationally hunted

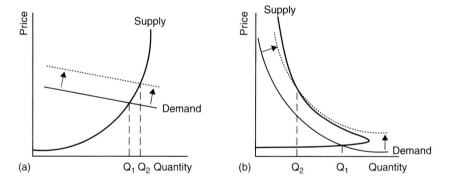

Figure 5.4 **Supply and demand curves for two scenarios, starting with the same quantity of hunting being supplied (Q_1) and showing the effect of a similar rise in income. The dashed line is the demand curve after the rise in income, and point Q_2 is the quantity supplied after this rise. (a) A sole owner of the hunted resource and an elastic demand for hunting. (b) An open access resource and a rarity-driven demand for hunting.**

species (Figures 5.4(a) and 5.4(b)). With a resource that is privately owned by many small operators and an elastic demand curve, an increase in hunter income leads to an increase in the price paid and the amount of the good supplied, and there is no issue with sustainability. This might be the scenario for privately owned game ranches. However, if the resource is *de facto* open access and there is a rarity value to hunting it, the same increase in income could change the situation from a relatively sustainable one to one in which very high prices are paid by hunters exerting large amounts of effort on a highly depleted population, with low catches resulting. This might be the situation for a species such as argali *Ovis ammon* in parts of Central Asia.

These are highly simplified supply and demand curves. Real-life situations soon get much more complicated, particularly if there are oligopolies involved (Bulte & Damania, 2005). However the analysis highlights the importance of considering issues of supply and demand, market structure, ownership and hunter motivations when determining whether a recreational hunting system is likely to be sustainable in the face of changes in factors such as hunter income.

Managing recreational hunting systems for sustainability

The final characteristic that distinguishes recreational hunting from other forms of hunting (other than highly commercial enterprises like fisheries) is that there is often the potential for management through formal or informal rule-setting. There are a number of prerequisites for effective management, and hence for keeping hunting within sustainable limits. Firstly, data are required on the size and productivity of the hunted population and on the levels of off take. Then appropriate rules need to be set and these rules must be followed by the hunters. Monitoring needs to be in place to assess the outcome of the chosen hunting strategy, and then there needs to be the institutional capacity available to act on this assessment and improve management. In this section we outline a case study of the effects of observation error in translating management strategies into a desired population size, and discuss ways of developing a more holistic approach to management that incorporates uncertainty into the management process.

The relationship between perception and reality in red deer management

It is notoriously difficult to obtain good estimates of population densities for most hunted species. Fisheries scientists are particularly aware of the problems in getting good stock size estimates (Hilborn & Walters, 1991), but the issue is also acute in tropical forests, where confidence intervals on estimates of the density of important bushmeat species such as duikers are routinely 30 per cent or more (Hart, 2000). The calculation of sustainable off take quotas is reliant on a time series of population size estimates but more often than not these estimates are infrequent, imprecise and biased. We illustrate the problems that can arise using the red deer in Scotland, a population subject to few of the problems that beset fisheries or bushmeat, being relatively visible, with well understood biology, located in a wealthy country and with a management infrastructure involving both private landowners and an overarching governing body.

Red deer in Scotland are routinely censused by the Deer Commission for Scotland, with a given estate being censused about once every seven years. Independent counts carried out at the same time suggest that these censuses underestimate population sizes by around 20 per cent on average and that the observation error in the censuses is also about 20 per cent (Milner-Gulland *et al.*, 2004). The implication of these three sources of uncertainty (infrequent, biased and imprecise counts) is illustrated in Figure 5.5. At a given target hunting mortality rate, say 14 per cent of the population, there are three different potential resulting hind densities – the low density that would result if the population was actually harvested at 14 per cent, the rather high density that results when the population is harvested at what the managers think is 14 per cent (but is actually more like 10 per cent, because the population size has been underestimated), and the middling density that the managers think the deer population is at, based on their censuses.

It is clear that observation error needs to be explicitly accounted for in management decision-making, or the chasm between reality and perceptions could get very large indeed. This could then lead to unexpected collapses of hunted systems as warning bells fail to ring.

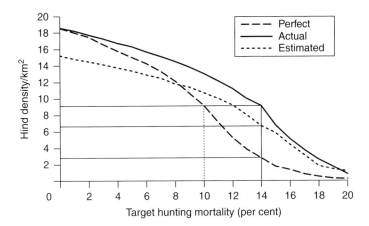

Figure 5.5 **The relationship between hind density and perceived hunting mortality. The line with long dashes ('perfect') shows the true relationship if the perceived hunting mortality was actually applied. The continuous black line ('actual') shows the relationship based on the target hunting mortality set based on observed hind densities. The line with short dashes 'estimated' shows the relationship that is observed. Taking the example of a hunting mortality of 14% (thin solid lines): if this mortality rate was actually imposed, hind density would be about 3 hinds/km^2. In actual fact, the hind density achieved by setting a target hunting mortality of 14% is about 9 hinds/km^2, which is perceived by the manager to be about 6.5 hinds/km^2. When the manager thinks that they are implementing a hunting mortality of 14%, what they are actually achieving (vertical dashed line) is about 10%.**

Accounting for uncertainty in management

In order to account for uncertainty in monitoring and implementation, it is necessary to consider explicitly not just the dynamics of the harvested system but the management system as well (Figure 5.6). Traditional wildlife management, of the type still used in many recreational hunting systems, focuses on the dynamics of the hunted population. However, as we have seen, the hunted population is set within a much wider system. The reactions of the hunters to changes in economic, social and political circumstances need to be considered. The management system takes in information from monitoring, and produces rules by which hunters are supposed to abide. There

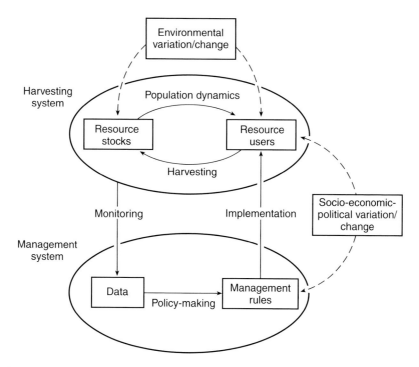

Figure 5.6 **A framework for management. Both the harvesting system and the management system need to be explicitly included when evaluating potential harvesting strategies. The two systems are linked by monitoring (collecting data about the harvested system) and implementation (putting harvest rules into practice), both of which are likely to involve uncertainty.**

is substantial uncertainty in the data collected by monitoring, as discussed above. There is also uncertainty surrounding implementation – will the rules be followed correctly, or will hunters circumvent them by killing the wrong age or sex animal, or by killing too many in the wrong place at the wrong time? How will these rules be enforced? The development of the rules themselves requires some sort of model for understanding the sustainability of harvesting, and this model may be inadequate or wrong (Hilborn & Mangel, 1997).

This broader approach to the management of natural resources is becoming normal best practice in fisheries science, where it is known as the operating model approach (McAllister *et al.*, 1999). The approach involves developing linked models of stock dynamics, the observation process and the management process, and then testing potential management strategies for their robustness and ability to meet management objectives. It was first developed by the International Whaling Commission in response to the particular need for any resumption of whaling to be demonstrably sustainable and precautionary (Kirkwood & Smith, 1996).

The natural next step after developing an operating model to test the sustainability of potential management options is to use it as a tool for ongoing management. Management strategies that are developed and implemented one year produce data that are fed into the next year's management plan. This new information then updates the manager's understanding of the system, and this improved understanding leads to a revision of the harvest rules that are imposed on hunters. This process is known as passive adaptive management (Walters, 1986; Parma & NCEAS Working Group on Population Management, 1998). It is also possible to set out specifically to experiment with the system with the aim of improving understanding of its dynamics. This is known as active adaptive management (Shea *et al.*, 2002), but it involves a substantial increase in complexity and potential risk to balance against the increased understanding obtained (Lee, 1999; Hauser & Possingham, 2008).

Although passive adaptive management is the way of the future for commercial species, what are its prospects for recreationally hunted species? This depends on the complexity of the management system, the value of the population involved, and the consequences of getting things wrong. If recreational hunting is small-scale or local, based upon common or replaceable species, and not particularly lucrative, monitoring and management may be best kept simple and low-level (cf. Hockley *et al.*, 2005). However, when species are of conservation concern or recreational hunting is of economic importance, then sustainability is of prime importance and the adaptive management approach is worth pursuing.

Few examples yet exist of formal adaptive management in the recreational hunting context, but one example is the management of recreational duck hunting by the US Fish and Wildlife Service (Johnson & Williams, 1999). This large-scale and long-standing management system is supported by the stakeholder groups such as the NGO, Ducks Unlimited, and demonstrates that when conditions are right, recreational hunting can be managed in a holistic and resilient manner.

Conclusions

The science of recreational hunting does not differ qualitatively from that of any other type of hunting. There are differences in emphasis that can usefully be highlighted, and it can also be useful to consider its similarities to other fields where innovative approaches have been tried.

One of the key features of recreational hunting is that it tends to be selective, targeting particular individuals on the basis of some phenotypic characteristic. There is both empirical and theoretical evidence that selective hunting can cause indirect evolutionary and ecological effects that could compromise sustainability in the long run. However it is still not clear how significant these effects are in the context of the direct effects of mortality and other perturbations to the ecosystem, both natural and anthropogenic. It could be that selective hunting is only of major concern under extreme conditions, when populations are isolated and under extreme demographic or selective pressure.

Recreational hunting, like all other forms of hunting, may be prone to problems of instability and over-exploitation under conditions of open access when demand for the products of hunting is inelastic. It is important for conservationists to consider these issues when assessing the sustainability of a hunting system, because even if the system seems sustainable at present, changes in the economic circumstances of hunters, ultimate consumers or resource owners may trigger a rapid decline in population size.

Biological processes, economic processes and the management system itself are interlinked. Uncertainty in monitoring and in our understanding of the biology of the system, imperfect compliance with rules and variability in the biological and sociopolitical environments interact to make hunting systems dynamic. This calls for an adaptive management system, in which managers can evaluate the degree of uncertainty and bias that affects their understanding of the world, and can build this evaluation into their procedures. By bringing our uncertainty and ignorance out into the open, we can act appropriately to deal with it on an ongoing basis. This is the key to sustainability.

References

Begon, M., Mortimer, M. & Thompson, D.J. (1996) *Population Ecology: A Unified Study of Animals and Plants.* Blackwell Science, Oxford.

Bulte, E.H. & Damania, R. (2005) An economic assessment of wildlife farming and conservation. *Conservation Biology*, 19, 1222–1233.

Cattadori, I.M., Haydon, D.T., Thirgood, S.J. & Hudson P.J. (2003) Are indirect measures of abundance a useful index of population density? The case of red grouse harvesting. *Oikos*, 100, 439–446.

Clark, C.W. (1990) *Mathematical Bioeconomics: The Optimal Management of Renewable Resources*. John Wiley, New York.

Clutton-Brock, T., Albon, S.D. & Guinness, F.E. (1982) *Red Deer: Behaviour and Ecology of Two Sexes*. University of Chicago Press, Chicago.

Davis, G. & Brown, D. (2007) *Bushmeat and Livelihoods: Challenges for the 21st Century*. Cambridge University Press, Cambridge.

Edwards-Jones, G., Hussain, S. & Davies, B. (2000) *Ecological Economics*. Blackwell Science, Oxford.

Hall, R.J., Milner-Gulland, E.J. & Courchamp, F. Endangering the endangered: the effect of perceived rarity on species extinction. *Conservation Letters*, 1(2), 75–81.

Hardin, G. (1968) The tragedy of the commons. *Science*, 162, 1243–1248.

Hart (2000) Impact and sustainability of indigenous hunting in the Ituri Forest, Congo-Zaire: a comparison of unhunted and hunted duiker populations. In *Hunting for Sustainability in Tropical Forests*, eds. J.R. Robinson & E.L. Bennett. Columbia University Press, New York.

Hauser, C. & Possingham, H. (2008) Experimental or precautionary: adaptive management over a range of time horizons. *Journal of Applied Ecology*, 45, 72–81.

Haydon, D.T., Shaw, D.J., Cattadori, I.M., Hudson, P.J. & Thirgood, S.J. (2002) Analysing noisy time-series: describing regional variation in the cyclic dynamics of red grouse. *Proceedings of the Royal Society of London*, B 269, 1609–1617.

Hilborn, R. & Mangel, M. (1997) *The Ecological Detective: Confronting Models with Data*. Princeton University Press, Princeton, New Jersey.

Hilborn, R. & Walters, C.J. (1991) *Quantitative Fisheries Stock Assessment: Choice, Dynamics and Uncertainty*. Kluwer Academic Publishers, Dordrecht.

Hirshfield, M.F. & Tinkle, D.W. (1975) Natural-selection and evolution of reproductive effort. *Proceedings of the National Academy of Sciences of the United States of America*, 72, 2227–2231.

Hockley, N.J., Jones, J.P.G. & Andriahajaina, F.B. *et al.* (2005) When should communities and conservationists monitor exploited resources? *Biodiversity and Conservation*, 14, 2795–2806.

Hudson, P.J. (1986) *The Biology and Management of a Wild Game Bird*. The Game Conservancy Trust, Fordingbridge.

Hudson, P.J. (1992) *Grouse in Space and Time: The Population Biology of a Managed Gamebird*. Game Conservancy Limited, Hampshire.

Hudson, P.J. & Dobson, A.P. (2001) Harvesting unstable populations: red grouse *Lagopus lagopus scoticus* (Lath.) in the United Kingdom. *Wildlife Biology*, 7, 189–195.

Hudson, P.J., Dobson, A.P. & Newborn, D. (1998) Prevention of population cycles by parasite removal. *Science*, 282, 2256–2258.

Hutchings, J.A. (2005) Life history consequences of overexploitation to population recovery in Northwest Atlantic cod (*Gadus morhua*). *Canadian Journal of Fisheries and Aquatic Sciences*, 62, 824–832.

Jennings, S., Reynolds, J.D. & Mills, S.C. (1998) Life history correlates of responses to fisheries exploitation. *Proceedings of the Royal Society of London Series B-Biological Sciences*, 265, 333–339.

Johnson, F. & Williams, K. (1999) Protocol and practice in the adaptive management of waterfowl harvests. *Conservation Ecology*, 3(1), 8. http://www.consecol.org/vol3/iss1/art8/

Kirkwood, G.P. & Smith, A.D.M. (1996) Assessing the precautionary nature of fisheries management strategies. In *FAO Fisheries Technical Paper*. No. 350, Part 2. FAO, Rome. http://www.fao.org/DOCREP/003/W1238E/W1238E07.htm

Kokko, H., Lindström, J. & Ranta, E. (2001) Life histories and sustainable harvesting. In *Conservation of Exploited Species*, eds. J.D. Reynolds, G.M. Mace, K.H. Redford & J.G. Robinson, pp. 301–322. Cambridge University Press, Cambridge.

Law, R. (2000) Fishing, selection, and phenotypic evolution. *ICES Journal of Marine Science*, 57, 659–668.

Law, R. & Grey, D.R. (1989) Evolution of yields from populations with age-specific cropping. *Evolutionary Ecology*, 3, 343–359.

Lee, K.N. (1999) Appraising adaptive management. *Conservation Ecology*, 3(2), 3. http://www.consecol.org/vol3/iss2/art3/

Ling, S. & Milner-Gulland, E.J. (2006) Assessment of the sustainability of bushmeat hunting based on dynamic bioeconomic models. *Conservation Biology*, 20, 1294–1299.

Loveridge, A., Reynolds, J. & Milner-Gulland, E.J. (2006) Does sport hunting benefit conservation? In *Key Topics in Conservation*, eds. D. Macdonald & K. Service. Blackwell Science, Oxford.

McAllister, M.K., Starr, P.J., Restrepo, V.R. & Kirkwood, G.P. (1999) Formulating quantitative methods to evaluate fishery-management systems: what fishery processes should be modelled and what trade-offs should be made? *ICES Journal of Marine Science*, 56, 900–916.

Milner-Gulland, E.J. & Mace, R. (1998) *Conservation of Biological Resources.* Blackwell Science, Oxford.

Milner-Gulland, E.J., Bukreeva, O.M., Coulson, T.N. *et al.* (2003) Reproductive collapse in saiga antelope harems. *Nature*, 422, 135.

Milner-Gulland, E.J., Coulson, T.N. & Clutton-Brock, T.H. (2004) Sex differences and data quality as determinants of income from hunting red deer. *Wildlife Biology*, 10, 187–201.

Parma, A.M. & The NCEAS Working Group on Population Management (1998) What can adaptive management do for our fish, forests, food and biodiversity? *Integrative Biology, Issues, News, and Reviews*, 1, 16–26.

Proaktor, G., Milner-Gulland, E.J. & Coulson, T. (2007) Evolutionary responses to harvesting in ungulates. *Journal of Animal Ecology* 76, 669–678.

Robinson, J.R. & Bennett, E.L. (2000) *Hunting for Sustainability in Tropical Forests.* Columbia University Press, New York.

Roldan, A.I. & Simonetti, J.A. (2001) Plant–mammal interactions in tropical Bolivian forests with different hunting pressures. *Conservation Biology*, 15, 617–623.

Saether, B.E., Solberg, E.J. & Heim, M. (2003) Effects of altering sex ratio structure on the demography of an isolated moose population. *Journal of Wildlife Management*, 67, 455–466.

Shea, K., Possingham, H.P., Murdoch, W.W. & Roush, R. (2002) Active adaptive management in insect pest and weed control: intervention with a plan for learning. *Ecological Applications*, 12, 927–936.

Sinclair, A.R.E., Fryxell, J. & Caughley, G. (2005) *Wildlife Ecology, Conservation and Management.* Blackwell Science, Oxford.

Walters, C. (1986) *Adaptive Management of Renewable Resources.* Macmillan, New York.

(6)

Guns, Sheep, and Genes: When and Why Trophy Hunting May Be a Selective Pressure

Marco Festa-Bianchet[1] and Ray Lee[2]

[1]Département de biologie, Université de Sherbrooke,
Québec, Canada
[2]Foundation for North American Wild Sheep, Wyoming, AZ, USA

Introduction

Recreational hunting is a lucrative economic activity that has been recognised as one tool to foster the conservation of biodiversity by directing some of the economic revenue it generates to conservation programs and by demonstrating to local populations the value of wildlife and habitat protection (Leader-Williams *et al.*, 2001). Recreational hunters are most interested in killing males with large secondary sexual characteristics (or trophies) such as horns, antlers or tusks. A hunter's willingness to pay for the hunt is typically correlated with the expectation of harvesting a male with large horns, antlers, or tusks. Because hunters are willing to pay more to harvest males with larger trophies, trophy hunting can produce a substantial amount of revenue, which can potentially be used for conservation (Harris & Pletscher, 2002; Hofer, 2002). For example, in North America many successful bighorn sheep *Ovis canadensis* reintroduction

Recreational Hunting, Conservation and Rural Livelihoods: Science and Practice, 1st edition.
Edited by B. Dickson, J. Hutton and W.M. Adams. © 2009 Blackwell Publishing,
ISBN 978-1-4051-6785-7 (pb) and 978-1-4051-9142-5 (hb).

programmes have been financed entirely with funds generated through the auction of trophy ram permits by the Foundation for North American Wild Sheep. In these auctions, the highest bids are reserved for permits from areas with a reputation for producing the largest rams.

Overall, there is little question that modern ungulate management, based on recreational hunting, is sustainable from a demographic viewpoint. In many countries where large predators have been eliminated, populations of ungulates are increasing despite high levels of recreational hunting, sometimes causing loss of biodiversity and economic damage (Côté *et al.*, 2004; Milner *et al.*, 2006). Economically successful trophy hunting programmes exist in many parts of the world (Leader-Williams *et al.*, 2001). One area of particular interest is Central Asia, where several mountain ungulates, but particularly wild sheep, are the object of increasing interest from trophy hunters and where both the ungulates and their habitat are threatened by poaching, overgrazing, and disease transmission by domestic livestock.

However, recreational hunting does not just have a numerical impact on exploited populations. Such hunting is typically selective, either in terms of which sex–age classes are killed, or in terms of the specific morphological attributes of the animals that are harvested. Sex- and age-specific mortality caused by recreational hunting is typically very different from that seen in non-hunted ungulate populations, and therefore it is likely to be different from the mortality regimes prevalent during most species' evolution. In particular, in most hunted populations, the mortality of males and of prime-aged females is much higher than in populations that are not hunted (Festa-Bianchet, 2003; Bender *et al.*, 2004; Milner *et al.*, 2006). If hunting is the main cause of adult mortality, as in many hunted populations of ungulates (Langvatn & Loison, 1999; Bender *et al.*, 2004; Milner *et al.*, 2006), one may expect recreational hunting to become a selective pressure on life-history patterns. Life-history evolution and sex- and age-specific survival probabilities are intimately linked, so that a change in mortality patterns is likely to change the fitness payoffs of different reproductive strategies. For example, in heavily hunted populations animals may benefit by investing as many resources as possible in a reduced number of breeding opportunities.

Recently, researchers have become increasingly aware that the selective effects of recreational hunting in general, and of trophy hunting in particular, may be important (Jachmann *et al.*, 1995; Harris *et al.*, 2002; Coltman *et al.*, 2003; Festa-Bianchet, 2003; Garel, 2006). In some circumstances, intense removal of trophy males may select for genetically small horns, antlers or tusks. This has been suggested for African elephants *Loxodonta africana* (Jachmann *et al.*, 1995), one population of bighorn sheep (Coltman *et al.*, 2003) and one of European mouflon *Ovis aries* (Garel, 2006). But overall, very few studies have examined the potential selective effects of trophy harvest, or of any other management strategy. Trophy hunting may also have negative demographic consequences, because males with large horns may also have other fitness-related genetic traits that could be selectively removed from a population. While the potential evolutionary impacts of trophy hunting are worthy of consideration, there is, again, currently not enough evidence to determine when they should be seen as a significant concern for conservation.

Any potential selective genetic effects of trophy harvest can be significantly decreased by lowering harvest pressure to ensure that some males with large trophies participate in breeding. Good wildlife management programmes – using reduced season lengths and limiting permits – are designed to regulate hunting pressure to ensure adequate survival of older age animals. There are a number of cases demonstrating that trophy size of bighorn sheep has not diminished through time when populations are properly managed. In this context, it is important to avoid human-made barriers to gene flow, such as fences or major habitat disruptions, that could prevent movement between hunted and protected areas. Gene flow from protected areas should reduce the long-term genetic effects of selective removal of large-horned rams through hunting.

In this chapter we consider under what circumstances hunting may be a selective pressure leading to changes in gene frequencies in hunted populations, examine current research needs and suggest management strategies to reduce the genetic impact of trophy hunting.

The case of Ram Mountain

Ram Mountain is an isolated outcrop in Alberta, Canada, 30 kilometres from the main range of the Rocky Mountains (Figure 6.1). Since 1971, it has been the site of numerous wild sheep studies. Some of the most recent studies have dealt with the potential for artificial selection, through trophy hunting, of bighorn sheep. This potential was increased by a number of factors, of which possibly the most important was that rapidly growing rams developed horns large enough to be considered trophies before those horns gave them a reproductive advantage (Hogg & Forbes, 1997; Coltman *et al.*, 2002) (Figure 6.2).

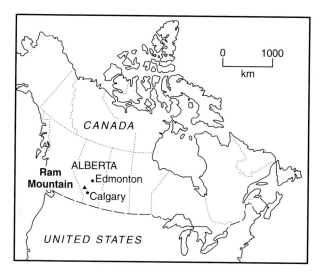

Figure 6.1 **Location of Ram Mountain, Alberta.**

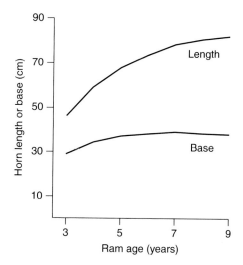

Figure 6.2 **Bighorn rams – most horn growth occurs before 5 years of age.**

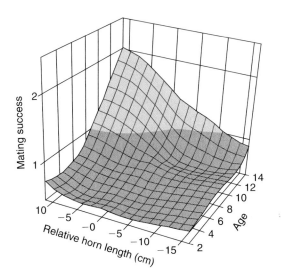

Figure 6.3 **Mating success in relation to horn length and age.**

Rapidly growing horns put a ram at risk of being harvested at four years of age, but do not give him a mating advantage unless he survives to seven to eight years (Figure 6.3). Rams with slow-growing horns that did not meet the legal definition of 'trophy', on the other hand, could expect to die of old age, after some of them had reached a high dominance status and obtained a high mating success. Coltman *et al.* (2002) also noted that younger rams, using a variety of mating strategies, sired about 50 per cent of the lambs on Ram Mountain.

Other characteristics of the bighorn sheep population made it especially susceptible to artificial selection. Harvest by Alberta residents was not limited by a quota, but only by a definition of which rams could be legally hunted in terms of horn shape and size. In theory, all rams could be shot as soon as their horns reached 4/5 of a curl (Figure 6.4). The population is small (it ranged from 26 to 152 adult sheep during the 30-year study) and isolated, therefore it is susceptible to genetic drift and cannot receive 'unselected' immigrants from protected areas (Hogg *et al.*, 2006). Genetic drift due to the small population size, however, cannot currently be distinguished from the selective effect attributed to hunting.

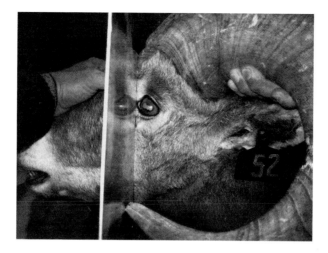

Figure 6.4   **A barely legal four-year-old ram from Alberta. Shot two days after leaving a protected area. The plexiglass is used to determine if the tip of the horn extends beyond an imaginary straight line drawn from the base of the horn to the tip of the eye – that makes the ram legal to harvest.**

Of course, many factors have been shown to affect the size of secondary sexual characteristics (horns and antlers). Hengeveld & Festa-Bianchet (2006) and Plensky (2006) stated that their findings suggest that trophy sheep management based on minimum horn curl criteria and unlimited entry hunts may over time favour rams with slow growing horns. Habitat degradation in the study area was also mentioned by these authors. Indeed, in earlier reports on Ram Mountain (Jorgenson *et al.*, 1993, 1998; Festa-Bianchet *et al.*, 1997; LeBlanc *et al.*, 2001) reductions in horn size were attributed to density related decreases in forage availability. Frisina & Frisina (2004) note that habitat conditions over ride genetic potential when it comes to horn size. While genetics certainly plays a role in horn size, nutrition typically dominates.

Reviewing some of the European literature, Geist (2004) noted that declines in trophy quality of antlers could be reversed with time and careful management. Like the authors mentioned above, Geist explains the apparent reductions in horn size primarily through nutrition. He felt that, contrary to the fears of Coltman *et al.* (2003), the apparent decline in horn and body size they observed was not permanent, and could be reversed.

Research needs

The reason why trophy hunting may be a selective pressure is that hunters are attracted by the same morphological character that may determine male mating success. Clearly, to understand the potential impacts of trophy hunting on the distribution of male mating success, we need to know the relationships between horn or antler size, mating success and interacting variables such as age, body mass, and population density. Much of the controversy that followed the publication of the paper by Coltman *et al.* (2003) is attributable to a lack of understanding of these relationships. They are likely to vary substantially across species, and possibly across populations as well. Sexual selection, favouring males with large horns or antlers, will counter the potential artificial selection of trophy hunting. The outcome should depend on the specific relationships between horn or antler size and mating success, the proportion of males killed by hunters according to their age and horn or antler size, and probably also on how changes in the age structure of the male population affect male–male competition (Clutton-Brock *et al.*, 1997).

In the case of bighorn sheep, rams develop large horns a few years before horns play an important role in both dominance status and mating success. Importantly, while horn size and mass (both of which are age-related and affect social rank) play an important role in affecting the mating success of rams near the top of the dominance hierarchy, there is little evidence that they affect the mating success of subordinate rams (Hogg & Forbes, 1997; Coltman *et al.*, 2002). The success of alternative mating strategies adopted by all rams other than the top two to three in the social hierarchy appears independent of horn size. Therefore, one cannot simply assume a linear relationship between horn size and mating success. Factors such as age, mass and the characteristics of competing rams must be taken into account. Similarly, in red deer *Cervus elaphus* both age and antler size play a role in mating success: average antler mass peaks at about ten years of age and remains high thereafter, but mating success peaks at eight to eleven years of age, then declines (Kruuk *et al.*, 2002). The genetic impacts of harvesting a large-antlered male when he is seven or when he is 12 are, therefore, very different.

Immigration from unselected areas, and the effects of barriers, corridors, and habitat fragmentation are also worthy of investigation, because immigration from protected areas should lessen the selective effect of trophy hunting (Hogg, 2000; Hogg *et al.*, 2006). In species with a high level of male mating skew such as fallow deer *Dama dama*, removal of a few highly successful males may have a selective effect, but may also increase the effective population size by allowing a larger number of males to mate, again depending on the particular mating system of each population (Apollonio *et al.*, 1989). In other species, such as white-tailed deer (*Odocoileus virginianus*), large-antlered males may not be able to monopolise a large proportion of estrous females, and therefore lower the potential for artificial selection through trophy hunting.

Finally, in some species, such as mountain goats *Oreamnos americanus*, the size of the horns may affect harvest decisions by trophy hunters, but may not be directly involved in mating success, for example if the latter was more dependent on male body mass. Therefore, several interesting avenues of research on the effects of selective harvests are open, with important implications for wildlife management in addition to our fundamental interest in understanding evolution in wild populations.

So far, unfortunately, very few studies of ungulates have quantified male reproductive success, and fewer still have been able to correlate individual mating success with physical and population attributes (Hogg & Forbes, 1997; McElligott *et al.*, 2001; Coltman *et al.*, 2002). Without information about the relationships between horn or antler size and mating success, it is difficult to assess the potential selective impact of various management strategies. Similarly, there has been little work on long-term changes in horn shape and size in hunted and non-hunted populations (Coltman *et al.*, 2003; Garel, 2006), and more is needed, particularly in situations where pedigrees are also available. Long-term monitoring of marked individuals, whose physical characteristics can be measured repeatedly, offers the most promising avenue to learn more about artificial selection through recreational hunting.

Management

Some wildlife managers have long considered the potential impacts of over utilisation of the renewable wildlife resources they are charged with managing. When the recreational harvest of wild sheep in the southwestern United States was first reopened (1952 in Nevada, 1953 in Arizona), after up to a century of closure, strict regulations were written – addressing season lengths, sex, age and horn length – to limit the harvest (Lee, 1989a).

In Arizona, harvest protocols were established to provide: (1) adequate ratios of young to old rams; (2) a desired ratio of rams to ewes; and

(3) a restriction of harvest to no more than ten per cent of the total rams, nor more than 25 per cent of the rams older than six years of age. Lee (2000) reported that Arizona's current sheep management plan calls for the harvest of 6 per cent of the estimated rams, or 12 per cent of the estimated number of older age (six and above) rams, whichever number is the lowest. The Boone and Crockett scoring system measures both the mass and the symmetry of the horns; it is a measurement of the total number of inches of total horn length and horn circumference at the base and at each of the quarters, with differences between the two horns deducted from the total score. Lee (1989b) showed the marked increases through time in 162 plus and 170 plus point rams in the harvest in Arizona. These increases were from 12 per cent of the harvest in 1953–1958 to 46 per cent in 1983–1988 for 162 plus point rams, and from two per cent of the harvest to 14 per cent for 170 plus point rams. Lee (1991) illustrates this increase in size for the 1431 rams harvested in the period 1953–1990. These results demonstrate the long-term sustainability of these harvest regimes.

In New Mexico's main populations of Rocky Mountain bighorn sheep on Wheeler Peak and in the Pecos Wilderness, where the season first opened in 1990, ram harvests were limited to 5 per cent and 8 per cent of the estimated ram numbers, respectively. Mean ages and scores increased with this level of harvest. Data from the Pecos Wilderness reports that for the 112 rams harvested since 1990, the mean age for the four top scoring rams increased from 8.4 to 9.6 years of age for the periods 1990–1997 and 1998–2006, respectively. Mean score also increased from 171.1 to 173.9 Boone and Crockett points.

Winkler (1987) listed the methodologies for determining harvest quotas for six US states and Mexico. These ranged from 8 per cent of the ram population in New Mexico, to 12 per cent of the rams observed in Utah, to 15 per cent of the mature rams observed in California. These numbers contrast with the harvest percentage at Ram Mountain of about 40 per cent of the rams legally (four-fifths of a curl or more) available for harvest each year (Coltman *et al.*, 2003). Of the 57 rams harvested during the study period, 45 were harvested before reaching eight years of age, and nine were shot as early as four years of age. This does not represent sustainable trophy hunting.

Organisations such as the Boone and Crockett Club have kept long-term records of trophies. Frisina & Frisina (2004) reviewed record books to determine the top 100 bighorn sheep entries. Of these entries, 47 per cent were taken in the ten decades from 1880 to 1970, while 53 per cent were taken in just the past two decades 1980s and 1990s. Due to the conservative hunt management strategies, the number of trophy animals is on the rise. While there are many factors that weigh upon this simple analysis, it is apparent that there have been no long-term, adverse impacts due to trophy harvest.

While it may theoretically be possible to produce genetic changes due to selective and intense levels of harvest, in well-managed populations, as described above, this is not a management issue of significant concern.

Management options

When faced with unclear results and complex relationships between variables, scientists will typically attempt to increase their sample size. Managers and conservationists, faced with immediate problems, cannot afford to wait for more data. They must take decisions based on available evidence. Currently, the evidence suggesting that trophy hunting has a deleterious effect is very limited. Nevertheless, while artificial selection is not the most pressing threat facing hunted populations of wild ungulates, that threat should not be ignored, especially when there are management options readily available that should prevent it. In this section we discuss some of those options in the light of the available biological evidence.

We know what needs to be avoided: a situation where males with large horns or antlers are shot before they can obtain the high breeding success that may derive from those large weapons. Consequently, harvest should be directed to the older age classes. In the case of bighorn sheep, that would mean harvesting rams aged at least seven to eight years. One practical problem is that the ability of hunters to assess age in the field varies from species to species: in ibex *Capra ibex*, a six-year-old and a ten-year-old look very different even from a distance, but in bighorn sheep they may not, and in chamois *Rupicapra rupicapra* or tahr *Hemitragus jemlahicus* they do not. On the other hand, long-term monitoring of the average age of harvested animals is an important tool for managers. For bighorn sheep, we suggest that a situation where the average age of harvested rams is less than six years should be a concern. If the average age is nine years or older, there is much less reason to worry.

The potential selective pressure caused by trophy hunting is likely to increase with the proportion of trophy males that are harvested. An unlimited-entry scheme based on a morphological definition of what can or cannot be shot may create a selective pressure favouring rams with slower growing horns. Therefore, a quota system based on estimates of available trophy males should be used to ensure that an adequate number of large males survive to an advanced age and hence to decrease the potential selective effects of hunting. In addition, some agencies are combining quotas with 'any ram' as the legal animal for harvest – this further reduces the pressure on the older-age-class males.

Research has shown the importance of connectivity among populations for the exchange of genetic material (Hogg, 2000; Epps *et al.*, 2005). Large protected areas such as national parks (there are large populations of wild sheep in Glacier, Grand Canyon, Rocky Mountain and Yellowstone national parks in the US, the Pinacate Reserve in Mexico, and Banff and Jasper in Canada), or remote areas with very low hunting pressure, provide potential sources of unselected immigrants that could limit any selective effects of trophy hunting. It is therefore important to maintain those protected areas and to ensure that movement routes are not blocked by barriers such as transportation corridors, housing developments or fences. The potential selective effects of hunting are likely to be greater in populations that are small and isolated, and where the selective pressure is applied over the entire range of occurrence. On the other hand, park managers should be concerned about the possible effects that high hunting pressure outside the protected area may have on the genetic structure of protected populations. Unidirectional gene flow from the protected area to outside may decrease the effective population size in the protected area (Hogg, 2000).

Finally, we caution against the facile remedy of resorting to artificial translocations to counter real or perceived 'genetic problems'. While in a few specific cases of very small and endangered populations genetic rescue is a valid option (Hogg *et al.*, 2006), systematic supplementation for trophy hunting purposes is not sustainable and may have deleterious consequences (Tallmon *et al.*, 2004). Transplanted individuals may break up local adaptations or introduce new pathogens or parasites. Attempts to artificially increase horn or antler size by introducing animals from populations with larger trophies are unjustified both from a conservation and an ethical viewpoint.

Conclusions

The conservation of biodiversity cannot ignore social, economic and cultural aspects, and trophy hunting offers a substantial potential for conservation benefits. Although the potential for artificial selection by trophy hunting may be a concern, it pales in comparison to the threats presented by habitat destruction, poaching and diseases. Research on artificial selection in harvested wildlife population is only beginning to develop. Meanwhile, managers should continue to consider the potential genetic consequences of different selective harvesting schemes. Hunter groups should be encouraged to place more emphasis on the conservation values of hunting and less on the size of secondary sexual characteristics of harvested animals. More importantly, however, the promise of trophy hunting as a conservation tool must be more fully realised, so that a greater share of the proceeds from recreational hunting in general, and trophy hunting in particular, finds its way directly into conservation activities.

As an example of the possibilities of the trophy hunting as a conservation tool, bighorn sheep numbers have more than tripled in North America during the past 30 years. This is due primarily to the funds raised by hunter/conservationists and spent on the conservation of wild sheep.

Recreational hunters have driven one of the most successful conservation programs in history.

References

Apollonio, M., Festa-Bianchet, M. & Mari, F. (1989) Effects of removal of successful males in a fallow deer lek. *Ethology*, 83, 320–325.

Bender, L., Schirato, G., Spencer, R., McAllister, K. & Murphie, B. (2004) Survival, cause-specific mortality, and harvesting of male black-tailed deer in Washington. *Journal of Wildlife Management*, 68, 870–878.

Clutton-Brock, T., Rose, K. & Guinness, F. (1997) Density-related changes in sexual selection in red deer. *Proceedings of the Royal Society of London B*, 264, 1509–1516.

Coltman, D., Festa-Bianchet, M., Jorgenson, J. & Strobeck, C. (2002) Age-dependent sexual selection in bighorn rams. *Proceedings of the Royal Society of London B*, 269, 165–172.

Coltman, D., O'Donoghue, P., Jorgenson, J., Hogg, J., Strobeck, C. & Festa-Bianchet, M. (2003) Undesirable consequences of trophy hunting. *Nature*, 426, 655–658.

Côté, S., Rooney, T., Tremblay, J., Dussault, C. & Waller, D. (2004) Ecological impacts of deer overabundance. *Annual Review of Ecology and Systematics*, 35, 113–147.

Epps, C., Palsboll, P., Wehausen, J., Roderick, G., Ramey, R. & McCullough, D. (2005) Highways block gene flow and cause a rapid decline in genetic diversity of desert bighorn sheep. *Ecology Letters*, 8, 1029–1038.

Festa-Bianchet, M. (2003) Exploitative wildlife management as a selective pressure for the life-history evolution of large mammals. In *Animal Behavior and Wildlife Conservation*, eds. M. Festa-Bianchet & M. Appollonio, pp. 191–207. Island Press, Washington.

Festa-Bianchet, M., Jorgenson, J., Berube, C., Portier, C. & Wishart, W. (1997) Body mass and survival of bighorn sheep. *Canadian Journal of Zoology*, 75, 1372–1379.

Frisina, R. & Frisina, M. (2004) Sport hunting: a model of bighorn success. *Proceedings of the Northern Wild Sheep and Goat Council*, 14, 195–199.

Garel, M. (2006) *Consequences de la chasse et des contraintes environnementales sur la demographie evolutive des populations d'ongules.* Ph.D. thesis, Universite Claude-Bernard Lyon, Lyon.

Geist, V. (2004) Trophy males as individuals of low fitness. *Proceedings of the Northern Wild Sheep and Goat Council*, 14, 200–204.

Harris, R. & Pletscher, D. (2002) Incentives toward conservation of argali *Ovis ammon*: a case study of trophy hunting in western China. *Oryx*, 36, 373–381.

Harris, R., Wall, W. & Allendorf, F. (2002) Genetic consequences of hunting: what do we know and what should we do? *Wildlife Society Bulletin*, 30, 634–643.

Hengeveld, P. & Festa-Bianchet, M. (2006) Ramifications of the hunt: horn growth, selection, and evolution in British Columbia. *Proceedings of the Northern Wild Sheep and Goat Council*, 15, 15 (abstract only).

Hofer, D. (2002) *The Lion's Share of the Hunt. Trophy Hunting and Conservation – A Review of the Legal Eurasian Tourist Hunting Market and Trophy Trade under CITES.* TRAFFIC Europe, Brussels.

Hogg, J. (2000) Mating systems and conservation at large spatial scales. In *Vertebrate Mating Systems*, eds. M. Apollonio, M. Festa-Bianchet & D. Mainardi, pp. 214–252. World Scientific, Singapore.

Hogg, J. & Forbes, S. (1997) Mating in bighorn sheep: frequent male reproduction via a high-risk "unconventional" tactic. *Behavioral Ecology and Sociobiology*, 41, 33–48.

Hogg, J., Forbes, S., Steele, B. & Luikart, G. (2006) Genetic rescue of an insular population of large mammals. *Proceedings of the Royal Society B – Biological Sciences*, 273, 1491–1499.

Jachmann, H., Berry, P. & Imae, H. (1995) Tusklessness in African elephants – a future trend. *African Journal of Ecology*, 33, 230–235.

Jorgenson, J., Festa-Bianchet, M. & Wishart, W. (1993) Harvesting bighorn ewes: consequences for population size and trophy ram production. *Journal of Wildlife Management*, 57, 429–435.

Jorgenson, J., Festa-Bianchet, M. & Wishart, W. (1998) Effects of population density on horn development in bighorn rams. *Journal of Wildlife Management*, 62, 1011–1020.

Kruuk, L., Slate, J., Pemberton, J., Brotherstone, S., Guiness, F. & Clutton-Brock, T. (2002) Antler size in red deer: heritability and selection but no evolution. *Evolution*, 56, 1683–1695.

Langvatn, R. & Loison, A. (1999) Consequences of harvesting on age structure, sex ratio and population dynamics of red deer *Cervus elaphus* in central Norway. *Wildlife Biology*, 5, 213–223.

Leader-Williams, N., Smith, R. & Walpole, M. (2001) Elephant hunting and conservation. *Science*, 293, 2203.

LeBlanc, M., Festa-Bianchet, M. & Jorgenson, J. (2001) Sexual size dimorphism in bighorn sheep (*Ovis canadensis*): effects of population density. *Canadian Journal of Zoology*, 79, 1661–1670.

Lee, R. (1989a) *The Desert Bighorn Sheep in Arizona*. Arizona Game and Fish Department, Phoenix, Arizona.

Lee, R. (1989b) Status of bighorn sheep in Arizona, 1988. *Desert Bighorn Council Transactions*, 33, 9–10.

Lee, R. (1991) Status of bighorn sheep in Arizona, 1990. *Desert Bighorn Council Transactions*, 35, 9–10.

Lee, R. (2000) A working hypothesis for desert bighorn sheep management. In *Transactions of the 2nd North American Wild Sheep Conference*, eds. A. Thomas & H. Thomas, pp. 67–72. Desert Bighorn Council and Northern Wild Sheep and Goat Council, Reno, Nevada.

McElligott, A., Gammell, M., Harty, H. *et al.* (2001) Sexual size dimorphism in fallow deer (*Dama dama*): do larger, heavier males gain greater mating success? *Behavioral Ecology and Sociobiology*, 49, 266–272.

Milner, J., Bonenfant, C., Mysterud, A., Gaillard, J., Csany S. & Stenseth, N. (2006) Temporal and spatial development of red deer harvesting in Europe: biological and cultural factors. *Journal of Applied Ecology*, 43, 721–734.

Plensky, D. (2006) Influence of trophy hunting and habitat degradation on horn growth in bighorn sheep. *Proceedings of the Northern Wild Sheep and Goat Council*, 15, 16 (abstract only).

Tallmon, D., Luikart, G. & Waples, R. (2004) The alluring simplicity and complex reality of genetic rescue. *Trends in Ecology and Evolution*, 19, 489–496.

Winkler, C. (1987) Desert bighorn sheep hunting regulations and methodology for determining harvest quotas. *Desert Bighorn Council Transactions* 31, 37–38.

Science and the Recreational Hunting of Lions

Andrew J. Loveridge[1], Craig Packer[2] and Adam Dutton[1]

[1]Wildlife Conservation Research Unit, Department of Zoology, University of Oxford, Tubney, UK
[2]Department of Ecology, Evolution and Behavior, University of Minnesota, St. Paul, MN, USA

Introduction

The African lion *Panthera leo* is an iconic species and one by which the fortunes of conservation efforts on the African continent can be gauged. Lions require large areas of relatively pristine habitat and sizeable populations of medium sized ungulate prey. They have the potential to act as an umbrella species in that protection of a population of African lions also protects biodiversity in the area over which it ranges. The widely held consensus is that the species has declined in both number and geographic range in recent decades. Although no quantitative historical estimates of lion numbers are available, it is thought that continental populations have declined from 1 to 200,000 (Nowell & Jackson, 1996) to between 23,000 and 47,000 over the last few decades (Chardonnet, 2002; Bauer & van der Merwe, 2004). Because of poor historical data, the extent to which the decline has occurred will probably never be known. However, the estimated 82 per cent decline in geographic range from over 22 million km^2 to just over three million km^2 (Ray *et al.*, 2005) resonates

Recreational Hunting, Conservation and Rural Livelihoods: Science and Practice, 1st edition.
Edited by B. Dickson, J. Hutton and W.M. Adams. © 2009 Blackwell Publishing,
ISBN 978-1-4051-6785-7 (pb) and 978-1-4051-9142-5 (hb).

with the proposed decline in number and suggests that estimates of numerical decline are of the right order of magnitude.

The African lion is currently listed as 'vulnerable' by the World Conservation Union (IUCN) (IUCN, 2000), and is listed in Appendix II of the Convention on International Trade in Endangered Species of Wild Fauna and Flora (CITES). The alarming decline in lion numbers prompted a proposal to upgrade lions from Appendix II to Appendix I at the 13th CITES conference of parties (CITES, 2004). While this proposal was withdrawn, it has focused the conservation community's concern over the decline in lion numbers and stimulated debate over the possible causes of decline. Among these is trophy hunting.

In this chapter we examine the extent to which trophy hunting of lions affects the conservation status of the species. We discuss the impact of trophy hunting on lion populations and the conditions under which this can be sustainable and highlight situations where trophy hunting may be less advisable.

Background to trophy hunting of lions

Hunting lions for sport has occurred for thousands of years with records of lion hunts going back to the time of Pharaoh Amenhoteb III in 1400 BC (Guggisberg, 1962) and medieval tapestries depict European nobles hunting lions. African people have a tradition of hunting lions as a test of manhood – notably the still-extant tradition of *olamaiyo* where young Maasai men prove their courage by killing a lion with only a spear (Kruuk, 2002). In colonial Africa, lions were viewed as vermin and killed wherever possible; this hunting was, at least partially, motivated by 'the thrill of the chase', particularly amongst the colonial elite hunting in the more remote areas of East and southern Africa. Early hunting expeditions to the Serengeti typically killed large numbers (sometimes hundreds) of lions (Turner, 1987) often in the mistaken belief that removal of carnivores would 'protect' ungulate populations, thereby providing improved opportunities for hunting. Similarly, misguided motivations saw 450 lions killed by rangers in the 1950s in Kruger National Park, South Africa (Smuts, 1978a; Frank & Woodroffe, 2001).

Lions are still a popular species amongst contemporary trophy hunters, although in general harvests of lions tend to be more restrained, perhaps because predators are no longer viewed as vermin (at least amongst Western sport hunters) and because there is an improved understanding of the role of

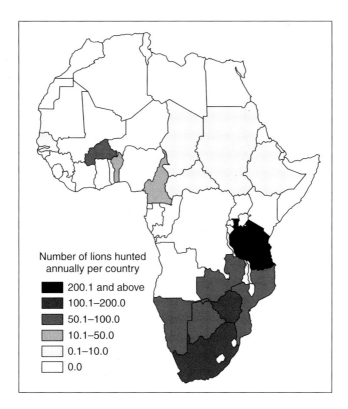

Figure 7.1   **Map of Africa showing the number of lions trophy hunted per year in the range states that allow lion trophy hunting. Number of lions hunted in west and central Africa from Chardonnet (2002). Number of lions trophy hunted in East and southern Africa calculated as mean of trophy exports 2000–2004 (UNEP-WCMC, 2006) for all countries except Botswana where pre hunting ban exports (1997–2000) where used and Zimbabwe, where trophy hunting offtake data from the WWF-PWMA NP9 database (1998–2002) were used.**

predators in ecosystem function. In sub-Saharan Africa, 15 of the 32 countries that have populations of lions allow trophy hunting of lions (Figure 7.1). In West Africa only Senegal, Burkina Faso and Benin permit lion hunting, with Burkina Faso apparently having had a sustainable offtake of around 12 lions per year over the last 20 years (Chardonnet, 2002). Three central African countries offer trophy hunting of lions, Chad, Cameroon and the Central African

Republic, of which Cameroon is reported to have stable offtakes of around nine lions per year.

As the majority of lions occur within the eastern and southern regions of the continent (especially Tanzania, Zimbabwe, Botswana, Zambia, Namibia, South Africa and Mozambique), it is not surprising that most lion trophy hunting occurs in these countries, with nine of the 16 countries in the region allowing trophy hunting of lions. By far the most lion trophy hunting on the continent occurs in Tanzania, with an average of 250 lions trophy hunted each year. However, since Tanzania is estimated to hold around half the lions present on the continent (14,432 individuals; Chardonnet, 2002) and when hunting offtakes are viewed as a proportion of the total population Tanzania's hunting offtake is only around 1.8 per cent of the population (Figure 7.2). Zimbabwe, Botswana, Zambia, Namibia, South Africa and Mozambique, are key lion range states in the Southern African region. Trophy hunters visiting Botswana, Zambia and Mozambique typically export between 13 and 18 lion trophies per year, while Zimbabwe and South Africa export around 76 and 170 trophies per year respectively (UNEP-WCMC, 2006).

Trophy hunting does not account for more than 2 to 4 per cent of the total lion population of most African countries each year (Packer *et al.*, 2006), Figure 7.2), and no African country exceeds the 6 per cent offtake that World Wide Fund for Nature (WWF) recommends as being a sustainable harvest from a lion population (WWF, 1997). This suggests that on a continental scale trophy hunting offtake is relatively conservative. Zimbabwe and South Africa stand out as exporting higher numbers of trophies in relation to wild populations than their neighbours. In the case of Zimbabwe this may be because of the practice of including female lions on hunting quotas: between 1998 and 2002, 29 per cent of lions trophy hunted were female (Figures 7.1 and 7.2). In the case of South Africa, many trophy exports are likely to have been derived from so called 'canned hunts' where captive-bred lions are hunted, frequently in small enclosures or under artificial conditions. Thus the offtake figures for South Africa are unlikely to reflect accurately the potential impact of trophy hunting on the country's wild populations. The controversial activity of hunting captive-bred lions does not pose a direct problem for the conservation of wild lion populations, unless wild lions are captured for the trade. It is unclear whether 'canned hunting' reduces the demand for trophy hunts of lions from wild populations. The ethics of this type of hunting are widely questioned and the very public controversy surrounding the issue (Michler, 2002) may colour public perceptions of trophy hunting of wild animals in operations that are less ethically controversial and contribute to conservation of wild populations.

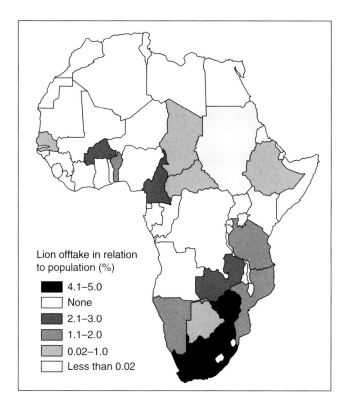

Figure 7.2 **Map of Africa showing trophy hunting offtake of lions in relation to estimated population size in each country. Population sizes taken from Chardonnet (2002), trophy exports are from the same sources as in Figure 7.1.**

Over the last decade, hunting of lions has been temporarily suspended in a number of southern African countries due to concerns over management. Zambia suspended all trophy hunting in 2002 while wildlife management institutions were restructured and Zimbabwe has suspended lion trophy hunting in western Zimbabwe from 2005 to 2008 due to concerns over excessive offtakes in previous years. Botswana imposed a ban from 2001 to 2004 due to concerns over management of problem animal control. In the light of Botswana's relatively conservative trophy hunting offtakes, this ban may have been unnecessary from a population management point of view. The local hunting industry in the region is estimated to have lost US$1.26 million during the first year of the ban (Botswana Wildlife Management Association (BWMA), 2001).

Does trophy hunting impact lion populations?

Lion populations can be incredibly resilient to perturbation, provided the social structure of the population remains relatively intact and particularly if immigration is possible from nearby populations. Experimental culling of prides in Kruger National Park in the 1970s showed that population numbers returned to their original levels in depleted areas within 15 months, despite some prides being culled by 60 per cent and others completely removed. Recovery was largely due to immigration from the surrounding undisturbed population and by increases in the reproductive rate (Smuts, 1978b). Similarly, relatively high offtakes in Matetsi Safari Area, Zimbabwe (0.83 males per 100 km^2) in the 1970s and 1980s may have depressed population numbers, but did not result in complete extirpation (Loveridge *et al.*, in preparation; Packer *et al.*, 2006). The lion population of Amboseli National Park, Kenya, was extirpated in the 1990s by pastoralists in the areas surrounding the park, but recolonisation occurred within a few years from population reservoirs close to the park (Frank *et al.*, 2006). Similar population recoveries have been observed after disease outbreaks. The Serengeti population declined by 30 per cent after a canine distemper virus outbreak in 1993–1994, however the population had fully recovered within a few years (Packer *et al.*, 1999).

There are examples of trophy hunting impacting lion populations. Yamazaki (1996) found that widespread removal of male lions by trophy hunters altered social behaviour in Luangwa National Park, Zambia. Similar social perturbations were found in Hwange National Park (HNP), Zimbabwe, where high offtakes of both adult sub-adult male lions in surrounding safari concessions altered adult sex ratios and male ranging behaviour (Loveridge *et al.*, 2007). In this case hunting quotas were exceptionally high in relation to population size. For instance in the Gwaai Intensive Conservation Area (ICA), adjacent to HNP, average quotas between 1999 and 2003 were 1.8–3.1 lions per 100 km^2, and offtakes of male lions were on average 0.6 per 100 km^2 (range 0.3–1.1 per 100 km^2). Recent surveys of the ICA (Davidson & Loveridge, 2006) suggest that lion densities in the area may only be around 0.7 lions per 100 km^2, despite a recovering population after suspended hunting of lions in 2004–2006. Loveridge *et al.* (2007) showed that hunting on the boundary of the park created a vacuum effect which drew male lions from the protection of the park into male-depleted hunting areas. It is likely that high trophy hunting harvests were only sustained because of the proximity of the national park.

Another example of social perturbation caused by trophy hunting was the introduction of female hunting quotas in nearby Matetsi Safari Area, Zimbabwe, in the early 1990s. This may have caused population declines. In turn, declining population sizes and a relatively high male lion harvest appear to have reduced trophy quality and the number of male lions harvested in the hunting area (Packer *et al.*, 2006; Loveridge *et al.*, in preparation). This may have been a direct consequence of introducing harvests of females. Regular removal of even small numbers (>3 per cent) of reproductive females has been shown to expose a lion population to decline (Hermann *et al.*, 2002; Van Vuuren *et al.*, 2005). Reproductive success in lions is closely related to pride size and prides of three or more adult females are significantly more successful at rearing cubs than smaller prides (Packer *et al.*, 1988). Apart from overall population reduction, one of the consequences of killing adult females is the reduction of the size and success of breeding units in the population. Populations of long-lived species have been shown to be relatively resilient to population disturbance if the reproductive female life stage remains intact, as with cheetahs *Acinonyx jubatus* (Crooks *et al.*, 1998) and loggerhead turtles *Caretta caretta* (Crouse *et al.*, 1987), and the same is likely to be true of lions.

Despite these examples there is a growing consensus amongst conservationists that lions can be trophy hunted on a sustainable basis, particularly if hunting targets only surplus males. This is supported by a model based on population parameters from long-term lion research in the Serengeti, Tanzania (Whitman *et al.*, 2004). This model shows that infanticidal behaviour by new males replacing territorial males shot by trophy hunters can exacerbate population decline, particularly if young males are harvested. However, if only males of six years or older are removed from the population, there is very little impact on the size and viability of the harvested population (Figure 7.3). This finding may form the basis for lion management in the future.

Quotas and quota setting

Traditionally, African trophy hunting has been managed on the basis of licences sold to shoot individual animals, and harvests were (and still are) often controlled by quotas limiting offtake. Quotas are usually set on the basis of the

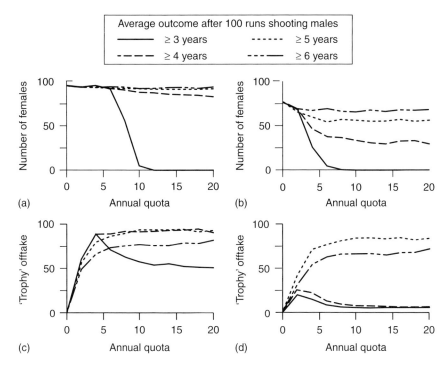

Figure 7.3   (a)–(d) Effects of trophy hunting as a function of quota size and male age. (a) Number of adult females after 30 years in hypothetical populations where males are non-infanticidal. (b) Number of females in infanticidal populations; note that infanticidal populations are smaller and more vulnerable to trophy hunting. (c) Total number of males harvested over 30 years in infanticidal populations. (d) Total number of 5–6 year old 'trophies' harvested in infanticidal populations. From Whitman *et al.* (2004).

population size, trends in trophy quality, and hunting success in relation to quotas in the previous season (WWF, 1997; Taylor, 2001). Where suitable data exist, changes in 'catch effort' can also be taken into account, although this measure can be confounded by changing demand for hunts between years or changing hunter effort.

Carnivores, including lions, are notoriously difficult to census (Loveridge *et al.*, 2001) which means that quotas are often arbitrary or set on the basis of historical precedent and may be unrealistic (e.g. Loveridge *et al.*, 2007). For example, quotas for male lions in Tanzanian hunting concessions were on average 0.38 lions per 100 km^2 (Whitman *et al.*, 2004), while quotas in concessions around Hwange National Park were 2.5 and 6.6 times higher (0.9–2.5 lions per 100 km^2), despite generally lower population densities compared to Tanzanian populations.

For lions a maximum offtake of 6 per cent of the population (WWF, 1997) or 5 to 10 per cent of adult males (Creel & Creel, 1997; Greene *et al.*, 1998) has been recommended. These harvest levels are consistent with recommended harvests of other carnivores, e.g. brown bears *Ursus arctos* (Swenson *et al.*, 1994). However, because accurate population numbers do not exist for most lion populations, and census methods are often time-consuming, expensive and require considerable expertise, it is often difficult to determine appropriate quotas based on percentage offtake in relation to total population size.

Indirect measures of whether quotas and harvest rates are suitable can also be used. These can be used independently or in conjunction with population estimates. Trends in trophy size (usually measured using the Safari Club International criteria of the sum of the length and breadth of the skull) can easily be monitored in hunting areas, provided trained staff are available to measure and record sizes. This can be a useful measure of whether or not harvest rates are appropriate, with declining trophy sizes potentially indicative of excessive offtake. Long-term records of trophy quality are useful in this regard. An analysis of long-term records undertaken for Matetsi Safari Area (Loveridge *et al.*, in preparation, Figure 7.4) revealed declines in the size of the largest trophies over a 20-year period, prompting a reassessment of hunting quotas in this hunting area (Grobbelaar & Masulani, 2003).

Similarly, where ages of hunted animals can be accurately estimated, this can provide a useful indicator of offtake. As noted above, Whitman *et al.* (2004) suggest that, because of the difficulty (and often arbitrary nature) of quota setting, it might be more effective to limit harvest on the basis of age, removing males that are six years or over (Figure 7.4). Assessment of age in the field (before the animal is shot) and verification once the trophy has been retrieved could lead to an effective means of managing and monitoring sustainable trophy hunting of lions.

Whitman *et al.* (2004) showed that the lions' noses become increasingly pigmented with age, and Whitman *et al.* (2007) showed that the 'six-year rule' could be achieved by restricting hunting to males with noses that are at least 60 per cent black. Although the tip of the nose may not be the easiest metric to evaluate in the field, it is a far more reliable indicator of age than the lion's

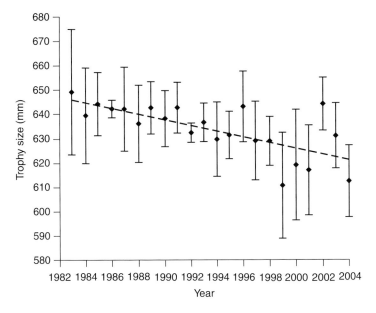

Figure 7.4 **Trophy sizes (based on Safari Club International measurements of length + breath of skull) of the largest five lions shot in Matetsi Safari Area 1983–2004. Trophy size declined over this period.** $r^2 = 51.9$, **Pearson's correlation** $= 72, P = 0.00.$

mane. Male lions show far greater variation in mane darkness, hair length or extent of mane at a given age. Further criteria are currently in development that will hopefully be easier to measure in field conditions, such as overall coat condition on the face, tooth coloration and wear, and the degree of discoloration on the backs of the legs of males that have just started breeding (Whitman & Packer, 2006). It is essential to verify the lion's age after the animal has been hunted. The pulp cavities of the lions' teeth fill in before five years of age, and wear patterns of the carnassial teeth also indicate the dead animal's age (Whitman & Packer, 2006).

To ensure the sustainability of harvested populations monitoring is required at some level, either of population size or of trophy quality and/or age. Population monitoring has the advantage that it can be done before animals are harvested, however it is expensive, time consuming and requires extensive expertise if robust estimates are to be obtained. Most wildlife management institutions in Africa do not have the capacity to undertake this kind of survey on a regular basis. For this reason monitoring trophy quality or trophy age in age-limited harvests may be more feasible, in that in most cases trophies for export must be submitted for inspection, veterinary controls (dipping) and for export paperwork to be provided. It may be relatively easy for the relevant measures or samples (e.g. teeth) to be taken in order to determine trophy quality (whether in terms of trophy size or age). Adaptive management has long been a feature of African conservation. Monitoring trophy quality may allow quotas to be increased or decreased on the basis of the age/trophy quality of harvested trophies. In addition it may be feasible to make submission of trophy measures/samples a prerequisite for exportation of a trophy.

Monitoring the behaviour of trophy hunters

While there are many advantages to replacing a quota system with an age-based harvest strategy, and several hunting organisations have formally adopted the six-year age minimum for 'trophy' lions (e.g. the Tanzanian Hunting Operators Association (TAHOA) and the Niassa Reserve in Mozambique) only the Niassa Reserve has adopted any sort of formal verification system. Professional hunters and their clients often lack a clear appreciation of how to estimate a lion's age (Whitman & Packer, 2006), and the Internet abounds with examples of hunting companies showing photographs of lions recently shot by their clients that are far younger than four years of age. The disadvantage of allowing harvests of lions to be based entirely on age as estimated by hunting operators is that if too many young males are shot (either inadvertently through inaccurate ageing, or because trophies and ages are not adequately monitored by managers) then harvests could have deleterious effects on populations, adding to existing threats such as habitat loss and control associated with human–lion conflict.

Following the example of the Forest Stewardship Council (FSC), efforts are currently under way by Packer to expand on the success of the Niassa Reserve in implementing transparent oversight on lion harvests by establishing a conservation certification system called Savannas Forever in Tanzania, Mozambique and Botswana. In this system, lion trophies would be inspected by an independent third party to estimate the animals' ages and to provide feedback to the companies on improving their harvest strategies in the future.

Trophy hunting as a problem animal control tool

There is some suggestion that trophy hunting could be used as a tool by which problem lions could be controlled, instead of relying on official Problem Animal Control (PAC) operations. This would serve the dual purpose of eliminating particular problem animals and earning revenue for local communities, potentially leading to greater tolerance of predators if financial rewards were forthcoming. While at face value this appears to be an attractive win–win situation, the practicalities of marketing specific problem animals to a largely overseas based clientele present considerable challenges. It is often difficult to predict the spatio-temporal occurrence of problem animals, which in turn makes the logistics of having hunting clients in the right place at the right time problematic. Furthermore, it is not always possible to identify the actual individual or group of individuals responsible for livestock predation or human attack after the fact. Independent verification of whether or not conflict is in fact occurring may be needed in order to prevent bogus claims being put forward about 'problem' animals in order to increase quotas. Furthermore, there is a danger that selling 'problem animal hunts' might provide a perverse incentive for poor animal husbandry, which is ultimately detrimental to lion conservation.

Nevertheless, hunting guides and operators and their clients do often respond to requests to deal with 'problem' animals on an *ad hoc* basis while hunting in an area, and this may improve local perceptions of both hunting and 'problem' animals. It is, however, questionable that trophy hunting on its own is the solution to problem animal control. For example, in the US it is often argued that trophy hunting cougars *Puma concolor* reduces the incidence of harmful encounters between this species and people. However there is little scientific evidence that trophy hunting of cougars mitigates this conflict (Cougar Management Guidelines Working Group, 2005), largely because hunting does not suppress cougar populations sufficiently to decrease the frequency with which people and cougars encounter one another, particularly when human settlement and leisure activities encroach on cougar habitat (Cougar Management Guidelines Working Group, 2005). In many cases quick resolution to conflict is advisable, particularly where people have been killed. In the case of cougars, and probably also in the case of lions, individual problem animals need to be removed expeditiously by a professional PAC unit rather than by paying trophy hunters.

Currently, trophy hunting can assign a monetary value to 'problem' species and this may increase tolerance, especially if revenue from trophy hunting can offset losses. One example of this is the rehabilitation of farmers' perceptions of wild cheetahs *Acinonyx jubatus* in Namibia. Here, recently introduced CITES export quotas have allowed trophy hunting to take place on private ranches – where previously, because of livestock predation, cheetahs had been viewed solely as a pest species (Marker *et al.*, 2003). The potential financial benefit of tolerating cheetahs (as well as strong advocacy on the part of local conservationists) has led to reduced levels of mortality and an increasing cheetah population. This is a particularly interesting case as trophy hunting has been linked specifically to control of problem individuals in that the Namibian Professional Hunters Association attempts to link potential trophy hunters directly with farmers who have problem cheetahs on their property (Leader-Williams & Hutton, 2005). This, in theory, enhances the general incentive to tolerate cheetahs with the added incentive of controlling specific problem animals rather than destroying them. If such a scheme could be made viable for problem lions (which of course offer much greater potential threat to human life), trophy hunting of problem lions could reap similar benefits.

Conclusions

Substantial areas of well-connected habitat with abundant natural prey populations are crucial for healthy, self-sustaining lion populations. Problem animal control, whether legal or illegal, together with loss of habitat, have significantly more impact on lion populations than legalised hunting. Hunting contributes substantially to protection of habitat, particularly in East and southern Africa (Loveridge *et al.*, 2007). Trophy hunting, when well regulated and well monitored, has small or negligible impact on the viability of lion populations, and hunting areas contribute substantially to the total area available for conservation of this species. In the face of expanding human and livestock populations, protecting habitat and prey populations is likely the most important single factor in the conservation of lions in Africa. The consensus is that well-managed hunting of lions can improve the fortunes of hunted populations by mitigating the effect of other anthropogenic factors, for example by creating revenue streams that reduce motivations to kill lions illegally and provide justification for protection of habitat, particularly in marginal areas unsuitable for other livelihood activities.

It is difficult and often expensive to monitor lion populations, which makes quota setting challenging. Age-based harvests of male lions and indirect monitoring of trophy age and quality may provide a more cost-effective and practical solution. However, verification of proper hunting practices through an independent certification process is essential not only to protect lions from unscrupulous practices by rogue hunting operators but also to improve public perception of hunting as a valid conservation tool.

However, there may be situations under which trophy hunting from a population may not be advisable. Small or declining populations may be more vulnerable to trophy hunting than current models predict. This is particularly true of populations that are also subject to high levels of other forms of mortality such as illegal killing of lions, where additional mortality caused by trophy hunting may exacerbate population decline.

References

Bauer, H. & van der Merwe, S. (2004) Inventory of free-ranging lions *Panthera leo* in Africa. *Oryx*, 38, 26–31.

Botswana Wildlife Management Association (BWMA) (2001) *Economic Analysis of Commercial Consumptive Use of Wildlife in Botswana*. Botswana Wildlife Management Association and ULG Northumbrian Ltd., Leamington Spa, UK.

Chardonnet, P. (2002) *Conservation of the African Lion: Contribution to a Status Survey*. International Foundation for the Conservation of Wildlife, France & Conservation Force, USA.

Convention on International Trade in Endangered Species of Wild Fauna and Flora (2004) *Consideration of Proposals for Amendment of Appendices I and II. Transfer of Panthera leo from Appendix II to Appendix I, CoP13 Prop. 6*. CITES Secretariat, Châtelaine-Geneva, Switzerland.

Cougar Management Guidelines Working Group (2005) *Cougar Management Guidelines*. WildFutures, Washington.

Creel, S. & Creel, N.M. (1997) Lion density and population structure in the Selous Game Reserve: evaluation of hunting quotas and offtake. *African Journal of Ecology*, 35, 83–93.

Crooks, K.R., Sanjayan, M.A. & Doak, D.F. (1998) New insights on cheetah conservation through demographic modeling. *Conservation Biology*, 12, 889–895.

Crouse, D.T., Crowder, L.B. & Caswell, H. (1987) A stage based population model for loggerhead sea turtles and implications for conservation. *Ecology*, 68, 1412–1423.

Davidson, Z. & Loveridge, A.J. (2006) *The Gwaai Conservancy. Lion Density and Quota Recommendations based on a Poor Density Estimation*. Wildlife Conservation Research Unit, Oxford, UK.

Frank, L., Maclennan, S., Hazzah, L., Bonham, R. & Hill, T. (2006) *Lion Killing in the Amboseli-Tsavo Ecosystem, 2001–2006, and its Implications for Kenya's Lion Population*. Kilimanjaro Lion Conservation Project & Predator Compensation Fund, Nanyuki & Nairobi, Kenya.

Frank, L.G. & Woodroffe, R. (2001) Behaviour of carnivores in exploited and controlled populations. In *Carnivore Conservation,* eds. J. Gittleman, S. Funk, D.W. Macdonald & R. Wayne, pp. 419–442. Cambridge University Press, Cambridge.

Greene, C., Umbanhowar, J., Mangel, M. & Caro, T. (1998) Animal breeding systems, hunter selectivity, and consumptive use in wildlife conservation. In *Behavioural Ecology and Conservation Biology* ed. T. Caro, pp. 271–305. Oxford University Press, Oxford and New York.

Grobbelaar, C. & Masulani, R. (2003) *Review of Offtake Quotas, Trophy Quality and 'Catch Effort' Across the Four Main Wildlife Species Elephant, Buffalo, Lion and Leopard*. WWF SARPO, Harare, Zimbabwe.

Guggisberg, C.A.W. (1962) *Simba, the Life of the Lion*. Bailey Bros & Swinfen, London.

Hermann, E., Van Vuuren, J.H. & Funston, P.J. (2002) Population dynamics of lions in the Kgalagadi Transfrontier Park: a preliminary model to investigate the effect of human-caused mortality. In *Lion Conservation Research. Workshop 2: Modelling Conflict,* eds. A.J. Loveridge, T. Lynam & D.W. Macdonald. Wildlife Conservation Research Unit, Oxford.

IUCN (World Conservation Union) (2000) *2000 IUCN Red List of Threatened Species*. IUCN, Gland, Switzerland.

Kruuk, H. (2002) *Hunter and Hunted. Relationships Between Carnivores and People*. Cambridge University Press, Cambridge, UK.

Leader-Williams, N. & Hutton, J.M. (2005). Does extractive use provide opportunities to offset conflicts between people and wildlife? In *People and Wildlife: Conflict or Co-existence?* eds. R. Woodroffe, S. Thirgood & A. Rabinowitz. Cambridge University Press, Cambridge.

Loveridge, A.J., Lynam, T. & Macdonald, D.W., eds. (2001) *Lion Conservation Research. Workshop 1: Survey Techniques*. Wildlife Conservation Research Unit, University of Oxford, Oxford, UK.

Loveridge, A.J., Reynolds, J.C. & Milner-Gulland, E.J. (2007) Does sport hunting benefit conservation? In *Key Topics in Conservation*, eds. D.W. Macdonald & K. Service, pp. 224–240. Blackwell Publishing, Oxford.

Loveridge, A.J., Searle, A.W., Murindagomo, F. & Macdonald, D.W. (2007) The impact of sport hunting on the population dynamics of an African lion population in a protected area. *Biological Conservation*, 134, 548–558.

Loveridge, A.J., Murindagomo, F., Moyo, G. & Macdonald, D.W. (in prep) *Impact of trophy hunting on lions in Matetsi Safari Area, Zimbabwe*. Paper not yet submitted.

Marker, L.L., Mills, M.G.L. & Macdonald, D.W. (2003) Factors influencing perceptions of conflict and tolerance towards cheetahs on Namibian farmlands. *Conservation Biology*, 17, 1290–1298.

Michler, I. (2002) To snap or snipe? *Africa Geographic*, 10(9). www.africageographic. com/archives [last accessed 15 August 2008].

Nowell, K. & Jackson, P. (1996) *Wild Cats. Status Survey and Conservation Action Plan.* IUCN/ SSC Cat Specialist Group, Gland, Switzerland.

Packer, C., Altizer, S., Appel, M. *et al.* (1999) Viruses of the Serengeti: patterns of infection and mortality in African lions. *Journal of Animal Ecology*, 68, 1161–1178.

Packer, C., Herbst, L., Pusey, A. *et al.* (1988) Reproductive success of lions. In *Reproductive Success: Studies of Individual Variation in Contrasting Breeding Systems*, ed. T.H. Clutton-Brock, pp. 363–383. University of Chicago Press, Chicago.

Packer, C., Whitman, K., Loveridge, A.J., Jackson, J. & Funston, P.J. (2006) Impacts of trophy hunting on lions in East and Southern Africa. Recent offtake and future recomendations. http://www.felidae.org/JOBURG/lion.htm

Ray, J.C., Hunter, L. & Zigouris, J. (2005) *Setting Conservation and Research Priorties for Larger African Carnivores*. Wildlife Conservation Society, New York.

Smuts, G.L. (1978a) Interrelations between predators and prey, and their environment. *BioScience*, 28, 316–320.

Smuts, G.L. (1978b) Effects of population reduction on the travels and reproduction of lions in Kruger National Park. *Carnivore*, 1, 61–72.

Swenson, J.E., Sandegren, F., Bjärvall, A., Söderberg, A., Wabakken, P. & Franzén, R. (1994) Size, trend, distribution and conservation of the brown bear *Ursus arctos* population in Sweden. *Biological Conservation*, 70, 9–17.

Taylor, R. (2001) Participatory natural resource monitoring and management. Implications for conservation. In *African Wildlife and Livelihoods. The Promise and Performance of Community Conservation* eds. D. Hulme & M. Murphree, pp. 267–279. James Currey, Oxford, UK.

Turner, M. (1987) *My Serengeti Years*. W.W. Norton & Company, New York.

United Nations Environment Programme – World Conservation Monitoring Centre (2006) *CITES Trade Database*, UNEP-World Conservation Monitoring Centre, Cambridge, UK.

Van Vuuren, J.H., Herrmann, E. & Funston, P.J. (2005) Lions in the Kgalagadi Transfrontier Park: modelling the effect of human-caused mortality. *International Transactions in Operational Research*, 12, 145–171.

Whitman, K. & Packer, C. (2006) *A Hunter's Guide to Aging Lions in Eastern and Southern Africa*. Safari Press, Brandon, Vermont, USA.

Whitman, K., Starfield, A.M., Quadling, H.S. & Packer, C. (2004) Sustainable trophy hunting of African lions. *Nature*, 428, 175–178.

Whitman, K., Starfield, A., Quadling, H. & Packer, C. (2007) Modeling the effects of trophy selection and environmental disturbance on a simulated population of African lions. *Conservation Biology*, 21(3), 591–601.

World Wide Fund for Nature (1997) *Quota Setting Manual*. World Wide Fund for Nature, Zimbabwe Programme Office, Harare.

Yamazaki, K. (1996) Social variation of lions in a male-depopulated area in Zambia. *Journal of Wildlife Management*, 60, 490–497.

Livelihoods

(8)

Sportsman's Shot, Poacher's Pot: Hunting, Local People and the History of Conservation

William M. Adams

Department of Geography, University of Cambridge, Cambridge, UK

Hunting and the history of conservation

The histories of sport or recreational hunting and conservation are closely entwined. Hunting, particularly the shooting of large mammals, had an important place in the history of wildlife conservation in the 20th century. A critical source of conservation sensibilities in the late Victorian era came from those who hunted 'big game' in the British Empire, such as the so-called 'penitent butchers' of the Fauna Preservation Society (Fitter & Scott, 1974). The modern patterns and practices of conservation that developed through the 20th century owed a great deal to the political organisation of elite hunters (for example in the British colonial conservation organisation the Society for the Preservation of the Wild Fauna of the Empire (SPWFE) (1905), or the US Boone and Crockett Club (MacKenzie, 1988; Prendergast & Adams, 2003; Adams, 2004).

There is a standard historical narrative of the way hunting stimulated and eventually gave way to conservation through the first half of the 20th century.

Recreational Hunting, Conservation and Rural Livelihoods: Science and Practice, 1st edition.
Edited by B. Dickson, J. Hutton and B. Adams. © 2009 Blackwell Publishing,
ISBN 978-1-4051-6785-7 (pb) and 978-1-4051-9142-5 (hb).

It draws especially on accounts of British colonial Africa (e.g. MacKenzie, 1988, 1997; Neumann, 1996, 1998), where commercial hunting for ivory and skins gave way to hunting as a subsidy for European advance, leading in turn to the development of ritualised and idealised hunting practice by a colonial elite obsessed with trophies, sportsmanship and other ideals of British boys' education (MacKenzie, 1988). Alongside these, hunting for subsistence was important among the indigenous population of all continents, before and after the advent of European traders, armies of annexation and colonial administrators. It was important too for those who came to settle on annexed land, settlers drawing on the same animal resources as those they displaced.

By the late 19th century, the depredations of colonial hunters in the African Cape had led to massive reductions in the populations of plains antelope and elephant. Game regulations were passed in the Cape in 1886, and subsequently in German East Africa (1896) and British East Africa (1897 in Uganda and the East African Protectorate; 1900 in Kenya: MacKenzie, 1988; Grove, 1987; Beinart & Coates, 1995).

The first members of the SPWFE, founded in London in 1903, included politicians and aristocracy, colonial administrators, businessmen and a scattering of scientists and naturalists. Many were hunters like the founder, Essex businessman and saviour of Epping Forest, Edward North Buxton (Prendergast & Adams, 2003). One early member commented 'its first year saw it in the shape of a modest and unpretentious group of gentlemen of wide experience of the outposts of Empire, and a common enthusiasm for the preservation from wanton destruction of many of its fauna' (Whitbread, 1907, p. 10). Around the same time, elite hunters in the United States were also becoming preservationists. From the 1860s recreational hunting and fishing had become an integral part of a romanticised 'frontier' experience enjoyed by leading east coast industrialists in the US (Jacoby, 2001). This era, and these sensibilities, are epitomised by Theodore Roosevelt. On a hunting trip in Dakota in 1887, he found the plains empty of game, the buffalo hunted out. In response he founded the Boone and Crockett Club in 1888, which became influential in conservation, both in the US and internationally. Other organisations soon followed, such as the New York Zoological Society (1895).

Given the prevalence of hunters among its ranks, it is interesting that the SPWFE remained distinct from purely hunting organisations such as the Shikar Club, founded in 1908. There was substantial cross-membership between the Societies, and the question of merger was discussed by the SPWFE in 1908, and again in 1925 (McKenzie, 2000). By that time the Society was far more concerned with the creation of national parks as inviolable sanctuaries for game than about the niceties of sportsmanship, however hunting was still important to a number of leading members of the society. Lord Onslow, the SPFE's President, said in 1928 'I do not think anybody believes that this Society will interfere in any way with reasonable and legitimate sport … we count on all big game sportsmen for their help'.[1] In that year the Shooting editor of the *Field* agreed to give prominence to the Society's work.

Hunting, development and tourism

In the early decades of the 20th century, conservationists justified the establishment of game reserves in African colonies (in the face of sceptical governments and disapproving settlers) on the basis that safari hunting provided valuable revenue to the territorial exchequer. Edward Buxton presented the business case for preservation to the British Colonial Office in 1907, arguing that a share of the revenues from hunting licences could help pay for effective game protection by well-qualified staff; indeed, he argued that reserves could even create a profit as well as providing an outlet for the energies of young officers in remote areas.[2] Henry Seton-Karr, hunter, traveller and Member of Parliament, claimed that wild fauna contributed to the material wealth and revenue of the Empire. He estimated that the direct annual revenue from licences in British East Africa was between £8000 and £10,000 a year, sums not to be despised in 'a young and sparsely populated portion of the Empire' (Seton-Karr, 1908).

The idea of game as a resource that could be managed to maximise human benefit received its strongest support in the US. It drew closely on the model of rational 'wise use' conservation developed in the US during the Progressive Era at the turn of the 19th and 20th centuries (Hays, 1959). Such thinking was extended to game management in the American Game Policy Committee in the 1920s, chaired by Aldo Leopold, whose 1933 book *Game Management* established a scientific discipline and underpinned the utilitarian approach to wild species managed for hunting.

Arguments about the economic benefits of sport hunting were supported by the growth of the safari industry in the 20th century. Elite hunters from Europe and North America beat a path to Africa, and a safari industry evolved to package their experiences and meet their needs for luxury and sport (Cameron, 1990): in the track of Theodore Roosevelt (1909) came the Duke and Duchess of York (1924–25), the film magnate George Eastman (1926 and 1928), the Prince of Wales (1928–30), and Ernest Hemingway (1934). Hemingway's stories *The Short Happy Life of Francis Macomber* and *The Snows of Kilimanjaro* became widely known and reprinted. Through this and other means, African safari hunting became a global cultural phenomenon, romantic, dangerous and irresistibly associated with the rich and powerful. As hunting holidays became more packaged, so such sport became more accessible, and the safari industry grew. Air flights and motor travel allowed speedier access to both Africa and game, although they also presented new threats, including the 'monstrous habit' of shooting from vehicles.[3]

However, by the 1930s, the safari industry was changing away from hunting. There was already a transition from bullets to film as a strategy for the capture of trophies, and the photo-safari was advocated by conservationists as more skilful and more humane. Walt Disney's *Bambi* was made in 1942, and the culture of hunting in the US began to be overlain for the growing urban middle class with a new humanitarian view of nature. At the same period in East Africa, the hunting safari began to play a secondary role to game viewing safari tourism. It ushered in an era where conservation, especially national parks, became associated not with hunting but with nature viewing. This trend was long-established in the US, where the first tourist lodge had been built at Yosemite as early as 1857. National parks such as Yellowstone in the US and Banff in Canada were central to the boosting of trans-continental railroads. By the end of World War I, the automobile had taken over: by 1918 seven times more people visited Yellowstone by car than rail. The model spread, for example among middle-class white people in South Africa following the creation of the Kruger National Park in 1926 (Carruthers, 1995). By the 1930s, the SPFE was advocating national parks in East Africa to support an international tourist industry.

By the second half of the 20th century, nature viewing had become an integral part of the automobile economy and urban middle-class lifestyles: even in the 1950s, 99 per cent of visitors to US national parks came by car. In the 1950s also, the advent of inter-continental air travel by jet airliner ushered in international wildlife tourism as a mass industry. The vast majority of these new tourists came to see the real-life equivalents of Bambi or Dumbo, or the new wild stars of TV documentaries made by Armand Denis or David

Attenborough and their myriad successors. The new industry was built on the idea of viewing game in areas where it was nurtured and protected, not on the generation of revenues by bringing in elite hunters to shoot it. Conservation, underpinning and justified by this industry, came to be understood in terms of the protection of the wild against human threat. Hunting, whether for recreation or for food, increasingly became part of that threat to nature, not part of the solution to its conservation.

Except for the first two or three decades of the 20th century, hunting for pleasure was therefore primarily viewed throughout the century as part of the essentially destructive human approach to wild species. Recreational hunting was relegated into the same broad category as the commercial trade in wild species, addressed by campaigns against whaling, fur, ivory and trade in endangered species (Adams, 2004). International hunting, particularly the shooting of large mammals and birds, persisted, but it no longer played a large part in public ideas about conservation. From the 1960s especially, popular campaigns to promote the protection of endangered species drew closer to the animal welfare movement in their anthropomorphism, their emphasis on the beauty and uniqueness of wild species, and their account of the barbarity of human hunting.

Hunting and exclusion

The influence of early hunters in the early 20th century left conservation a range of legacies. The most significant was the idea of exclusion, and the game reserve, as a specific zone where hunting was strictly controlled, with fixed and patrolled boundaries. In many countries, private recreational hunting concentrated on land appropriated by the state as hunting reserves alongside other areas appropriated as national parks. Both tend to exclude local people. The creation of protected areas involves not only the partition of space (between that for nature and that for development) but also the designation of rights – the determination of who has the right to enter, to enjoy wildlife, or to kill it.

The colonial enthusiasm for game reserves drew closely on established European traditions for the regulation of hunting. The idea of parks as royal or aristocratic hunting grounds was long established, reflecting ancient traditions of the Middle East and practices across Eurasia. The critical element in this tradition was its social exclusivity: bounded hunting parks and estates demarcated where hunting might be done, but also who could do it. These were pleasure grounds for an aristocratic elite. Access was physically but also socially restricted. Mammals, birds and fish considered attractive quarry were classified as 'game', and unlicensed killing of game by rural people was strictly prevented. Those who transgressed were punished severely (Thompson, 1975). Formal hunting in Victorian Britain involved a complex social infrastructure of gamekeepers, stalkers and beaters to raise game and drive it to hunters' guns, to stalk deer on cleared Scottish estates, or to pursue foxes on horseback. In the British Empire, modified forms of such practices came to include shooting tigers, sticking pigs, and even hunting jackals and other relatives of the fox with packs of local hounds.

The idea of protected areas was fundamental to the thinking of early hunting conservationists. The trigger for the formation of the SPWFE was a British Colonial Office proposal to degazette game reserve in the Sudan. Between 1905 and 1909, the SPWFE constituted three delegations to the British Secretary of State for the Colonies to argue for stronger hunting regulations and for game reserves. The direct influence of the SPWFE on policy waned in subsequent decades, but its more general impact was maintained between the two world wars, with the Earl of Onslow as its Secretary and (from 1929) the Prince of Wales (later King Edward VIII) as Patron (Neumann, 1996). Their demand from British colonial government was simple: the enforcement of regulations to restrict hunting throughout colonial territories to a select group of licensed sportsmen, and to provide reserves for game where numbers could be replenished. Neumann (1996) accused the SPFE of recreating Edwardian sporting estates (in catastrophic decline at home after the World War I) in the African bush, with rural Africans enrolled as agricultural peasantry, barred from hunting.

Game reserves, set aside on land alienated by the crown from their inhabitants, allowed aristocratic styles of hunting to continue and made them accessible to the new gentlemen managers of Empire. Increasingly, this reservation of land was seen also to be meeting wider purposes in the preservation of game herds for posterity. Buxton believed that conservation was part of the responsibility of empire, arguing 'the nation has "pegged out claims", and must bear charges' (SPWFE, 1905, p. 13). By 1907, it was noted with satisfaction that 'our great administrators are alive to the Imperial obligation of guarding from wanton destruction the marvellous varieties of life which are still to be found within the circumference of His Majesty's Dominions beyond the seas' (Whitbread, 1907, p. 10).

The sporting hunter and the poacher

Historically, formal sport hunting was practised by those who respected certain moral codes of sportsmanship, and indeed could claim the breeding and financial independence to be true 'sportsmen'. Hunting for subsistence, whether by poor settlers or indigenous people, has usually been frowned upon by both recreational hunters and conservationists. In conservation writing, subsistence hunters have conventionally been portrayed as cruel in their methods, greedy in the hunting and lacking in an ability to judge or achieve sustainability.

A key theme in early 20th century conservation writing about Africa was the idea of the 'reckless' hunter. In 1902, Buxton blamed the decline of game in Africa on: 'reckless shooting', bloodthirstiness, the failure of 'true sportsmanship', and excessive killing. He deplored circumstances where 'an otherwise sane man runs amuck' (Buxton, 1902). An editorial in the London *Saturday Review* in 1908 criticised 'rich and irresponsible young Englishmen' who amassed large game bags.

In response, hunting conservationists urged a model of the 'true sportsman'. Henry Seton-Karr maintained that 'British sportsmen, as a class, have done nothing in any wild country to reduce or wipe out any kind of wild big game' (Seton-Karr, 1908). William Hornaday claimed that 'the great mass of worthwhile sportsmen are true protectors and conservators' (Hornaday, 1921, p. 56). This argument remained central to the SPWFE's work: in promoting national parks in East Africa, Richard Hingston maintained 'The sportsman does not obliterate wild life. True, he kills. But seldom is his killing wholesale or indiscriminate' (Hingston, 1931, p. 404). The brute fact of the sportsman's kill was deemed acceptable because of the manner of the hunt (on foot, traversing hard country and facing the wild animal at close range), and the expert cleanness of the killing shot, and the fact that hunters 'sincerely desire the perpetuation of game and hunting sport, and the conservation of the rights of posterity therein' (Hornaday, 1921, p. 56).

The figure whom Hingston, and many before him, referred to as 'the native hunter' was regarded very differently. Colonial ideas about hunting by local people built longstanding opposition to subsistence hunting among British and other European landowners. To some Europeans, poaching had a certain romance, being a product of nostalgia for a bucolic rural past in Victorian Britain, including the myth of Robin Hood and of the skilful lone poacher, outwitting the blundering forces of the law to put meat on the table (for example Richard Jefferies's *The Amateur Poacher c.* 1890s) and John Buchan's sporting gentlemen poaching salmon and stags from hapless neighbours in *John Macnab* (1925). Poachers in colonial Africa were also sometimes regarded with paternalistic tolerance. Much more often, African subsistence hunting was judged haphazard, inefficient, wasteful and cruel. Worst, perhaps, it undermined colonial revenue generation by allowing rural people to survive without entering into gainful employment in cash crop production or wage labour in mines or settler farms. Local hunters not only broke the law, they were too independent. The 'native hunter', Hingston said, 'cares nothing about species or trophies or sex nor does he hunt for the fun of the thing. What the native wants is as many animals as possible for the purpose of either of meat or barter' (Hingston, 1931, p. 404). Romanticism was incompatible with efficiency in killing, and especially with the use of rifles by local hunters.

Early conservationists often claimed that there was a 'colonial duty' to control native hunting (Buxton, 1902). To Buxton, the traditional balance between ill-armed indigenous hunters and abundant populations of prey had been upset by colonialism: '*Pax Britannica*' had created 'new opportunities for killing game'. There were Kamba hunters, 'at every water hole' on the Athi Plains in Kenya because the Maasai were not there to keep them away, and 'everything that walked was killed with poison arrows'. It was a clear colonial duty to prevent indiscriminate hunting: 'as we allow the natives to kill game to a certain extent by preventing fighting among them, we should also prevent their trapping and killing on a large scale' (SPWFE, 1905, p. 17). The SPFE deputation to the Colonial Office in 1930 urged 'a close watch' on native hunting 'to prevent indiscriminate slaughter of game by natives'.

Sport hunting was deemed acceptable, but hunting for subsistence or trade was not. This was true not only of Africa, but also the US and elsewhere. Local indigenous people, including 'Boer farmers' in South and East Africa and poor white farmers in the US, were all regarded as direct threats to game and targeted assiduously by state conservationists as hunting outlaws (Jacoby, 2001). These hunters were not 'sportsmen'. Those who hunted for food, or for trade, were seen by definition to lack any sense of 'sportsmanship'. This was demonstrated by their methods, which were deemed arbitrary and cruel, with low-technology traps and weapons. There could be no 'clean kill', no selection of a male animal or a good trophy, no self-regulated limits on 'greed'. Market forces were driven by direct utility and precluded a higher ethical code.

Conservationists in British colonial Africa were caught on the horns of a dilemma. Local hunting that used modern weapons that killed efficiently was unacceptable. The pressure of such hunting was everywhere thought to be on the increase as rural African populations grew and conservationists feared that without science or a sporting code, game would be wiped out. On the other hand, inefficient local techniques, using traps, simple point weapons or poison, were also unacceptable, and such methods were increasingly regarded as cruel. Opposition to native hunting was expressed in (and to an extent stemmed from) a sense that the methods used showed an unacceptable lack of concern at animal suffering. In 1936, Dr. A.H.B. Kirkman of the University of London Animal Welfare Society proposed that a resolution be sent to the secretary of State for the Colonies about West Africa, deploring the 'uninterrupted persecution of the fauna by barbarous native methods and the danger of extermination of many of Africa's game animals'.[4]

However, sport hunters themselves were also being caught out (as they are today) by growth in humanitarian concern for their prey. The idea that conservation involved safeguarding animals for shooting was itself potentially repugnant to many. By the 1930s, these concerns were being widely expressed in London and elsewhere. Indeed, in their efforts to increase membership, the SPFE was itself extending its appeal to the 'many in this country to whom cruelty and senseless slaughter are abhorrent' (Hobley, 1938, p. 47). That slaughter was epitomised by the poacher, but even the sporting hunter's claim to moral and scientific integrity was increasingly viewed with scepticism.

As the preservationist international conservation model spread after World War II, poaching became a scourge of newly created national parks. The transparent cruelty of wire snares and the catastrophic decline in charismatic species such as the African elephant reinforced these stereotypes of unsporting hunting. The lack of a scientific basis for setting levels of harvest under informal hunting added the charge of unsustainability. The growing power of the idea of nature as 'wilderness' made hunting of all kinds, but especially uncontrolled informal (and illegal) hunting increasingly unacceptable. Poachers not only transgressed the sportsman's moral code, they destroyed the stock of game and violated nature's wildness. Through most of the 20th century, informal hunting was anathema to conservationists.

Hunting as a conservation strategy

From the 1940s, recreational hunting increasingly lost its central position in justifications for conservation in Africa, and hunters ceased to control, lead or dominate conservation organisations. Hunting continued, but at arm's length from conservation. However, in the closing decades of the 20th century, recreational hunting began to undergo a rehabilitation in the eyes of international conservation planners. This came about because of its potential role in strategies to generate economic benefits from conserved lands. Such strategies have been remarkably successful in areas of large commercial farms that are too dry for productive agriculture, for example in the Zimbabwean 'low veld' and northern South Africa. Changes to the law to allow landowners rights to exploit wildlife on their land fostered the creation of a hunting industry on private farms that appears highly profitable and compatible with effective conservation of many wild animals, including the target species (Duffy, 2000; Wolmer, 2007). Such legal arrangements mirror those of countries like the UK where private rights to hunt on private land are an important element in the rural economy and the livelihoods of some households.

More important, perhaps, is the radical suggestion that recreational hunting can also contribute to the welfare of poor as well as rich rural people in these environments. The idea that conservation should be 'community-based' became a central feature of conservation strategies (Adams & Hulme, 2001; Western et al., 1994; Hulme & Murphree, 2001). In the 1980s and 1990s much debate focused on strategies to bring benefits to people living as neighbours of protected areas (strategies of park outreach), or to people living in other wildlife-rich areas (community-based natural resource management, CBNRM; Barrow & Murphree, 2001). The key context for this revolution in thinking was southern Africa, and particularly Zimbabwe, where the CAMPFIRE movement captured the spirit of the moment in offering CBNRM as a means for impoverished African communities consigned to arid and infertile lowlands in remote regions to generate revenue from wildlife (Murombedzi, 1999; Duffy, 2000). At the stroke of a pen, wild animals such as elephants or lions could be transformed from dangerous pests into walking assets. In the same happy alchemy, wealthy foreign big game hunters (paying tens of thousands of US dollars for a hunt, culminating in the shooting of one or more of the 'big five') became the sources of potentially huge sums in hard currency that could be shared among the rural poor. Moreover, this money was not the result of some passive charitable handout, but it was earned through private sector enterprise. Big game hunting, in its new guise, fitted the neo-liberal agenda of the Washington consensus (http://en.wikipedia.org/wiki/Washington_Consensus): commercially-based, targeted at the poor and (with the right science in place) potentially environmentally sustainable. The positioning of recreational hunting, and the private industries that supported it, at the apex of the new strategies to combine conservation and poverty alleviation was a turnaround in image, fortunes and arguably practice to equal the boldest corporate re-branding of the same era.

The argument that recreational hunting can in part be justified by its socio-economic benefits, because of its capacity to generate income for development and poverty alleviation, is novel. The question lies at the heart of this book. Recreational hunters and their supporters face significant challenges in achieving this outcome. It is clear that wealthy recreational hunters will pay handsomely for their sport, and that their quarry persists chiefly in marginal areas of countries with substantial and entrenched rural poverty. However, it is not enough simply to point to the juxtaposition of wealthy hunters and poor rural people. Major political issues of power and control are deeply embedded in such schemes. How is this revenue to reach the hands of the poor? What institutional arrangements can ensure that it moves effectively and fairly? What if different people stand to benefit from hunting revenues to different degrees – who compensates those who gain least, or who suffer depredations from animals most severely? How can threats to life and limb be compensated? How much revenue actually reaches the hands of the poor, and how much gets stuck in the hands of the safari operator and travel industry? How much is captured to meet the urgent needs of local government and the wider area? How much is captured by the kleptocracy that characterises governance in many developing countries? Who, among local communities, has any power to determine what animals are hunted, and how the money is shared out, and how can such arrangements be made transparent and fair? Can hunting revenue deliver sustained livelihoods for the rural poor or simply provide short-term cash subsidies for an impoverished life?

Evidence from CAMPFIRE in the 1990s showed that safari revenues yielded a significant income to rural households only in a few instances where there were plentiful populations of desirable species (elephant, buffalo, lion), and strong local social capital (Murombedzi, 1999; Bond, 2001). What happens in areas where wildlife is more scarce? What happens in even favoured areas if rural population densities rise: how long can safari revenues out-compete other land uses?

Conclusion

There is some research, and a great deal of advocacy, relevant to the role of recreational hunting in conservation, especially in developing countries. Many questions remained to be answered, and the argument about hunting's role in conservation is one that will inevitably evolve. A key issue for the future is undoubtedly the issue of rights to land and resources. Recreational hunting is of proven value as an economic strategy for those holding large dry farms and ranches in contemporary Africa. It is also a useful source of government revenue in extensive state-owned safari areas in countries like Zambia, Zimbabwe and Tanzania. Such farms and protected areas have, however, almost always been created through the displacement of rural Africans. Those displacements are increasingly controversial, and contested as an infringement of human rights (Chatty & Colchester, 2002; Cernea & Schmidt-Soltau, 2003; Cernea, 2006). Can hunting contribute to conservation solutions that overcome the political and economic inequities of the colonial past in areas such as Africa? If 'poachers' are to be genuine proprietors with recreational hunters as their clients, key elements of the models of hunting and society that dominated in the nineteenth and twentieth centuries need to be reversed. That is a challenging agenda indeed.

Notes

1 Minutes of meeting SPFE 17 February 1928, *Journal of the Society for the Preservation of the Fauna of the Empire* 1927: 20–27. The SPWFE changed its name to the Society for the Preservation of the Fauna of the Empire (SPFE) in 1921.
2 SPWFE (1907) Minutes of proceedings at a deputation of the Society for the Preservation of the Wild Fauna of the Empire to the right Hon. The Earl of Elgin, His Majesty's Secretary of State for the Colonies). *Journal of the Society for the Preservation of the Wild Fauna of the Empire*, III: 20–32.
3 General Meeting SPFE London Zoo, 3 March 1930, *Journal of the Society for the Preservation of the Fauna of the Empire* 1930: 5–11 (p. 9).
4 Meeting of the Society for the Preservation of the Fauna of the Empire 27 January 1936, *Journal of the Society for the Preservation of the Fauna of the Empire* 28: 7–8.

References

Adams, W.M. (2004) *Against Extinction: the Story of Conservation.* Earthscan, London.

Adams, W.M. & Hulme, D. (2001) Conservation and communities: changing narratives, policies and practices in african conservation. In *African Wildlife and Livelihoods: The Promise and Performance of Community Conservation*, eds. D. Hulme & M. Murphree, pp. 9–23. James Currey, Oxford.

Barrow, E. & Murphree, M. (2001) Community conservation: from concept to practice. In *African Wildlife and Livelihoods: The Promise and Performance of Community Conservation*, eds. D. Hulme & M. Murphree. James Currey, Oxford.

Beinart, W. & Coates, P. (1995) *Environment and History: the Taming of Nature in the USA and South Africa*. Routledge, London.

Bond, I. (2001) CAMPFIRE and the incentives for institutional change. In *African Wildlife and Livelihoods: The Promise and Performance of Community Conservation*, eds. D. Hulme & M. Murphree, pp. 227–243. James Currey, Oxford.

Buxton, E.N. (1902) *Two African Trips: With Notes and Suggestions on Big Game Preservation*. E. Stanford, London.

Cameron, K.M. (1990) *Into Africa: the Story of the East African Safari*. Constable, London.

Carruthers, J. (1995) *The Kruger National Park: a Social and Political History*. Natal University Press, Durban.

Cernea, M. & Schmidt-Soltau, K. (2003) The end of forcible displacement? Conservation must not impoverish people. *Policy Matters*, 12, 42–51.

Cernea, M.M. (2006) Population displacement inside protected areas: a redefinition of concepts in conservation politics. *Policy Matters*, 14, 8–26.

Chatty, D. & Colchester, M. (ed.) (2002) *Conservation and Mobile Indigenous Peoples: Displacement, Forced Resettlement and Sustainable Development*. Berghahn Press, New York.

Duffy, R. (2000) *Killing for Conservation: Wildlife Policy in Zimbabwe*. James Currey, Oxford.

Fitter, R. & Scott, P. (1974) *The Penitent Butchers: 75 Years of Wildlife Conservation*. Fauna Preservation Society, London.

Grove, R.H. (1987) Early themes in African conservation: the Cape in the nineteenth century. In *Conservation in Africa: People, Policies and Practice*, eds. D.M. Anderson & R.H. Grove, pp. 21–40. Cambridge University Press, Cambridge.

Hays, S.P. (1959) *Conservation and the Gospel of Efficiency: the Progressive Conservation Movement 1890–1920*. Harvard University Press, Cambridge, MA.

Hingston, R.W.G. (1931) Proposed British National Parks for Africa. *Geographical Journal*, 77, 401–428.

Hobley, C.W. (1938) The conservation of wild life: retrospect and prospect, Part II. *Journal of the Society for the Preservation of the Fauna of the Empire*, 33, 39–49.

Hornaday, W.T. (1921) Post-war game conditions in America. *Journal of the Society for the Preservation of the Fauna of the Empire*, 1, 53–58.

Hulme, D. & Murphree, M. (ed.) (2001) *African Wildlife and Livelihoods: The Promise and Performance of Community Conservation*. James Currey, Oxford.

Jacoby, K. (2001) *Crimes Against Nature: Squatters, Poachers, Thieves, and the Hidden History of American Conservation*. University of California Press, Berkeley.

MacKenzie, J. (1988) *The Empire of Nature: Hunting, Conservation and British Imperialism*. Manchester University Press, Manchester.

MacKenzie, J.M. (1997) Empire and the ecological apocalypse: the historiography of the imperial environment. In *Ecology and Empire: Environmental History of Settler Societies*, eds. T. Griffiths & L. Robin, pp. 215–228. Keele University Press, Edinburgh.

McKenzie, C. (2000) The British big-game hunting tradition: masculinity and fraternalism with particular reference to 'the Shikar Club'. *The Sports Historian*, 20(1), 70–96.

Murombedzi, J.S. (1999) Devolution and stewardship in Zimbabwe's CAMPFIRE programme. *Journal of International Development*, 11, 287–293.

Neumann, R.P. (1996) Dukes, Earls and ersatz Edens: aristocratic nature preservationists in colonial Africa. *Environment and Planning D: Society and Space*, 14, 79–98.

Neumann, R.P. (1998) *Imposing Wilderness: Struggles Over Livelihood and Nature Preservation in Africa*. University of California Press, Berkeley.

Prendergast, D.K. & Adams, W.M. (2003) Colonial wildlife conservation and the origins of the Society for the Preservation of the Wild Fauna of the Empire (1903–1914). *Oryx*, 37, 251–260.

Seton-Karr, H. (1908) The preservation of big game. *Journal of the Society for the Preservation of the Wild Fauna of the Empire*, 4, 26–28.

Society for the Preservation of the Wild Fauna of the Empire (SPWFE) (1905) Minutes of proceedings at a deputation from the Society for the Preservation of the Fauna of the Empire to the Right Hon. Alfred Lyttelton (His Majesty's Secretary for the Colonies), February 2, 1905. *Journal of the Society for the Preservation of the Wild Fauna of the Empire*, 2, 9–18.

Thompson, E.P. (1975) *Whigs and Hunters: the Origin of the Black Act*. Allen Lane, London.

Western, D., Wright, M. & Strum, S. (ed.) (1994) *Natural Connections: Perspectives in Community-based Conservation*. Island Press, Washington, DC.

Whitbread, S.H. (1907) The year. *SPWFE Journal*, 1907, 10–13.

Wolmer, W. (2007) *From Wilderness Vision to Farm Invasions: Conservation & Development in Zimbabwe's South-East Lowveld*. James Currey, Oxford.

9

Exploitation Prevents Extinction: Case Study of Endangered Himalayan Sheep and Goats

Michael R. Frisina[1] and Sardar Naseer A. Tareen[2]

[1]Department of Animal and Range Sciences,
Montana State University, Bozeman, MT, USA
[2]SUSG-CAsia, BRSP House, Quetta, Balochistan, Pakistan

Introduction

When leaders at Torghar (black mountains or hills in Pushtoo), in the Toba Kakar mountain range of Balochistan Province, Pakistan, decided to act to conserve the rapidly dwindling populations of Suleiman markhor *Capra falconeri jerdoni* and Afghan urial *Ovis orientalis cycloceros* on their Jazalai tribal lands, they had to overcome a number of obstacles (Figure 9.1). Among them were funding and gaining acceptance by the local rural community who own and have occupied the area for centuries. Adequate funding and acceptance by local people are common problems confronting conservation in developing countries (Lewis *et al.*, 1990; Bodmer *et al.*, 1997). Local acceptance is especially important when people depend on natural resources for subsistence, as at Torghar.

Torghar' managers pursued an innovative approach new to the region, the implementation of a program by which funds raised by limited trophy

Recreational Hunting, Conservation and Rural Livelihoods: Science and Practice, 1st edition.
Edited by B. Dickson, J. Hutton and W.M. Adams. © 2009 Blackwell Publishing,
ISBN 978-1-4051-6785-7 (pb) and 978-1-4051-9142-5 (hb).

Figure 9.1   **The Torghar Conservation Project was initiated to conserve populations of Suleiman markhor** *Capra falconeri jerdoni*, **left and Afghan urial** *Ovis
orientalis cycloceros*, **right.**

hunting of markhor and urial would be used to fund conservation of these
species while simultaneously benefiting the local community (Khan, 2002).
They initiated the programme by informally designating most of the Torghar
area as the Torghar Conservation Project (TCP). The purpose was to conserve
the area's biodiversity while enhancing community stability. The TCP can
be considered a success in that there is clear evidence that it has achieved its
objectives of conserving biodiversity while improving the lives of local families
(Johnson, 1997a). Here we draw upon our first hand knowledge of the project
and the written record to describe how the TCP, initiated in 1985, progressed
over the subsequent 21 years, emphasising aspects that are potentially applicable to solving similar problems in other areas. The first author has made four
trips to Pakistan since 1997 to work closely with project managers in population monitoring and habitat management aspects of the program. The second
author is a Pushtoon tribal leader.

Study area

Geographic setting

The approximately 1000 km^2 Torghar Conservation Project lies within the
Torghar hills, Toba Kakar range, of Balochistan Province in west central

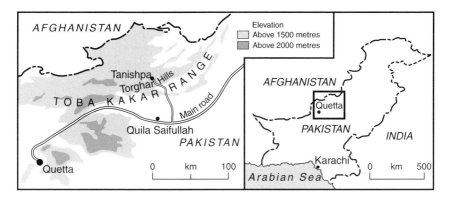

Figure 9.2 **Location of the Torghar area within Pakistan.**

Pakistan near the north central border with Afghanistan (Figure 9.2). The Torghar area is a series of rugged upturned sandstone ridges. The ridges are approximately 90 km long and vary from about 15 to 30 km in width. Johnson (1997b) described the TCP as predominantly a series of three parallel ridges separated by two north-east-running stream drainages. The southernmost ridge has a north-facing slope that gradually rises to an elevation of about 2800 m, and is dissected by several deeply incised drainages. The south-facing slopes drop precipitously from the crest forming a series of step-like cliffs to the Khaisore valley; the northern ridges consist of steeply upturned rock layers resembling a series of parallel, jagged-toothed combs (Figure 9.3).

The climate is dry, with cold winters and warm summers. Violent duststorms and thunderstorms occur during summer months (SGPC, 1991). During July and August, the mean temperature is about 26°C; strong winds are common. Total annual precipitation within the region varies from 18 to 27 cm. Most precipitation occurs between December and March. Periodic droughts are common and may last for several years at a time. According to local tribesmen, the years 1997 to 2001 constituted an unusually harsh drought cycle (Woodford *et al.*, 2004), with drought conditions continuing through 2004, but moderating somewhat in 2005.

Balochistan is characterised by a diverse flora, typically Persian in character (Burkill, 1969). The TCP lies within the Balochistan juniper and pistachio scrub forest and dry subtropical semi-evergreen scrub forest zones described by Roberts (1997). Shrub–steppe plant communities dominate the semidesert landscape of the Torghar area.

Figure 9.3   **Much of the Torghar landscape consists of steeply upturned rock layers resembling a series of parallel jagged-tooth combs.**

Human setting

The human population numbers about 4000 tribal people, predominantly Jazalais, a Pathan tribe. The Pathans still practice an ancient code of honour based on tribal rules (Caroe, 1958; Spain, 1962; Quddus, 1987). The people of Torghar are further divided into clans, each occupying its own specific area of the mountain or valley. Individual residents are dispersed throughout the mountains with some centred in five villages: Kundar, Tanishpa, Khaisore, Torghberg and Tubli. Tanishpa is the largest village with about 25 house-holds. The Jazalais are a semi-nomadic people, tending large flocks of sheep and goats. In early spring some families move north out of the mountains to the plains of Kakar Kahorasan taking their families and flocks with them, and returning to the Torghar area in early autumn. Other tribes in surrounding areas live similarly. The TCP lies across one of the traditional migration routes followed by these tribes. Nearly 20 tribal groups, said to number in the thousands, with large numbers of sheep and goats, pass through the area twice a year (Woodford *et al.*, 2004). In early spring these tribes travel to

Khorsan and beyond the Durand Line (political boundary between Pakistan and Afghanistan) (Hayat Khan, 2000) into Afghanistan to summer in their tribal territories there. Although cultivable land and perennial water are limited, some families have developed fruit and nut tree orchards in valley bottoms near human settlements. Extreme poverty is a characteristic of human society at Torghar.

Wildlife

Suleiman markhor and Afghan urial are the only large wild ungulates inhabiting the TCP. Suleiman markhor occur in low numbers and have a limited distribution in Pakistan, including the rugged mountains of western Pakistan. Afghan urial are more widespread and common than Suleiman markhor, but are not abundant (Roberts, 1997). Mitchell (1989) and Johnson (1997a, 1997b) concluded that by the early 1980s Suleiman markhor and Afghan urial populations at Torghar were at very low levels. Both species are listed in the Third Schedule of the Balochistan Wildlife Protection Act of 1974 as animals which can only be hunted under specific circumstances (Johnson, 1997b). Suleiman markhor are listed as 'Endangered' under the US Endangered Species Act (FWS, 1997) and are listed in Appendix I of the Convention on International Trade in Endangered Species of Wild Fauna and Flora (CITES) which effectively halts legal trophy hunting by foreign hunters (CITES, 2006). However, markhor are legally hunted at Torghar through a special CITES exemption for Pakistan that was approved at the tenth CITES meeting of the Conference of Parties in 1997 (Shackleton, 2001). The Afghan urial is not listed under the US Endangered Species Act, but is listed in Appendix II of CITES.

Conservation hunting programme

In the early 1980s tribal leaders Nawab Taimur Shah Jogezai and Sardar Naseer Tareen were alarmed at what appeared to be a dramatic decline in Suleiman markhor and Afghan urial in the Torghar area (Johnson, 1997b; Khan, 2002; Mitchell & Frisina, 2007). The decline in markhor and urial was probably due to increased poaching as the result of the influx of weapons, ammunition, and millions of refugees into the area during the Soviet–Afghanistan war (Johnson, 1997b; Khan, 2002; Mitchell & Frisina, 2007). The war began in 1979, ending in 1992.

Tribal leaders sought assistance from the United States Fish and Wildlife Service (USFWS) in early 1984, and from this association the TCP developed. They agreed that a game guard programme with limited trophy hunting was essential to save Torghar's urial and markhor populations from extinction. Trophy hunting could provide funding to maintain the conservation programme.

The people of Torghar and surrounding areas have a centuries-long tradition of hunting in their own areas. Hunting was an established right and norm for the tribesmen. Banning of hunting by local tribesmen was a major hurdle for tribal leaders to overcome before the conservation programme could begin in earnest. After careful consideration, the tribesmen agreed to accept a ban on hunting in exchange for the potential employment opportunities and economic benefits associated with a conservation programme. In 1985, the TCP began by purchasing equipment and paying the salaries of seven game guards to protect markhor and urial, utilising a US$10,000 loan from the Chisholm Foundation. The loan was repaid using trophy hunting income.

The employment of game guards was a complex issue. Important social considerations included access to resources, opportunities for development, and right of representation in the decision-making process among the various tribal affiliations. Discussions and negotiations were centred on perceptions by some that employment was a 'right' rather than a position based on merit or need (T. Rasheed, personal communication, 2006). The decision was to distribute the employment of game guards among tribal clans. The distribution of benefits based on need and skills versus 'rights' is an ongoing issue; the challenge has been to find ways to accommodate both viewpoints (Bellon, 2005). Game guards were responsible for protection of wildlife within areas belonging to their clan. Initially game guards were placed at points of entry into the area protected and were faced with the task of informing approaching migrating tribesmen that hunting was banned. To avoid jeopardising safe passage through Jazalai territory, people dependent on seasonal migration through Torghar agreed to respect the hunting ban. This was a major accomplishment for the conservation effort because the culture is strictly tribal and it would be difficult, if not impossible, to enforce Government writ without tribal acceptance. Local people feel themselves bound only by tribal law.

In 1986 the first foreign hunters harvested one markhor and four urial for a fee of US$6500 per urial and a US$20,000 fee for markhor, making the TCP self-sufficient. Local leaders maintained the TCP on an informal basis up to 1993. Each year additional game guards were employed. In 1994 the TCP evolved into the Society for Torghar Environmental Protection (STEP), an officially registered nongovernmental organisation under Pakistani law. The primary goal of STEP is the conservation of urial and markhor; however, since all wildlife species in the area are protected, a diversity of species, including many non-hunted species, benefits. Trophy hunting is not a goal of STEP, rather a means to fund the programme (Khan, 2002). During the 2005–2006 hunting season five markhor and five urial were harvested. The fees are now US$35,000 per markhor and US$11,000 per urial. Until 2000, 25 per cent of the hunting fees went to the government and 75 per cent to the community-based hunting programme where the hunt occurred. In 2000 this ratio was changed to a 20/80 (Shackleton, 2001).

Each year a limited number of urial and markhor hunters visit the area to pursue trophies. Their hunting fees employ local tribesmen, who refrain from hunting in exchange for employment as game guards in charge of preventing poaching in the TCP. STEP currently employs 82 game guards (T. Rasheed, personal communication, 2006). The TCP hunting program is the oldest community-controlled program in Pakistan (Shackleton, 2001).

All hunting within the TCP is administered by STEP following Pakistani law and rules formulated by the local Torghar community. Trophy hunting of markhor and urial is open to anyone willing to pay the associated fee. STEP locates hunters through a booking agent and they are guided by STEP personnel.

Biodiversity influences

The population status of markhor and urial were monitored from 1994 to 2005 through four surveys following field protocols described by Johnson (1997b) and Frisina et al. (1998). During each autumn survey, observations were made from the ground by experienced observers within the same predefined observation areas. The number of markhor and urial observed, and the size of each survey area, were the basis for calculating an observed density for each species. The core portion of Torghar (Johnson, 1997b) was divided into observation areas and each area was classified as high or low quality habitat for markhor and urial. Observed densities were then applied to the entire TCP to formulate an estimated population. The results of the first three surveys were reported by Johnson (1997b), Frisina et al. (1998) and Frisina (2000). During a fourth survey (using the same field protocols and calculations for population

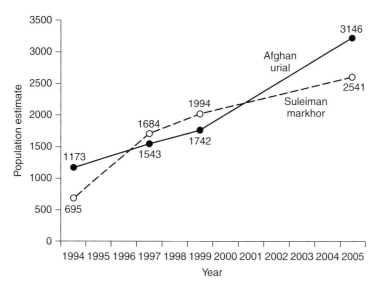

Figure 9.4 **Population trend of Suleiman markhor and Afghan urial in the Torghar Conservation Project. Data from Johnson (1997a), Frisina *et al.* (1998), Frisina (2000), and Shafique (2006).**

estimation) conducted in November 2005, the population was estimated to be 2541 markhor and 3146 urial (Shafique, 2006).

The trend for both markhor and urial within the TCP has been up for several years (Figure 9.4). While the survey protocols formulated by Johnson (1997b) do not lend themselves to robust statistical interpretation, they do provide conservative and comparable population estimates. The estimates are conservative because the protocol assumes that all markhor or urial within the observation areas were observed, which is unlikely to be the case. Even aerial surveys underestimate population density (Pollock & Kendall, 1987). When conducting autumn surveys utilising a helicopter to count sheep, one can only expect to observe 20–50 per cent of the population (Remington & Walsh, 1993). Because of visibility limitations in the TCP due to extremely rugged topography (Figure 9.3), observers saw a smaller proportion of the population from the ground than would be expected using a helicopter. Johnson (1997a) concluded that the increases in markhor and urial populations were real and attributable to the effectiveness of the game guard programme in curtailing poaching.

The population estimates serve as a basis for estimating potential trophy quotas for the TCP. After examining the literature for similar polygynous species and populations, Harris (1993) concluded that annual harvests of trophy males in numbers equivalent to 1 or 2 per cent of the total population size can be maintained without negative consequences. Since markhor and urial have polygynous mating systems, the population's overall reproductive rate would be little influenced by loss of a small number of males (Caughley, 1977). Using this as a guide, potential maximum quotas have ranged from six to 17 for markhor and 11 to 13 for urial (Johnson, 1997a; Frisina *et al.*, 1998; Frisina, 2000). However, actual quotas for markhor and urial have been more conservative than this. Since the trophy hunting programme began in 1986, hunters have taken 45 urial and 35 markhor. The harvest has averaged two markhor and urial per year with annual harvests ranging from zero to five for each taxon. Hunters from 13 countries have hunted at Torghar; the majority of markhor hunters are Europeans while most of the urial hunters are North Americans (Table 9.1). Harvest levels have actually been within the range of one to two per cent of the total number of animals observed during population surveys (Frisina, 2000; Shafique, 2006). Thus, STEP has practised the precautionary principle (Cooney, 2004; Rosser *et al.*, 2005) through very conservative harvest quotas. It is critical to maintain sufficient mature males in the population for normal reproduction to be achieved and to not jeopardise long-term survival of the population (Shackleton, 2001). The success of this approach is self evident by the increasing populations of markhor and urial in the face trophy of hunting (Figure 9.4). The application of the precautionary principle is further substantiated by the high proportion of trophy and adult males observed in the population during the years of trophy hunting (Table 9.2).

Benefits to the local community

Since 1986 the trophy harvest has brought in a total income of US$1,332,000, less US$191,000 paid to the Government of Balochistan. A portion of the funds earned through the hunting program is used to provide health care for the local people. STEP has also undertaken measures to increase the number of jobs and improve education, roads, communications and agriculture in the TCP.

In addition to health care, the conservation effort is not only benefiting a few people by employing them as game guards, but is bringing prosperity

Table 9.1 **A summary of markhor and urial harvest at Torghar by hunter country of origin 1986–2006. (Figures are percentages of total harvest.)**

Country	Markhor	Urial
Austria	2.9	4.4
Belgium	5.7	4.4
Canada	2.9	4.4
Denmark	14.3	11.1
France	14.3	6.7
Germany	8.6	2.2
Hungary	2.9	0
Italy	11.4	0
Mexico	8.6	11.1
Spain	14.3	2.2
Switzerland	2.9	0
UK	2.9	2.2
USA	8.6	51.1
Unknown	0	4.4
North America (total)	20.1	62.2
Europe (total)	79.9	37.8

Table 9.2 **Proportion of trophy and adult males observed in the population during the years of trophy hunting.[a]**

Species	Trophy males			Adult males			
	≥6 years	Range	SD	≥2.5 years	Range	SD	N[b]
Markhor	14%	10–17	2.9	24%	13–29	7.2	4
Urial	16%	10–25	6.5	32%	24–49	11.8	4

[a]Data from Johnson (1997a), Frisina et al. (1998), Frisina (2000), and Shafique (2006).
[b]N = the number of annual autumn (fall) census surveys.

Figure 9.5 **The Society for Torghar Environmental Protection, with support from the United Nations Development Program, has constructed water storage structures that benefit the human community and wildlife.**

to the entire populace through projects designed to improve water supply (Figure 9.5). For example, with joint support from the United Nations Development Program, STEP has developed water tanks, wells, channels and storage dams at suitable locations. Among these was a water storage dam in the Khaisor Valley to reduce livestock grazing pressure within portions of the TCP. Water tanks were also developed at Tanishpa. STEP has assisted local people to develop agricultural fields and has supplied them with fruit and firewood sapling trees for their orchards. Ten technicians were trained to provide basic veterinary services to local herdsmen. Funds have been used to repair roads and trails to improve transportation for people living the Torghar area. The newly available water has allowed local people to increase production of their orchards and may reduce the sole dependence on raising livestock. STEP is being careful to place any new water sources at locations that will not increase competition between domestic and wild animals.

Future

STEP began a new phase in its development in 2000 by adding habitat maintenance and protection to its programme. Following several years of severe drought local tribesmen came to the realisation that the conservation programme was more important to their future than the long-established traditions of raising domestic livestock. Local tribesmen have asked STEP leaders for assistance in developing a habitat protection programme (Frisina *et al.*, 2002, 2006). STEP is in the process of developing a rotational grazing strategy for managing domestic sheep and goats in a manner harmonious with the habitat requirements of markhor and urial. The new strategy for range management will include the reintroduction of *pargor* a local and traditional form of rotational grazing that was abandoned several years ago (Woodford *et al.*, 2004). This traditional system will be adapted to maintain the soil and vegetation in harmony with the wild and domestic ungulates (Frisina *et al.*, 2002). Local tribesmen also requested that the plan include measures for protecting urial and markhor from diseases associated with domestic livestock. During the 1997–2001 drought years local tribesmen lost about 16,000 sheep and goats (Frisina *et al.*, 2002). By 2005, when the drought began to moderate, total sheep and goat numbers were reduced to about 20,000. Local tribesman intend on maintaining sheep and goat numbers at a maximum of 25,000 in future years (Frisina *et al.*, 2002). They believe revenue from urial and markhor populations will provide a more stable economic source than the more traditional livestock. The needs of people living within the TCP are great and STEP will continue to develop wells and to train people in improved livestock management and agricultural practices. Providing community health care is an ongoing effort.

Conclusions

The success of biodiversity conservation in the TCP has more to do with effectively managing for unique social characteristics and needs of the tribal society than the application of modern wildlife science. Over time, local people, in increasing numbers, have gained ownership of the conservation effort and have made it their own by sharing in the decision-making processes and the benefits of the programme. During years of severe drought salaries received through the program were an important source of income for some of the residents. Thus the community has come to realise that their survival and the successful conservation of biological resources are interdependent.

The increase of Suleiman markhor to an estimated 2541 in November 2005 (Shafique, 2006) is particularly noteworthy when one considers that Schaller (1977) estimated a total world wide range population of 2000 for this taxon in the early 1970s. The TCP population is now obviously the largest in existence and the only one with an apparently secure future.

Trophy hunting is a critical part of the conservation effort, generating revenue necessary to support the game guard programme and impressing upon the local people that their economic well-being is directly tied to the abundance of markhor and urial. STEP emphasises to the local community that the primary purpose of the hunting programme is not the generation of maximum funds, but instead the conservation of wildlife and wildlife habitats. The TCP developed using a limited dependence on external donor funds. Revenues obtained through hunting have made the programme financially self sustainable, and have served to encourage a sense of community ownership. Reliance on internally generated funds from trophy hunting can make the programme vulnerable to natural disasters, social unrest, and geopolitical instability typical of the region. On several occasions, geopolitical tensions have discouraged foreign hunters from participating. However, prior to funds derived from the hunting programme, the community had little in the way of reserve resources to use during difficult times. The current annual harvest quota level of four markhor and five urial meets the financial needs of conservation and provides funding for community needs. STEP considers the current harvest quotas adequate for maintaining the programme.

Most of the wildlife habitat in Balochistan Province occurs in tribal areas. It is well known that the Balochistan wildlife authorities cannot enforce the laws in these areas; wildlife is at the mercy of local tribesmen. STEP has proven that with the support and involvement of local people, wildlife conservation and management plans can succeed. Lessons learned from the TCP serve as an important practical example for conserving biodiversity under difficult circumstances that are not unusual to Central Asia. The value of the programme is verified by the fact that, the IUCN Central Asian Sustainable Use Group recently began applying the TCP approach to several other areas in Balochistan.

Acknowledgements

We thank David Ferguson, formerly of the United States Fish and Wildlife Service Division of International Affairs for his long-term commitment to STEP and wildlife conservation in Pakistan. Tahir Rasheed and R. Margaret Frisina reviewed early drafts of the manuscript.

References

Bellon, L. (2005) From hunting to sustainable use: a Pashtun tribal group's innovations, northern Balochistan, Pakistan. In *Innovative Communities: People-centered Approaches to Environmental Management in the Asia-Pacific Region,* eds. J. Velasquez, M. Yashiro, S. Yashimura, & I. Ono, pp. 109–130. United Nations Press, New York.

Bodmer, R.E., Penn, J.W., Puertas, P. & Fang, T.G. (1997) Linking conservation and local people through sustainable use of natural resources: community-based management in the Peruvian Amazon. In *Harvesting Wild Species Implications for Biodiversity Conservation,* ed. C. H. Freese, pp. 315–358. John Hopkins University Press, Baltimore, MD.

Burkill, I.H. (1969) *A Working List of the Flowering Plants of Baluchistan.* West Pakistan Government Press, Karachi.

Caroe, O. (1958) *The Pathans Within an Epilogue of Asia.* Oxford University Press, Karachi, Pakistan.

Caughley, G. (1977) *Analysis of Vertebrate Populations.* John Wiley and Sons, New York, NY.

Convention on International Trade in Endangered Species of Wild Fauna and Flora. (2006) *Appendices.* http://www.cites.org/eng/app/index.shtml

Cooney, R. (2004) *The Precautionary Principle in Biodiversity and Natural Resource Management. An Issues Paper for Policy-makers, Researchers, and Practitioners.* IUCN, Gland, Switzerland.

Fish and Wildlife Service (1997) Endangered and threatened wildlife and plants, 50CFR 17.11 & 17.2. August 20, 1994. US Department of the Interior, Washington, DC.

Frisina, M.R. (2000) *Suleiman Markhor (Capra falconeri jerdoni) and Afghan Urial (Ovis orientalis cycloceros) Population Status in the Torghar Hills, Balochistan Province, Pakistan.* A report to the US Fish and Wildlife Service and Society for Torghar Environmental Protection, Quetta, Pakistan.

Frisina, M.R., Woods, C. & Woodford, M.H. (1998) *Population trend of Suleiman markhor (Capra falconeri jerdoni) and Afghan urial (Ovis orientalis cycoceros) with reference to habitat conditions, Torghar hills, Baluchistan Province, Pakistan.* A report to the US Fish and Wildlife Service and Society for Torghar Environmental Protection, Quetta, Pakistan.

Frisina, M.R., Woodford, M.H. & Awan G.A. (2002) *Habitat and Disease Issues of Concern to Management of Straight-horned Markhor and Afghan Urial in the Torghar Hills, Balochistan, Pakistan.* A report to the US Fish and Wildlife Service and Society for Torghar Environmental Protection, Quetta, Pakistan.

Frisina, M.R., Woodford, M.H. & Awan G.A. (2006) *Management of Straight horned Markhor and Afghan Urial Habitat in the Torghar Hills, Balochistan, Pakistan.* A report to the Society for Torghar Environmental Protection, Quetta.

Harris, R.B. (1993) *Wildlife Conservation in Yeniquo, Qinghai, China: Executive Summary.* Doctoral dissertation, University of Montana, Missoula.

Hayat Khan, A. (2000) *The Durand Line. Its Geo-strategic Importance.* Area Study Centre, Preshwar, University of Preshwar, Pakistan.

Johnson, K.A. (1997a) Trophy hunting as a conservation tool for Caprinae in Pakistan. In *Harvesting Wild Species. Implications for Biodiversity Conservation,* ed. C. H. Freese, pp. 393–423. John Hopkins University Press, Baltimore, MD.

Johnson, K.A. (1997b) Status of Suleiman markhor and Afghan urial populations in the Torghar hills, Balochistan Province, Pakistan. In *Biodiversity of Pakistan,* eds. S.A. Mufti, C. A. Woods, & S. A, Hasan, pp. 469–483. Pakistan Museum of Natural History, Islamabad, Pakistan.

Khan, R. (2002) Tribal leadership paves the way, Sardar Naseer A. Tareen. In *Green Pioneers, Stories from the Grass Roots,* ed. M. Abidi, pp. 112–121. City Press, Karachi, Pakistan.

Lewis, D., Kawecha, G.B. & Mwenya, A. (1990) Wildlife conservation outside protected areas: lessons from an experiment outside Zambia. *Conservation Biology,* 4, 171–180.

Mitchell, R. (1989) *Status of Large Mammals in the Torghar Hills, Balochistan.* Report to the US Fish and Wildlife Service, Washington, DC.

Mitchell, R.M. & Frisina, M.R. (2007) *From the Himalayas to the Rockies. Retracing the Great Arc of Wild Sheep.* Safari Press Inc., Long Beach, CA.

Pollock, K. H. & Kendall, W.L. (1987) Visibility bias in aerial surveys: a review of estimation procedures. *Journal of Wildlife Management,* 51, 502–510.

Quddus, S. A. (1987) *The Pathans.* Feroszsons (dot.) Ltd., Karachi, Pakistan.

Remington, R. & Walsh, G. (1993) Surveying bighorn sheep. In *The Desert Sheep in Arizona,* ed. R.M. Lee, pp. 63–81. Arizona Game and Fish Department, Phoenix, AZ.

Roberts, T. J. (1997) *The Mammals of Pakistan.* Oxford University Press, Oxford, UK.

Rosser, A. M., Tareen, N. & Leader-Williams, N. (2005) The precautionary principle, uncertainty and trophy hunting: a review of the Torghar population of central Asian markhor (*Capra falconeri*). In *Biodiversity and the Precautionary Principle. Risk and Uncertainty in Conservation and Sustainable Use,* eds. R. Cooney & B. Dickson, pp. 55–72. Earthscan, London, UK.

Schaller, G.B. (1977) *Mountain Monarchs. Wild Sheep and Goats of the Himalaya.* The University of Chicago Press, Chicago, IL.

Shackleton, D.M. (2001) *A Review of Community-based Trophy Hunting Programs in Pakistan.* IUCN-Pakistan, Islamabad.

Shafique, C.M. (2006) *Status of Suleiman markhor (Capra falconeri jerdoni) and Afghan Urial (Ovis vignei cyloceros) with Relation to Community-Based Trophy Hunting in Torghar Conservancy, Balochistan Province, Pakistan.* A report to the Society for Torghar Environmental Protection, Quetta, Pakistan.

Spain, J.W. (1962) *The Way of the Pathans.* Oxford University Press, Karachi, Pakistan.

Superintendent of Government Printing, Calcutta (1991) *A Gazetteer of Baluchistan.* A reprint of the second edition, 1989. First published 1908. Vintage Books, Haryana, India.

Woodford, M. H., Frisina, M.R. & Awan, G.A. (2004) The Torghar conservation project: management of livestock, Suleiman markhor (*Capra falconeri*) and Afghan urial (*Ovis orientalis*) in the Torghar hills, Pakistan. *Game and Wildlife Science,* 21, 177–187.

10

Community Benefits from Safari Hunting and Related Activities in Southern Africa

Brian T.B. Jones

Environment and Development Consultant

Introduction

Some community-based natural resource management (CBNRM) pro-grammes rest on the assumption that recreational hunting, often combined with other forms of wildlife use such as ecotourism, can generate sufficient benefits to provide an incentive for rural communities to conserve wildlife and habitat over the long term. Sophisticated CBNRM programmes have been developed in southern Africa with donor support for more than 15 years (e.g. Gibson & Marks, 1995; Wainwright & Wehrmeyer, 1998; Twyman, 2000; Bond, 2001; Jones, 2001; Murombedzi, 2003). This chapter considers the benefits and costs to local people of such programmes in Botswana, Namibia, Zambia and Zimbabwe, and their impacts on local livelihoods.

It should be noted that widely diverging goals and contexts, and changes over time, make it difficult to generalise about CBNRM across southern Africa (Fabricius *et al.*, 2001). Few CBNRM projects have systematically collected data on livelihood impacts (Jones, 2004; Arntzen *et al.*, 2007b), or reviewed projects

Recreational Hunting, Conservation and Rural Livelihoods: Science and Practice, 1st edition.
Edited by B. Dickson, J. Hutton and W.M. Adams. © 2009 Blackwell Publishing,
ISBN 978-1-4051-6785-7 (pb) and 978-1-4051-9142-5 (hb).

over periods longer than 18 months. Data collection methods often differ, and the benefits measured are often different and do not necessarily occur in all communities. This chapter therefore does not aim to provide definitive conclusions for southern Africa, but provides some examples of how CBNRM is benefiting communities in some countries while also considering examples of negative impacts.

CBNRM in southern Africa

CBNRM involves the creation of enabling conditions for the establishment of a communal property regime, i.e. a regime in which a defined group collectively manages and exploits wildlife and wildlife habitat as a common property resource within a defined jurisdiction (Jones & Murphree, 2004). This involves the devolution of use rights and decision-making authority by government to a community-based organisation (CBO) with a committee to represent the local community that usually has a constitution, a defined area of authority, and a defined membership or group of resource users.

Once the CBO receives the rights over wildlife from government it is able to enter into contracts with hunting and photographic tourism companies to develop enterprises based on different forms of wildlife use. CBOs may also run their own enterprises such as camp sites and fishing camps. CBO committees need to account to local residents for income received and expenditure, and all residents need to agree on how the profit from the CBO contracts and enterprises will be used for community benefit.

There are differences in CBNRM across the region. In Zambia, Zimbabwe and Mozambique there are different forms of revenue sharing between government and the communities. In Zimbabwe, Mozambique, Zambia and Namibia community use rights are entrenched in legislation while in Botswana they are applied through policy directives. In Namibia and Botswana all income from community use rights over wildlife goes directly to the communities concerned.

In Botswana, a USAID-funded natural resource management programme (NRMP) ran from 1989 to 1999 (Rozemeijer, 2003). Local communities formed trusts, gained quotas from the wildlife department and entered into joint venture agreements for trophy hunting or photographic tourism with the private sector. By 2003, after 13 years, 96 registered trusts covered more than 100 villages (Arntzen *et al.*, 2007a). A new CBNRM Policy was approved in 2007 (Government of Botswana, 2007), partly responding to allegations of widespread mismanagement of funds in community trusts, which required trusts to surrender 65 per cent of their income from natural resource concessions and hunting quotas to a National Environmental Fund to which they may apply for the funds for community projects. The Government may vary the percentages depending on the circumstances and needs of a particular trust.

In Zimbabwe, the CAMPFIRE programme devolved 'appropriate authority' over wildlife to Rural District Councils (RDCs) under existing legislation. Policy guidelines provided for further devolution to sub-district administrative units called wards. Although there have been attempts to diversify the suite of resources managed by communities under CAMPFIRE, the programme remains heavily reliant on RDCs entering into contracts with private hunting companies, and to some extent with tourism lodge operators, to generate income for community development. RDCs have tended to hold on to large portions of the income from wildlife and tourism and have not devolved rights to the ward level. Despite recent economic collapse and political turbulence in Zimbabwe, the hunting industry and some CAMPFIRE institutions have shown remarkable resilience and continue to function using trophy hunting revenue (Child *et al.*, 2003; Taylor & Murphree, 2007). New government guidelines require that communities use revenue for social projects and not for household dividends (Taylor & Murphree, 2007).

In Namibia, planners learned from CAMPFIRE and developed legislation that gave rights over wildlife directly to self-identifying communities called 'conservancies' (Jones & Murphree, 2004). Conservancies are able to hunt certain game species without quotas or permits, to obtain a trophy hunting quota from government, to sell game and to control tourism within their boundaries (Figure 10.1). Conservancies retain all the income earned from wildlife and tourism and can use this to provide income or benefits to residents. Since the first conservancy was registered in 1998 there are now 50 conservancies, covering more than 11 million hectares (NACSO, 2006).

In Zambia, two main wildlife-based CBNRM programmes emerged (Jones, 2004). The ADMADE (Administrative Design Programme) programme was initiated in the mid-1980s as a national programme but has focused mainly on Game Management Areas (GMAs) in the Luangwa valley, around Kafue national park and in the lower Zambezi valley. Revenue is shared according to a set formula. The Luangwa Integrated Rural Development Project (LIRDP) was initiated in the Lupande GMA in the Luangwa valley in 1988 to link

Figure 10.1 **A meeting in northwest Namibia between the Purros Conservancy, Ministry of Environment and Tourism, and support NGOs to present the annual wildlife utilisation report-back and review the hunting quota. November 2007 (with permission, Greenwill Matongo).**

wildlife revenues with integrated rural development. From 1996, the project focused on wildlife and gave a greater share of income to communities and increased village-level decision-making (Child, 2004).

Benefits

CBNRM generates direct income that communities can use at their discretion for local development or other purposes and other more intangible benefits such as infrastructure, support for vulnerable groups, employment and game meat (Ashley, 1998; Fabricius *et al.*, 2001; Jones, 2004; Arntzen *et al.*, 2007b).

Direct financial benefits and community projects

Income from hunting and tourism can be substantial e.g. US$590,000 in 2005 for the Sankuyo community in Botswana (Arntzen *et al.*, 2007a) and US$114,000 in 2006 for the Nyae Nyae community in Namibia (Jones & Mosimane, 2007). In Botswana and Namibia large portions of this income have been used for recurrent project expenditure such as staff salaries, vehicles, and sitting allowances for committees, leaving small amounts for community benefits. In Namibia, communities started demanding greater accountability from their committees, and with the support of NGOs and Government have begun to budget specifically for community benefits (Davis & Kasaona, 2007).

Most reviews of CBNRM in the region conclude that direct financial benefits to rural households (usually in the form of dividend payments or jobs) are generally low, particularly in very large communities where income was shared between many people (Fabricius *et al.*, 2001; Arntzen *et al.*, 2007b). However, where small numbers of people coexist with abundant or high-value resources the dividends to households may be significant. Moreover, in Namibia there is evidence to suggest that job creation by conservancies enjoys widespread community support and is not simply being used by committees to establish patronage systems and extend their control over access to benefits (Jones & Mazambani, 2007).

Cash 'dividends' paid to households are small (Table 10.1), but it is important to compare the size of benefits against people's options and other local income opportunities (Fabricius *et al.*, 2001). For example, the 1989 household dividend of Z$200 (US$45 in 1990 terms) in the Masoka area of Zimbabwe represented a 56 per cent increase on household income from cotton (Fabricius *et al.*, 2001). According to Long (2004) the pay out of N$630 (US$63 in 2003 terms) to each member of the Torra Conservancy in north-west Namibia in 2003 could cover basic grocery costs for a local household for three months, was almost equivalent to the average amount raised annually from the sale of live goats and is equivalent to 14 per cent of the average annual income for individuals in the region and 8 per cent of the average annual income of households. The timing of cash dividends to households can be important: the most common use of cash in Torra Conservancy was for school fees as the payout occurred in January prior to the start of the new school year. In a Zimbabwe case study, net income from cotton provided an average of Z$2102 for each of the 81 cotton-growing households (with over 60 per cent of households earning less than Z$1000 and only 7 per cent earning over Z$3000),

Table 10.1   **Examples of cash 'dividends' from formal CBNRM programmes in southern Africa.**

Country		Year	Amount in local currency	Approx. US$ equivalent in reference year
Botswana (Sankuyo Tshwaragano Management Trust)	Arntzen *et al.* (2003 vol. II)	2003	P300 per family	75
Namibia (Torra & Nyae Nyae conservancies)	Murphy & Roe (2004)	2003	N$600 per member	73
Zambia (Lupande Game Manangement Area)	Child (2004)	2003	ZK5370 per adult	5.37
Zimbabwe (Nyaminyami District)	Sibanda (2004)	1996	Z$55 (mean household cash income)	5.55

while total income from wildlife was Z$526,593 and the 140 households on the membership register received Z$1000 each (Taylor & Murphree, 2007). As a cash crop, the total income from wildlife was several times that of cotton, and income from wildlife was far more equitably distributed across households than income from cotton.

Unfortunately, the surveys that provided this data were not replicated so it is not possible to track what happened in subsequent years regarding comparative cotton and wildlife incomes. Lack of data or surveys that track changes over time make it hard to arrive at broad conclusions about the scale and significance of cash dividends to households from CBNRM programmes.

In Namibia (and also Botswana), communities spend their income in diverse ways (Table 10.2). It is difficult to measure the impacts of these benefits on livelihoods. However, in remote areas (for example on the margins of the Namib Desert), conservancies have facilitated the provision of education and health facilities. At Puros, the conservancy has facilitated the construction of a school by one of its joint venture partners, and provides food and a cook

Table 10.2 **Benefits provided to members in four conservancies in north-west Namibia, August 2007.**

Benefit	Anabeb	Marienfluss	Puros	Sesfontein
Borehole for domestic water use		✓	✓	Planned
Lifts to clinics and in other emergencies with conservancy vehicle		✓	✓	
Employment in tourism lodges/activities linked to the conservancy such as joint ventures	8	14 (2 part-time)	35 (2 part-time)	No tourism joint ventures as yet
Conservancy employment (including conservancy owned campsites)	11	13	23	(data not obtained)
Employment in hunting	4 part-time			5 part-time
Support to local school (food and/or cash)	✓	✓	✓	✓
Own use hunting quota divided among local institutions			✓	
Transport children to and from schools in Sesfontein and Opuwo		✓	✓	
Donation of game meat for meetings and festivals	✓	✓	✓	✓
Meat from own use and other hunting to households	✓	✓	✓	✓
Income from craft sales to individuals			✓	
Loans to individuals			✓	
Social fund to help people in emergencies		✓		
Financial support for school children/prizes, bursaries for tertiary education		✓		✓
Financial support to provide teachers				✓

(Continued)

Table 10.2 **Continued.**

Benefit	Anabeb	Marienfluss	Puros	Sesfontein
Acquisition of conservancy property to house members on visits to town		✓		
Contribution to payments to members to offset livestock losses caused by predators	✓		Planned	✓
Provision of infrastructure (e.g. room for mobile clinic, classroom, etc.)		✓	✓	
Payments/cash support to Traditional Authority	✓			✓
Support to local sports teams/ tournaments	✓			✓

Source: Adapted from Jones and Mazambani (2007).

(Jones & Mazambani, 2007). This enables local children to stay in the village and avoid hostel fees in the village of Sesfontein, about 100 km away on a rough bush track. Marienfluss Conservancy has assisted in the provision of a mobile school and a mobile clinic, is negotiating with joint venture partners to bring flying doctors to the area, and provides transport (or cash to pay for transport) to the nearest clinic, 200 km away, where it has bought a house so families have accommodation while waiting for relatives to be treated. In conservancies with small populations (e.g. Puros 260; Marienfluss 300), the only jobs available are in the conservancy and related wildlife and tourism activities, and these have a significant impact on households and the local economy (Jones & Mazambani, 2007).

Bandyopadhyay *et al.* (2004) analyse data from a 2002 survey of 1192 households in seven conservancies in north-west and north-east Namibia, undertaken by the WILD (Wildlife Integration for Livelihood Diversification) Project and the Environmental Economics Unit of the Directorate of Environmental Affairs in the Ministry of Environment and Tourism (Long, 2004). Surveys were not undertaken in areas without conservancies, but comparison of welfare indicators from well-established conservancies (earning steady income) and newly established conservancies (earning little income) suggested that they concluded that conservancies overall had a beneficial effect on household welfare. The improved welfare effects were poverty-neutral in Kunene Region in the north-west and pro-poor in the north-eastern Caprivi Region (Bandyopadhyay *et al.*, 2004).

This survey did not include the Nyae Nyae Conservancy in north-eastern Namibia where the 2000 predominantly San residents are among the poorest and most marginalised in the country. In 2003, this conservancy directly provided 28 per cent of the jobs in the area and approximately 35 per cent of the cash income of residents (Weaver & Skyer, 2003). The conservancy also provided game meat from trophy hunting, supported the maintenance of village and wildlife water points and paid for local teachers.

In Botswana, the Sankuyo Tshwaragano Management Trust has increased its income from wildlife and tourism from approx. US$74,200 in 2000 to US$590,000 in 2005, and in each year except 2001 had a surplus after expenditure on operating costs and community benefits (Arntzen *et al.*, 2007a). The Trust has provided a wide range of community level and household benefits (Table 10.3). Household dividend varies year by year depending on trust income and decisions made by residents voting at the annual general meeting. An additional benefit is meat distribution from trophy hunting, some of which is given to poor people while the rest is auctioned to help provide an income for the Trust (Arntzen *et al.*, 2007a).

The four Namibian conservancies in Table 10.1, and the Sankuyo Trust in Botswana, are examples of community wildlife institutions which are well managed and have good natural resource endowments and tourism attractions, enabling them to earn high levels of income. They are also able under national policy and legislation to retain all income earned from hunting and photographic tourism, although it is not clear how Sankuyo's ability to retain all its revenue will be affected by the new CBNRM policy in Botswana referred to earlier. Within Namibia and Botswana there are many other communities that have adopted some form of CBNRM that do not have the same resource endowments as those considered above. For example, in Botswana there are 96 registered community trusts of which only 35 are generating income (Arntzen *et al.*, 2007a). In some cases improvements in governance, in internal management and in natural resource management may bring some modest increases in revenue, but by and large significant benefits in terms of direct household

Table 10.3   **Types of benefits provided by the Sankuyo Tshwaragano Management Trust, 1997–2005.**

Year	Household benefit	Community benefit
1997	Household dividends; 51 jobs	Built small community shop which ceased to function in 1999
1998	62 jobs	Built community hall and trust offices; purchased TV, solar panels and batteries; purchased vehicle for trust business and operations
1999	Household dividends; 62 jobs	Grant to football team; built Enviro Loo toilets which never worked due to poor construction
2000	62 jobs	Football team sponsorship; Built Shandereka cultural village and Kazikini campsite
2001	Funeral assistance to residents; 22 jobs	Football team sponsorship
2002	Household dividends; funeral assistance to residents; allowance to destitutes; 91 jobs	Football team sponsorship; Satellite TV subscription and installation
2003	Scholarship grants; 95 jobs	Satellite TV subscription.
2004	104 jobs	Donation to national disaster fund; training of staff to operate community owned tourism lodge
2005	Household dividends; funeral assistance; old age pension fund; housing for destitutes; micro credit scheme; 101 jobs	Football team sponsorship; support to local school

Note: Jobs throughout the table include those created by the Trust and by its hunting and photographic tourism joint venture partner

Source: Adapted from Arntzen *et al.* (2007a).

income and benefits from social projects will be few where the potential for wildlife use and tourism development is low.

Intangible benefits

Arntzen *et al.* (2007b) note the following major non-material benefits to communities from CBNRM:

- Empowerment of local population, encouraging self-esteem and pride, and reduced dependency on government. For example, Sankuyo in Botswana decided to operate a tourism lodge itself instead of entering into a joint venture agreement with a private sector operator.
- Exposure to commercial partners and business approaches for CBOs that are involved in joint ventures. Evidence from Namibia and Zimbabwe suggests that genuine joint-venture partnerships offer significant long-term benefits to communities in the areas of business skills and operation and specialised marketing.
- Development of a better understanding and working relationships with government, NGOs, and the private sector.

Arntzen *et al.* (2007b) suggest that non-material benefits of CBNRM in Zimbabwe, Namibia and Botswana may be more important to communities than the currently limited material benefits. Jones (2001) notes the importance of intrinsic values of wildlife to rural residents in the Kunene region of Namibia. Indeed, the well-documented current increases of wildlife on communal land in Kunene (e.g. NACSO, 2006), despite the fact that cash dividends to households are low or non-existent and jobs are not sufficient to provide for all community members, does provide some evidence that local people place importance on non-material benefits and on the existence value of wildlife.

Overall, the poverty reduction impact of CBNRM is limited, but in certain cases and conditions it can have positive impacts on livelihoods and community well-being.

Equity and power relations

A number of reviews of CBNRM have pointed to inequitable distribution of benefits in CBNRM projects. The poor lose out in CBNRM in two ways

(Turner, 2004). Firstly they are the worst equipped to capture benefits because they are less likely to compete for jobs in tourism or to benefit from expanded institutional functions or leadership opportunities. Secondly the very poor, who are most likely to be dependent on hunting or gathering, are disadvantaged by increased restrictions on resource harvesting (this issue is discussed in more detail below). Benefits are often appropriated by elites, often kin-related groups, who monopolise control of projects (Fabricius, 2004). In Botswana (Sankuyo), Boggs (2004) suggests that an elite can undermine *a conservancy* trust board and interfere with the democratic process. Power relations were disrupted when elders were forced to give up power to the young and educated. But unequal division of benefits created new layers of privilege in the community and caused conflict (Boggs, 2004).

In the early LIRDP project in Zambia, chiefs tended to appropriate income allocated to the community or directed it mostly in their own interests (Gibson, 1999; Child, 2004). This situation started to change with the introduction of a new project approach based on village-level participatory decision-making (Child, 2004). In this case, CBNRM activities resulted in shifting power away from the chiefs to local villagers.

Vaughan *et al.* (2004) suggested that the institutional structure of Namibian conservancies have tended to concentrate power in the hands of a few (community game guards and conservancy committees), allowing elites to benefit more than others. On the other hand, Bandyopadhyay *et al.* (2004) concluded that, even if the Namibian conservancies were not pro-poor, their benefits were not completely captured by the elite.

Murombedzi (2003) suggests there are also power imbalances between safari operators and local communities. In Zimbabwe, the safari industry is dominated by Europeans, as a result of the appropriation of rights from the local people by the colonial state. As a result of their privileged and powerful position in relation to communities, safari operators allow very little participation by Africans as skilled workers, and even the local trackers on which the hunt depends are treated as unskilled labourers. Operators impose restrictions on local activities that might interfere with the image of a pristine wilderness and ignore agreements that allow community members to monitor the hunts.

Costs of CBNRM to households

CBNRM programmes can impose costs on communities, in making the trade-off between sport hunting and local hunting for subsistence or cultural purposes, and there can be costs from living with wildlife from human–wildlife conflict.

Trade-offs between sport hunting and local hunting

The introduction of CBNRM can reduce access to game meat by some community members, including the poor, because hunting by community members is restricted to ensure that wildlife is available for sport hunting by rich foreigners. Thus Murombedzi (2003) found that communities entered into arrangements that guaranteed safari operators access to wildlife but restricted all local access to the same resources; local community members were called upon to regulate and police this arrangement (Murombedzi, 2003).

Prior to the introduction of the CBNRM programme in Botswana, poor people could obtain special game licences from the Department of Wildlife and National Parks to hunt for meat. In the Chobe Enclave community, the !Xo San group are traditional hunters. When the community wildlife management body, the Chobe Enclave Conservation Trust, gained the wildlife hunting quota for the enclave, special licences were no longer issued. Practically the entire quota was sold to a safari company, and few animals were available for allocation for local hunting. This severely disadvantaged the !Xo who could not afford to purchase meat from the Trust, and were also the group least involved in the Trust's decision-making structures (Jones, 2002).

In Zambia, Gibson (1999), using data from 1991 to 1994, found that hunting was important for the survival of rural residents as a source of meat and cash for consumer goods, and to pay school fees. Access to (illegally hunted) game meat was restricted by the anti-poaching activities of CBNRM projects, yet the direct economic benefit to rural residents of Zambia's CBNRM projects were not sufficient to replace the meat and cash from hunting. Individual returns from hunting far outweighed a resident's share in the benefits flowing from ADMADE and LIRDP (Gibson, 1999). Illegal hunting did not stop, but shifted to using snares and not firearms and targeting smaller mammals. Illegal hunters gained social status by continuing to provide their families and fellow villagers with game meat (Gibson, 1999). Changes in the approach to LIRDP in 1995 and 1998 aimed to place income directly in the hands of the local communities and their elected representative, and to create more attractive and equitable incentive structures which would reward individuals directly. Child

(2004) suggests that after four years of change the community had begun to appreciate its wildlife and poaching had been reduced.

In the Mahenye community in Zimbabwe, Gordon (1993) found that economic incentives (in the form of the annual dividend from safari hunting) also did not lead to a decline in poaching. Illegal hunting was being carried out by specific residents with the tacit support of the community because it was important to Shangaan cultural identity.

In Namibia, wildlife provided a safety net for some residents in conservancies in Kunene and Caprivi regions, in the form of food and to a lesser extent income security (Vaughan *et al.*, 2004). Wildlife users tended to be poorer or less secure households. In the Namibian conservancies, access to wildlife was restricted by the activities of community game guards and the poor lost access to the wildlife resource. Vaughan *et al.* (2004) recommended that the Namibian conservancies should take advantage of existing legislation that allows them to set their own quotas for certain game species so that local hunting could take place allowing poor people to get access to game meat.

Despite the introduction of community-based wildlife management initiatives, poor people may be disadvantaged through increased controls over wildlife use or the loss of access to wildlife that is allocated for use by foreign hunters. In some cases, hunting is likely to continue for subsistence purposes because benefits to households are low and are not sufficient to replace the loss of game meat. Further, hunting may continue for the expression of cultural identity, suggesting that economic incentives may be a necessary, but not a sufficient, condition for promoting conservation at community level, and there is a need to consider socio-cultural perspectives. The case of Zambia suggests that where programmes can be responsive to community needs, increasing benefits and local decision-making, there can be more positive outcomes. The Namibian case illustrates that not all programmes restrict wildlife use to sport hunting by foreigners, and through appropriate policy and legal provisions may enable local hunting to take place. As yet no Namibian communities have chosen to take advantage of the opportunity to regulate and allow hunting by local residents. Namibian conservancies do carry out their own hunts using local hunters, but the meat and by-products are used for community benefit. The case study from Canada presented in Box 10.1 reinforces arguments that policy and legislation should enable communities to make their own trade-offs regarding different uses of wildlife and whether they wish to allow sport hunting by outsiders.

> ## Box 10.1 Sport hunting and culture in Nunavut, Canada
>
> Polar bear hunting by the Inuit people in Nunavut in northern Canada presents interesting parallels with issues concerning sport hunting and communities in southern Africa. Four Nunavut Inuit communities receive quotas for hunting polar bears from either the Provincial Government or the National Government and use part of their quotas for sport hunting by foreigners. The hunting brings considerable economic benefits. In one of the communities a local hunting guide can earn up to C$7250 per hunt and will probably carry out two hunts in the short season. In one of the other communities, the sport hunt provides about 2000 kg of fresh meat. The hunts are carried out by local outfitters, so a large part of the income from the foreign hunter stays in the community. Overall income to Inuit from sport hunting is about C$1.5 million in a season. There are conflicts between the Inuit and Canadian regulatory authorities over lack of local involvement in deciding on quota restrictions and the Inuit are subject to external embargos imposed by the US Government on imports of polar bear hides from parts of Nunavut. There is also internal cultural conflict over sport hunting with some Inuit arguing that such hunting and other forms of conservation management are antithetical to maintaining an appropriate ethical relationship with polar bears. Partly as a result, the Inuit allocate only 25 per cent of their total quota to sport hunting, suggesting that cultural aspects of their relationship with wildlife are more important than the economic.
>
> *Source:* Wenzel and Dowsley (2005).

Human–wildlife conflict

Rural residents under CBNRM are expected to live alongside wildlife. Proponents of CBNRM believe that wildlife and tourism can generate sufficient income to cover the resulting costs of human–wildlife conflict (HWC). These can in some instances be high. The WILD project in Namibia estimated that crop losses to wildlife in Caprivi comprised 18 per cent of average annual household incomes for the region in Mayuni conservancy and 22 per cent in Kwandu conservancy (Murphy *et al.*, 2004). In the dry Kunene Region in north-west Namibia, major HWC costs are livestock losses to predators, and damage to water infrastructure and to gardens by elephants *Loxodonta africana*. Murphy *et al.* (2004) suggest that these types of impacts can contribute to poverty by reducing household food security and options for generating cash, and can be particularly problematic for households that have little security from the outset.

Across the region a number of attempts have been made to deal with the negative impacts of wildlife on livelihoods. In Namibia, conservancies in Kunene and Caprivi regions have begun compensating farmers for stock losses to predators. This scheme is administered by the conservancies in conjunction with the government and a Namibian NGO, Integrated Rural Development and Nature Conservation. Clear rules and guidelines have been established for the value of livestock and the grounds on which compensation will be paid. As conservancies are able to cover the costs from their own income, external funding is reduced and eventually terminated. In Caprivi the scheme is being extended to cover crop losses. Across the region, CBNRM funding is being used with varying degrees of success by communities to keep elephants away from crops by erecting a variety of barriers such as electric fences and chilli pepper fences.

In Namibia, wildlife numbers, including those of elephants and predators, are increasing in the Kunene and Caprivi regions partly due to the success of CBNRM (NACSO, 2006; Jones & Barnes, 2007) and in Caprivi because of the movement of elephants into the region from northern Botswana. These increases are placing heightened pressure on the CBNRM programme particularly in Caprivi where farmers are complaining of increased crop losses. Until recently, residents of conservancies have been tolerant of wildlife including elephants on their land, but generally household cash benefits are not sufficient to offset losses to elephants and the conservancy compensation scheme for crop losses is in its infancy. Unless income from wildlife increases (e.g. through hunting of elephants), local tolerance of crop losses is likely to be limited (Martin, 2005).

Conclusions

CBNRM can provide a range of benefits to local communities, including both material and non-material benefits. The level of benefit available varies with local context. Resource-rich communities will earn more income than others that have little potential for wildlife use or tourism development. The actual livelihood or poverty impacts of benefits are difficult to measure because often they are in the form of social projects. Actual cash benefits in the form of household 'dividends' are rare and when they are paid out are usually small amounts. However, the significance of the benefits available differs with the needs of the communities themselves. Depending on the local context even small amounts of money can have significant impacts and can be used for covering school fees or other livelihood contributions.

In some cases benefits are unevenly distributed within communities and local elites control CBNRM processes and benefits. However, in other cases adjustments to project approaches or demands for accountability by local residents are able to counter this tendency. There is evidence that poorer residents can be disadvantaged by restrictions on hunting and other resource use that previously they depended on. Initiatives based on privileging foreign sport hunters at the expense of local hunting are therefore unlikely to succeed. As a result policy and legislation need to provide more decision-making authority to local communities enabling them to decide for themselves how they want to use their wildlife. Further, there are costs to living with wildlife that CBNRM incomes do not necessarily cover. However, there are other intangible benefits that can also be important to residents and these will affect the way in which local community members evaluate for themselves whether they wish to live with wildlife. There is potential for resource-rich communities to increase income further through the development of additional enterprises but more attention needs to be given to the equitable distribution of benefits and the management of income.

National policy and legislation plays a role in determining the scale and nature of benefits to communities. In Zambia and Zimbabwe the state retains portions of revenue from wildlife and tourism, reducing the amount going to communities. In Namibia, and until recently in Botswana, communities retain all the income generated from hunting and photographic tourism concessions. Further Namibian legislation enables conservancies to gain income from a wide variety of wildlife uses, not only trophy hunting and a market exists for the sale of wildlife by conservancies. Community rights to benefit are better protected where they are entrenched in legislation rather than contained in policy guidelines or directives that can be changed at the whim of officials and the stroke of a keyboard. The disturbing trend of governments increasingly dictating how benefits can be spent by communities is disempowering, removes choice and reduces the flexibility of communities to use their income to respond to local needs. It remains to be seen how communities in Zimbabwe and Botswana will respond to the new government diktats.

References

Arntzen, J., Buzwani, B., Setlhogile, T., Kgathi, D.L. & Motsholapheko, M.R. (2007a) *Community-Based Resource Management, Rural Livelihoods, and Environmental Sustainability.* Phase three Botswana Country Report prepared for IUCN-South Africa and USAID FRAME. International Resources Group, Washington, DC.

Arntzen, J., Setlhogile, T. & Barnes, J. (2007b) *Rural Livelihoods, Poverty Reduction, and Food Security in Southern Africa: Is CBNRM the Answer?* International Resources Group, Washington, DC.

Arntzen, J.W., Molokomme, D.L., Terry, E.M., Moleele, N., Tshosa, O.D. & Mazambani, D. (2003) *Final Report of the Review of Community-Based Natural Resource Management in Botswana.* National CBNRM Forum, Gaborone.

Ashley, C. (1998) *Intangibles Matter: Non Financial Dividends of Community-based Natural Resource Management in Namibia.* Report for the World Wildlife Fund Living in a Finite Environment (LIFE) Programme. WWF-LIFE, Windhoek.

Bandyopadhyay, S., Shyamsundar, P., Wang, L. & Humavindu, M.N. (2004) *Do Households Gain from Community-based Natural Resource Management? An Evaluation of Community Conservancies in Namibia. (DEA Research Discussion Paper; no. 68).* Directorate of Environmental Affairs, Windhoek.

Boggs, L. (2004) Community-based natural resource management in the Okavango Delta. In *Rights, Resources and Rural Development: Community-based Natural Resource Management in Southern Africa,* eds. C. Fabricius & E. Koch with H. Magome & S. Turner, pp. 147–159. Earthscan, London.

Bond, I. (2001) CAMPFIRE and the incentives for institutional change. In *African Wildlife and Livelihoods,* eds. D. Hulme & M. Murphree, pp. 227–243. James Currey, Oxford.

Child, B. (2004) The Luangwa Integrated Rural Development Project, Zambia. In *Rights, Resources and Rural Development: Community-based Natural Resource Management in Southern Africa,* eds. C. Fabricius & E. Koch with H. Magome & S. Turner, pp. 235–247. Earthscan, London.

Child, B., Jones, B., Mazambani, D., Mlalazi, A. & Moinuddin, H. (2003) *Final Evaluation Report: Zimbabwe Natural Resources Management Program – USAID/Zimbabwe Strategic Objective No. 1. CAMPFIRE, Communal Areas Management Programme for Indigenous Resources.* United States Agency for International Development, Harare.

Davis, A. & Kasaona, J.K. (2007) Supporting the management of conservancy-generated income – achievements and challenges for IRDNC Kunene. Concept Paper prepared for the IRDNC mid-term evaluation, August 2007. Integrated Rural Development and Nature Conservation, Windhoek.

Fabricius, C. (2004) The fundamentals of community-based natural resource management. In *Rights, Resources and Rural Development: Community-based Natural Resource Management in Southern Africa*, eds. C. Fabricius & E. Koch with H. Magome & S. Turner, pp. 3–43. Earthscan, London.

Fabricius, C., Koch, E. & Magome, H. (2001) *Community Wildlife Management in Southern Africa: Challenging the Assumptions of Eden.* IIED Evaluating Eden Series No. 6. International Institute for Environment and Development, London.

Gibson, C.C. (1999) *Politicians and Poachers: The Political Economy of Wildlife Policy in Africa.* Cambridge University Press, Cambridge.

Gibson, C.C. & Marks, S.A. (1995) Transforming rural hunters into conservationists? An assessment of community-based wildlife management programs in Africa. *World Development*, 23, 941–957.

Gordon, D. (1993) *From Marginalisation to Centre-Stage? A Community's Perspective on Zimbabwe's CAMPFIRE Programme.* Major Paper submitted to the Faculty of Environment Studies in partial fulfillment of the requirements of the degree in Master of Environmental Studies, York University, North York, Ontario, Canada.

Government of Botswana (2007) *Community Based Natural Resources Management Policy.* Ministry of Environment, Wildlife and Tourism, Gaborone.

Jones, B. (2001) The evolution of community based approach to wildlife management in Kunene, Namibia. In *African Wildlife and Livelihoods: The Promise and Performance of Community Conservation*, eds. D. Hulme & M. Murphree, pp. 160–176. David Phillips, Cape Town.

Jones, B. (2002) *Chobe Enclave: Lessons Learned.* CBNRM Occasional Paper No. 7. National CBNRM Forum, Gaborne.

Jones, B.T.B. (2004) *CBNRM, Poverty Reduction and Sustainable Livelihoods: Developing Criteria for Evaluating the Contribution of CBNRM to Poverty Reduction and Alleviation in Southern Africa.* Commons Southern Africa Occasional Paper No. 7. Centre for Applied Social Sciences, University of Zimbabwe, Harare and the Programme for Land and Agrarian Studies, University of the Western Cape, Cape Town.

Jones, B.T.B. & Barnes, J.I. (2007) *WWF Human Wildlife Conflict Study: Namibian Case Study.* WWF Macroeconomics Programme Office and WWF Global Species Programme, Washington, DC and Gland.

Jones, B.T.B. & Mosimane, A. (2007) *Promoting Integrated Natural Resource Management as a Means to Combat Desertification: The LIFE Project and Namibian CBNRM.* International Resources Group. Washington, DC.

Jones, B.T.B. & Mazambani, D. (2007) *Managing Growth and Sustainability*. Mid-term Evaluation of IRDNC's Community-based Natural Resource Management Programme in Kunene Region and Caprivi Region, Namibia. WWF Programme Number NA000404/NA0001500. WWF UK, Godalming.

Jones, B.T.B. & Murphree, M.W. (2004) Community-based natural resource management as a conservation mechanism: lessons and directions. In *Parks in Transition: Biodiversity, Rural Development and the Bottom Line*, ed. B. Child, pp. 63–103. Earthscan and IUCN South Africa. London.

Long, S.A. (2004) Livelihoods in the conservancy study areas. In *Livelihoods and CBNRM in Namibia: The Findings of the WILD Project. Final Technical Report of the Wildlife Integration for Livelihood Diversification Project (WILD)*, ed. S.A. Long, pp. 55–80. Ministry of Environment and Tourism, Windhoek.

Martin, R.B. (2005) *Transboundary Species Project Background Study: Elephants*. Transboundary Mammal Project of the Ministry of Environment and Tourism, Windhoek.

Murombedzi, J. (2003) Devolving expropriation of nature: the devolution of wildlife management in southern Africa. In *Decolonizing Nature: Strategies for Conservation in a Post-colonial Era*, eds. W.M. Adams & M. Mulligan, pp. 135–151. Earthscan, London.

Murphy, C. & Roe, D. (2004) Livelihoods and tourism in communal area conservancies. In *Livelihoods and CBNRM in Namibia: The Findings of the WILD Project. Final Technical Report of the Wildlife Integration for Livelihood Diversification Project (WILD)*, ed. S.A. Long, pp. 119–138. Ministry of Environment and Tourism, Windhoek.

Murphy, C., Vaughan, C., Katjiua, J., Mulonga, S. & Long S.A. (2004) The costs of living with wildlife. In *Livelihoods and CBNRM in Namibia: The Findings of the WILD Project. Final Technical Report of the Wildlife Integration for Livelihood Diversification Project (WILD)*, ed. S.A. Long, pp. 105–117. Ministry of Environment and Tourism, Windhoek.

NACSO (2006) *Namibia's Communal Conservancies: A Review of Progress and Challenges in 2005*. Namibian Association of CBNRM Support Organisations, Windhoek.

Rozemeijer, N. (2003). CBNRM in Botswana. Chapter prepared for *Parks in Transition: Conservation, Development and the Bottom Line, Vol II*. IUCN Southern African Sustainable Use Specialist Group. IUCN South Africa, Pretoria.

Sibanda, B. (2004) Community wildlife management in Zimbabwe: the case of CAMPFIRE in the Zambezi Valley. In *Rights, Resources and Rural Development: Community-based Natural Resource Management in Southern Africa*, eds. C. Fabricius & E. Koch with H. Magome & S. Turner, pp. 248–258. Earthscan, London.

Taylor, R.D. & Murphree, M.W. (2007) *Case Studies on Successful Southern African NRM Initiatives and their Impact on Poverty and Governance. Zimbabwe: Masoka and Gairezi.* International Resources Group, Washington, DC.

Turner, S. (2004) Community-based natural resource management and rural livelihoods. In *Rights, Resources and Rural Development: Community-based Natural Resource Management in Southern Africa,* eds. C. Fabricius & E. Koch with H. Magome & S. Turner, pp. 44–65. Earthscan, London.

Twyman, C. (2000) Livelihood opportunity and diversity in Kalahari Wildlife Management Areas, Botswana: rethinking community resource management. *Journal of Southern African Studies,* 4, 783–806.

Vaughan, C., Long, S.A., Katjiua, J., Mulonga, S. & Murphy, C. (2004). Wildlife use and livelihoods. In *Livelihoods and CBNRM in Namibia: The Findings of the WILD Project. Final Technical Report of the Wildlife Integration for Livelihood Diversification Project (WILD),* ed. S.A. Long, pp. 81–104. Ministry of Environment and Tourism, Windhoek.

Wainwright, C. & Wehrmeyer, W. (1998) Success in integrating conservation and development? A study from Zambia. *World Development,* 26, 933–944.

Weaver, L.C. & Skyer, P. (2003) *Conservancies: Integrating Wildlife Land-use Options into the Livelihood, Development and Conservation Strategies of Namibian Communities.* Paper presented at the Vth World Parks Congress of IUCN to the Animal Health and Development (AHEAD) Forum, Durban, September 8–17.

Wenzel, G. & Dowsley, M. (2005) Economic and cultural aspects of polar bear sport hunting in Nunavut, Canada. In *Conservation Hunting: People and Wildlife in Canada's North: Papers from a Conference Titled: People, Wildlife and Hunting: Emerging Conservation Paradigms,* eds. M.M.R. Freeman, R.J. Hudson, & A.L. Foote. CCI Press, Edmonton.

Policy and Practice

Conservation Values from Falconry

Robert E. Kenward

Anatrack Ltd and IUCN-SSC European Sustainable
Use Specialist Group, Wareham, UK

Introduction

Falconry is a type of recreational hunting. This chapter considers the conservation issues surrounding this practice. It provides a historical background and then discusses how falconry's role in conservation has developed and how it could grow in the future.

Falconry, as defined by the International Association for Falconry and Conservation of Birds of Prey (IAF), is the hunting art of taking quarry in its natural state and habitat with birds of prey. Species commonly used for hunting include eagles of the genera *Aquila* and *Hieraëtus*, other 'broad-winged' members of the *Accipitrinae* including the more aggressive buzzards and their relatives, 'short-winged' hawks of the genus *Accipiter* and 'long-winged' falcons (genus *Falco*).

Falconers occur in more than 60 countries worldwide, mostly in North America, the Middle East, Europe, Central Asia, Japan and southern Africa. Of these countries, 48 are members of the IAF. In the European Union falconry

Recreational Hunting, Conservation and Rural Livelihoods: Science and Practice, 1st edition.
Edited by B. Dickson, J. Hutton and W.M. Adams. © 2009 Blackwell Publishing,
ISBN 978-1-4051-6785-7 (pb) and 978-1-4051-9142-5 (hb).

is regulated under the Wild Birds Directive (79/409/EEC) and in the United States by the Migratory Birds Act. Some countries with few falconers have no legal provisions. Falconry is recognised in international conventions: CITES has a system that allows individual raptors owned by falconers to be moved across international borders.[1]

A concise history of falconry

Falconry is probably 2–3000 years old. Raptor bones are frequent in the burial kurgans of Scythian tribes and the earliest indisputable evidence comes from a Chinese description that could be as early as 700 BC (Xaodie, 2005). A claim in the Shahnamei epic of Ferdowsi, that falconry in Iran predated Zoroaster (*c.* 4000 years BC), was written after the conquest by Muslims. If falconry was that old, there should have been signs in the relics of Persepolis (Yazdani, 2005) and in writings from Egypt and Greece. A bas-relief possibly depicting a falconer was found in the ruins of Khosabad (*c.*1700 BC), but Lindner (1973) concluded from many mosaics and writings that raptors were used by Greek and early Roman citizens for fowling (e.g. to attract mobbing birds down to nets or twigs covered with bird-lime, and in Thrace by flying raptors to frighten birds down into nets at ground level) but not for falconry.

Falconry reached Japan in the 3rd century AD and Europe with the Vandals in the fourth (Lindner, 1973). It thrived in early Muslim culture; the first Arabic treatise is from the eighth or 9th century AD (Allen, 1980). Trained raptors have been widely used across Asia, from Turkey, Iran, Mongolia and China in the north, to Arabia and the Indian subcontinent in the south, with extension into North Africa as far as Morocco. Sparrowhawks *Accipiter nisus* are still trapped widely on migration in the eastern parts of their range, for flying at migratory prey such as Eurasian quail *Coturnix coturnix*, before release in spring.

Owing to the levy value of hawks in Britain (tax could be paid as a hawk or £8–10 in lieu), the Domesday Book (11th century) records 24 nesting areas, presumed to be of goshawks *Accipiter gentilis*, for the county of Cheshire. This gives an early raptor density estimate of 0.9 pairs per 100 km² (Yalden, 1987). Within 200 years, Emperor Frederich II of Hohenstaufen was writing *De Arte Venandi cum Avibus* (von Hohenstaufen, 1248), for which he has been called the father of ornithology. His principle of testing hypotheses, for instance by sending a trusted servant to the north to see whether barnacles really metamorphosed into geese, was an important step in the development of modern science.

The Boke of St Albans (Berners, 1486) indicates that falcons were probably flown mainly by nobles, offering spectacular flights on excursions to large open spaces with a stable of reliable horses, whereas a goshawk was 'for ayeoman', being better for keeping a larder stocked with small-game. *An Approved Treatise on Hawks and Hawking* (Bert, 1619) reveals how sophisticated the veterinary treatment of trained raptors had become by the 17th century (Cooper, 1979). In Britain, falconry lost popularity after the English Civil War with the development of effective sporting guns and Land Enclosure Acts (which restricted access to good hawking land). By the late 18th century, the practice of falconry was restricted to a few landowners who formed a series of clubs until the present British Falconers' Club (BFC) was founded in 1927 (Upton, 1980).

Falconry was responsible for early laws to protect raptors. In England, goshawks were protected by Henry VII (1457–1509) 'in pain of a year and a day's imprisonment, and to incur a fine' (Cooper, 1981). When loss of interest in falconry was followed by the persecution of raptors to conserve game, Morant (1875) wrote in scorn of an 1873 committee on bird preservation 'No doubt, beside certain naturalists, it is our falconers who are anxious to make birds of prey more numerous'. Early in the 20th century, the BFC and the Royal Society for the Protection of Birds pioneered a bounty scheme for landowners who preserved raptor nests.

A small number of people kept falconry alive in most European countries and helped to establish it in North America between the two World Wars. The subsequent renaissance in Western falconry, illustrated by the increasing membership of the BFC (Figure 11.1), was stimulated by books, films, game fairs and journalists rediscovering a 'lost art'. It coincided with increasing general interest in wildlife, with falconers responsible for early quantitative studies of raptor predation (e.g. Craighead & Craighead, 1956; Brüll, 1964).

The early stage of the falconry renaissance also coincided in the 1950s and 1960s with some steep raptor population declines. Research eventually attributed these to the agricultural use of organochlorine pesticides (Ratcliffe, 1980; Newton, 1986), but not before falconry had been blamed. The removal of young from the last wild peregrines *Falco peregrinus* in Denmark and Schleswig-Holstein created fears that falconry was a threat to bird-of-prey populations. Laws banned hunting with raptors in countries with little history of falconry, including Sweden and (for recreation)

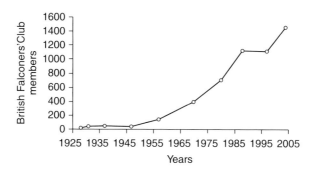

Figure 11.1   **Growth in membership of the British Falconers' Club from 1925 to 2005.**

Australia. However, most countries were content to tighten controls on the ownership of raptors.

Once aware of the pesticide problem, BFC members voluntarily restricted their licence applications for wild British peregrines. In the UK and North America, falconers created raptor conservation bodies (Hawk Trust, Raptor Research Foundation and Peregrine Fund) and began trying to breed peregrines, which had vanished in parts of Europe and the Americas. Peregrines had first been bred in Germany in 1943 (Waller, 1982) and isolated successes were achieved again in Germany and the US during 1970–1972. From a production of about 20 large falcons in 1972 (10 in Germany), the number bred annually rose to more than 200 in 1975 (Kenward, 1976). Falconers ran six of the seven major release projects for peregrines (in Germany, Poland and the US), and were heavily involved in the release programmes for Mauritius kestrels *Falco punctatus* and California Condors *Gymnogyps californianus* (Cade, 1986; Saar, 1988; Jones *et al.*, 1994; Trommer *et al.*, 2000; Wallace, 2001; Cade & Burnham, 2003).

Sources of raptors for falconry

Traditionally, raptors for falconry were obtained as 'eyasses' from wild nests or trapped after fledging. However, most are now domestic bred. Falconers seldom train wild adults ('haggards'), which have had longer in the wild to accumulate latent diseases or learn behaviour that hinders training.

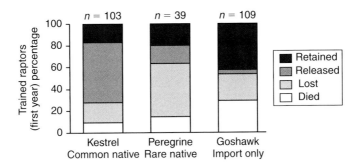

Figure 11.2 **Kestrels and peregrines, obtained from the wild under licence by members of the British Falconers' Club before 1970, were more often lost or released in their first year of life than Goshawks, which were relatively expensive imports. Data from Kenward (1974).**

Prior to 1970, raptors used in the UK were all native or potentially native species, and most were lost or released back to the wild in their first year (Figure 11.2, Kenward, 1974). Goshawks, which were relatively expensive imports, had fewest lost or released in their first year, but 52 per cent were lost or released eventually, which successfully re-established a native Goshawk population (Kenward, 2006).

In the UK, the supply of licences to obtain native or imported raptors was gradually restricted during the 1970s, with the result that the value of raptors rose to a level supportive of commercial breeding. However, different species are not equally easy to breed. The species bred earliest in Britain were Eurasian kestrels *Falco tinnunculus* and sparrowhawks, for which production peaked in 1987–1988, at more than 1000 and around 600 respectively (Fox, 1995). Domestic breeding was slower to develop for Goshawks than for falcons. Breeding of Harris hawks *Parabuteo unicinctus* from North America, which are a social raptor that is relatively easy to train, developed fastest of all the large raptors in Britain (Figure 11.3). Overall, the breeding of peregrines, goshawks and Harris hawks, representing the main species flown by falconers, rose from 100 in 1980 to 800 in 1991 and supported the growth in the numbers of falconers, many of whom were not registered in Figure 11.1 as members of the BFC. Production of pure species flown by falconers in Britain tended to plateau during the 1990s, as supply met demand. Prices for large raptors reduced from maxima of around £1000 to an average of perhaps £500 (€750)

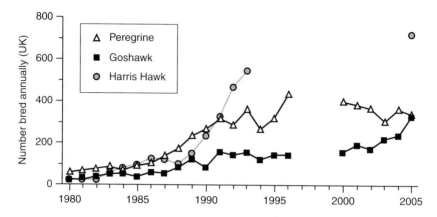

Figure 11.3 **During the period of government registration of all domestic breeding of raptors in the UK, production of goshawks developed more slowly than for other species favoured by falconers. Data, from Hawk Board (1988), Fox (1995) and British Falconers' Club, originated in the Government Environment Department and the Independent Bird Register (after 1993, Government data were kept only for rare native species).**

in 2006. Breeding of pure peregrines has tended to decline since the late 1990s (Figure 11.3).

Development in Britain of 'DNA fingerprinting' (Jeffreys et al., 1985) as a parentage test for raptors (Parkin, 1987) had proved a very strong deterrent against 'laundering' of wild birds. Initial tests showed a minority of breeding claims to be false and random survey of 20 domestic raptor broods in 1995–1996 found no illegality (Williams & Evans, 2000).

A survey for the European Commission recorded only 88 wild raptors (including 61 goshawks and eight peregrines) licensed for falconry in 2005. This is tiny in proportion to an estimated domestic production of about 10,000. Although nine states in the European Union (EU) still permit use of wild raptors for falconry, the proportion allowed probably exceeds five per cent of domestic production in only three.

The use of captive-bred birds in the UK and the rest of Europe contrasts with the situation in South Africa and the US. The US Fish and Wildlife Service permits a harvest of up to five per cent of wild raptor production and records the use of 800–900 wild raptors annually (Millsap & Allen, 2006).

In the Middle East there has also been greater reliance on birds captured from the wild and over-harvesting has had negative effects on both the raptors and the prey animals they are then used against. The tradition had been to use saker falcons *Falco cherrug*, trapped on migration between their Central Asian breeding grounds and wintering areas in Africa, to hunt migratory prey, especially the houbara bustard *Chlamydotis undulata* of which many make a similar migration to winter in the Middle East (Allen, 1980). Houbara are challenging prey for sakers, which normally take smaller species and breed best where there is abundance of mammals (Pfeffer, 1987) that do not require great speed. In the early 1990s, some 2750 wild sakers were being trapped annually (Riddle & Remple, 1994), and oil wealth had driven prices in the Middle East to an average US\$5000 (€7500) per bird. Whereas the difficulty of providing such birds with fresh meat had formerly provided a strong incentive to release them after nomadic hunting, a wealthier and settled population with refrigerators could keep more falcons. This put unsustainable pressure on houbara stocks and resulted in an unsustainable harvest of wild sakers in some areas, as political changes made Central Asian breeding areas more accessible (Fox, 2001; Kenward *et al.*, 2007).

Along with the use of captive bred birds has come the development of hybrid birds. In 1971 a female saker falcon and 'tiercel' (male) peregrine, which each courted other falcons but lacked conspecific partners, were put in a breeding enclosure in Ireland and reared two young from five eggs (Morris & Stevens, 1971). From this beginning grew a fashion for breeding falcon hybrids, initially partly as novelties (and proof of domestic parentage rather than laundering in days before DNA forensics) and subsequently because of advantageous traits in hunting particular quarry. This latter consideration applied especially in the Middle East where sakers have been bred with gyr falcons *Falco rusticolus* to be larger than pure sakers, and with peregrines to be faster.

Combined data from the surveys showed a very strong tendency for few hybrids to be produced, or used, in countries that permitted enough wild raptors for more than 3 per cent of their falconers annually (Figure 11.4). The reason for this relationship is uncertain. Only where countries are permitting falconers to acquire a new wild raptor every three to four years would supply approach demand and reduce domestic production. A plausible explanation is that where falconers were obliged to depend on domestic progeny by early restriction of access to wild stocks, early development of commercial breeding gave producers experience and competitive incentives to develop a fashion for hybrids.

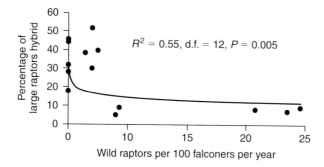

Figure 11.4   **The proportion of hybrids among large raptors produced or flown (whichever was greater in the IAF survey of 2000 and the EC survey of 2006) did not exceed 10% in countries where more than three wild raptors were permitted per 100 falconers each year. Data from Kenward (2004) and the European Commission survey.**

Contributions to conservation

There are a number of ways in which falconry is valuable for conservation. To enable eyasses to gain flight skills as they would in the wild, falconers 'hack' them by providing food at an artificial nest site to mimic the natural post-fledging period before recovery for training. Hacking has since become a highly efficient conservation technique for soft-release to restock raptor populations (Cade, 2000), modify nesting behaviour and rehabilitate wild birds after veterinary treatment. Young peregrine falcons 'hacked' from platforms on buildings readily adopted such sites for breeding (Tordoff et al., 1998). In Australia, incapacitated wild hawks maintained weight best after release if flown with falconry techniques (Holz et al., 2006).

At hack or when flown free after training, raptors wear location aids, traditionally as 1–2-cm-long closed bells on the legs or tail. Early use of wildlife radio tags on eagles (Southern, 1964) was followed by a commercial RB-4 receiver for falconry, named after Robert Berry who first bred goshawks by artificial insemination (Berry, 1972), which became the first reliable receiver widely used in wildlife research (the LA-12). The large falconry market for radio-tags will drive further developments.

Falconers are a small but elite proportion of the hunting community. A survey in the US showed that 83 per cent of falconers had tertiary education,

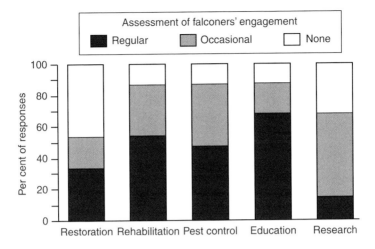

Figure 11.5   **The engagement of falconers in conservation activities as assessed by delegates of national authorities in 15 countries of the European Union. Data from the European Commission survey.**

compared with 47 per cent for other hunters and 57 per cent for wildlife-watchers. They spent twice as much time on their passion as hunters and six times that of watchers (Peyton *et al.*, 1995). US falconers also had a remarkably high engagement as volunteers in rehabilitation of wild raptors (57 per cent), conservation education projects (47 per cent) and raptor reintroduction work (35 per cent).

Similar engagement in Europe was recognised by delegates to the European Commission of national authorities responsible for the Wild Birds Directive. From 15 states with falconry, all but one recorded engagement in some aspect of raptor conservation. There was regular engagement in at least one activity in 12 cases, most often in education and awareness-raising, less in rehabilitation or use of raptors for biological control of nuisance species (typically on airfields) and least in research (Figure 11.5). Falconers in eight countries were engaged in conservation breeding or release work.

Falconers can also be important for monitoring wild raptors. Their per-egrine nest maps in the mid-20th century were crucial for survey and restoration work (Cade & Burnham, 2003). Marking at nests by falconers and others was combined with subsequent trapping, for mark–recapture surveys of goshawks (Kenward, 2006) and sakers. The IUCN World Conservation

Table 11.1 **Harvest rates and values of hunted red grouse, estimated for shooting by staff of the Game Conservancy Trust and measured for falconers at one estate in northern Scotland.**

Harvest parameters	Shooting (driven grouse)	Shooting (walking-up)	Falconry
Grouse/hunter/day	c. 25	c. 10	0.3
Payment/grouse killed	€ 100–200	€ 27–40	€ 65
Total value/grouse killed	€ 100–200	€ 27–40	€ 325

Congress (WCC2 was in Amman in 2000) called on saker range states and falconers to work with others to, *inter alia*, monitor populations and estimate sustainable yields (IUCN Resolution WCC 2.74).

Finally, there is potential for conservation in hunting with trained raptors, because it can bring high value to local communities through prolonged presence yet does not require a high density of quarry species. A shooter party may take 30–75 times as many grouse as a falconer daily, while the falconer is worth from two to ten times as much to the estate per grouse bagged (Table 11.1, from Kenward & Gage, in press). Falconry can also control pests where guns are undesirable (Saar *et al.*, 1999).

Conservation problems and responses in modern falconry

The recent survey for the European Commission addressed three concerns about falconry, namely the risk of introduction of exotic species; illegal procurement from wild populations; and the risk from introgression of genes through hybridisation. Of these, the introduction of exotic species was considered of least importance in Europe.

As has been seen there is little procurement of raptors from the wild in Europe. Indeed, wild peregrine populations tend to be highest in countries with most falconers (Figure 11.6), a relationship that remains even when the differing area of countries is taken into account ($P < 0.01$). There is therefore no evidence that falconers in Europe are reducing the numbers of wild peregrines. Overall, the number of raptors used in falconry in Europe remains

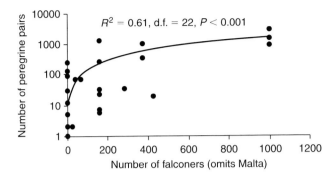

Figure 11.6 **High numbers of falconers in EU states were where BirdLife International recorded large peregrine populations. Data from Burfield & van Bommel (2004) and the European Commission survey.**

relatively low compared to the numbers breeding in the wild. Summary statistics gathered by Birdlife International in 2004 indicate a total of about 8400 pairs of peregrine falcons and more than 55,000 pairs of goshawks (Burfield & van Bommel, 2004) in the European Union.

The situation of the saker falcon in parts of Central Asia where trappers can easily access breeding areas is less healthy, with numbers reduced to a tenth of their former abundance. High productivity of this species gives it the potential for an unusually high yield of juveniles but is coupled with low survival in the first year to make it unusually sensitive to the harvest of adults (Kenward *et al.*, 2007). Nevertheless, saker populations are growing in Europe, remain large in west Kazakhstan and Mongolia (Dixon, 2005) and restocking has already started from domestic breeding with local genetic stock in depleted parts of southern Kazakhstan. It is essential for falconers to accept marking to certify the legal origin of sakers, so that value is low for birds outside official quota schemes. If necessary, marking can be associated by the banking of a feather or other genetic material as a 'mark-and-bank' security against marker tampering (Kenward, 2004). This could be linked to the development of a pay-to-use system, although this is challenging for a migratory species such as saker, because funds from harvesting in migration and wintering areas need to be transferred to ensure the conservation of breeding habitats. This might require an international agreement, perhaps under the Convention on Migratory Species.

The greatest concern in the European Commission survey was about hybrid falcons, with cases of hybrid breeding in the wild from four countries. However, it is important to recognise that natural hybridisation has been recorded between peregrine and prairie falcon *Falco mexicanus* in North America (Oliphant, 1991). It is also now clear from observation and genetic analyses that hybridization is not uncommon in zones where sakers encounter lanner falcons *Falco biarmicus* or gyr falcons (Wink & Sauer-Gürth, 2004; Nittinger *et al.*, 2006). For the species involved to have maintained their phenotypic identity for millennia in the face of natural hybridization, which represents a lack of behavioural isolation mechanisms, suggests that there are strong selection pressures against survival of intermediate phenotypes. This suggests that the *occasional* loss of artificial hybrids may not do great harm to healthy wild falcon populations.

Future directions

What is the future direction of falconry? Although severe action against falconry may be a tempting option for European governments because it will reduce lobbying pressure and the costs of regulation, it risks loss of special raptor management skills which take years to acquire.

Falconers certainly need to be cautious in their use of hybrids and thoughtful about exotic species, but governments need to recall that these are the product of restricted access to wild populations. When raptor populations are threatened, encouraging domestic breeding makes good sense. However, it may be that falconry in Europe would contribute more to conservation if more raptors were harvested from the wild. Perhaps the greatest cost to conservation associated with falconry in Europe now arises from obliging falconers to depend on domestic production, instead of using their funding and volunteer effort to help conserve wild stocks of popular species, like peregrines and goshawks, which are not at risk in the wild but need monitoring to ensure that populations remain healthy.

At an average €750 per bird, the domestic production of raptors in Europe is worth €7.5m. A one per cent harvest, of 1000 wild goshawks, would cover current demand in Europe and provide payments to landowners to compensate the predatory impact of this species as well as to fund monitoring. Although falconers in the US have recently regained access to peregrine populations that they helped to restore, harvesting peregrines more widely in Europe would require complex agreements to overcome entrenched positions and even five per cent (700 birds) would not meet the demand for large falcons. Domestic breeding would therefore remain important for species that have proved more vulnerable than the goshawk to unexpected human impacts. Maintenance of expertise and capacity in domestic breeding is important too, for providing insurance against problems that are detected so late in wild raptor populations that few are left, as in the case of Mauritius kestrels, Californian condors and the *Gyps* vultures in southern Asia.

Acknowledgements

I am grateful to the International Association for Falconry and Conservation of Birds of Prey and the Commission of the European Community for permission to use unpublished survey data. I thank Graham Irving and Barney Dickson for comments and Matt Gage for permission to use material in press.

Note

1 For more information on practical and ecological aspects of falconry, see www.i-a-f.org

References

Allen, M. (1980) *Falconry in Arabia*. Orbis, London.

Berners, J. (1486) *The Boke of St. Albans*. Schoolmaster Printer, St. Albans, UK.

Berry, R.B. (1972) Reproduction by artificial insemination in captive American Goshawks. *Journal of Wildlife Management*, 36, 1283–1288.

Bert, E. (1619) *An Approved Treatise of Hawks and Hawking*. Richard Moore, London.

Brüll, H. (1964) *Das Leben deutscher Greifvögel*. Fischer, Stuttgart, Germany.

Burfield, I. & van Bommel, F. (ed.) (2004) Birds in Europe – Population Estimates, Trends and Conservation Status. Birdlife International, Cambridge, UK.

Cade, T.J. (1986) Using science and technology to re-establish species lost in nature. In *Biodiversity*, ed. E.O. Wilson, pp. 279–288. National Academy Press, Washington, DC.

Cade, T.J. (2000) Progress in translocation of diurnal raptors. In *Raptors at Risk*, eds. R.D. Chancellor & B.-U. Meyburg, pp. 343–372. World Working Group on Birds of Prey and Owls, Berlin.

Cade, T.J. & Burnham, W. (ed.) (2003) *Return of the Peregrine – a North American Saga of Tenacity and Teamwork.* The Peregrine Fund, Boise, Idaho.

Cooper, J.E. (1979) The history of hawk medicine. *Veterinary History,* 1, 11–18.

Cooper, J.E. (1981) A historical review of goshawk training and disease. In *Understanding the Goshawk,* eds. R.E. Kenward & I.M. Lindsay, pp. 175–184. International Association of Falconry and Conservation of Birds of Prey, Oxford, UK.

Craighead, J.J. & Craighead, F.C. (1956) *Hawks, Owls and Wildlife.* Wildlife Management Institute, Washington, DC.

Dixon, A. (2005) Falcon population estimates: how necessary and accurate are they? *Falco,* 25/26, 5–8.

Fox, N.C. (1995) *Understanding the Bird of Prey.* Hancock House, Blaine, USA.

Fox, N.C. (2001) Future trends, captive breeding, trade controls or market forces? In *Saker Falcon in Mongolia: Research and Conservation,* eds. E. Potapov, S. Banzragch, N. Fox & N. Barton, pp. 211–214. Mongolian Academy of Sciences, Ulanbaatar, Mongolia.

Holz, P.H., Naisbitt, R. & Mansell, P. (2006) Fitness level as a determining factor in the survival of rehabilitated peregrine falcons (*Falco peregrinus*) and brown goshawks (*Accipiter fasciatus*) released back into the wild. *Journal of Avian Medicine and Surgery,* 20, 15–20.

Jeffreys, A.J., Wilson, V. & Thein, S.L. (1985) Individual-specific 'finger-prints' of human DNA. *Nature,* 316, 76–79.

Jones, C.G., Heck, W., Lewis, R.E., Mungroo, Y., Slade G. & Cade, T. (1994) The restoration of the Mauritius kestrel *Falco punctatus* population. *Ibis,* 137, 173–180.

Kenward, R.E. (1974) Mortality and fate of trained birds of prey. *Journal of Wildlife Management,* 38, 751–756.

Kenward, R.E. (1976) Captive breeding: a contribution of falconers to the preservation of Falconiformes. In *Proceedings of the World Conference on Birds of Prey, Vienna 1975,* ed. R.D. Chancellor, pp. 378–381. International Council for Bird Preservation, Cambridge, UK.

Kenward, R.E. (2004) Management tools for raptors. In *Raptors Worldwide,* eds. R.D. Chancellor & B.-U. Meyburg, pp. 329–339. World Working Group on Birds of Prey and Owls, Berlin.

Kenward, R.E. (2006) *The Goshawk.* T. & A.D. Poyser/A. & C. Black, London.

Kenward, R.E. and Gage, M.J.G. (in press) Opportunities in falconry for conservation through sustainable use. In *Peregrine Falcon Populations – Status and Perspectives in the 21st Century,* ed. J. Sielicki. EPFWG/Turul, Warsaw.

Kenward, R., Katzner, T., Wink, M. *et al.* (2007) Rapid sustainability modelling for raptors with radio-tags and DNA-fingerprints. *Journal of Wildlife Management,* 71, 238–245.

Lindner, K. (1973) *Beiträge zur Voglefang und Falknerei im Altertum.* Walter de Gruyter, Berlin.

Millsap, B.A. & Allen, G.T. (2006). Effects of falconry harvest on wild raptor populations in the United States: theoretical considerations and management recommendations. *Wildlife Society Bulletin*, 34, 1392–1400.

Morant, G.F. (1875) *Game Preservers and Bird Preservers*. Longmans, Green and Co., London.

Morris, J. & Stevens, R. (1971) Successful cross-breeding of a peregrine tiercel and a saker falcon. *Captive Breeding of Diurnal Birds of Prey*, 2, 5–7.

Newton, I. (1986) *The Sparrowhawk*. Poyser, Calton, UK.

Nittinger, F., Haring, E., Pinsker, W. & Gamauf, A. (2006) Are escaped hybrid falcons a threat to the Pannonian population of the Saker Falcon (*Falco cherrug*)? *Greifvögel & Eulen in Österreich*, 2006, 20–26.

Oliphant, L.W. (1991) Hybridization between a Peregrine Falcon and a Prairie Falcon in the wild. *Journal of Raptor Research*, 25, 36–39.

Parkin, D.T. (1987) The value of genetic fingerprinting to the breeding and conservation of birds of prey. In *Breeding and Management in Birds of Prey*, ed. D.J. Hill, pp. 81–86. University Press, Bristol.

Peyton, R.B., Vorro, J., Grise, L., Tobin, R. & Eberhardt, R. (1995) A profile of falconers in the United States: falconry practises, attitudes and conservation behaviours. *Transactions of the 60th North American Wildlife and Natural Resources Conference*, 181–192.

Pfeffer, R.H. (1987) *Saker Falcon*. Kainar, Almaty, Kazakhstan.

Ratcliffe, D.A. (1980) *The Peregrine Falcon*. Poyser, Berkhamsted, UK.

Riddle, K.E. & Remple, J.D. (1994). Use of the Saker and other large falcons in Middle East falconry. In *Raptor Conservation Today*, eds. B.-U. Meyburg & R.D. Chancellor, pp. 415–420. World Working Group on Birds of Prey and Owls, Berlin, Germany.

Saar, C. (1988) Reintroduction of the peregrine falcon in Germany. In *Peregrine Falcon Populations, Their Management and Recovery*, eds. T.J. Cade, J.H. Enderson, C.G. Thelander & C.M. White, pp. 629–635. Peregrine Fund, Boise, Idaho.

Saar, C., Henckell, T. & Trumf, J. (1999) Bekämpfung einer Kaninchenplage mit Beizhabichten. *Greifvögel und Falknerei*, 1998, 160–166.

Southern, W.E. (1964) Additional observations on winter bald eagle populations: including remarks on biotelemetry techniques and immature plumages. *Wilson Bulletin*, 76, 222–237.

Tordoff, H.B., Martell, M.S. & Redig, P.T. (1998) Effect of fledge site on choice of nest site by Midwestern Peregrine Falcons. *Loon*, 70, 127–129.

Trommer, G., Sielicki, S. & Wieland, P. (2000) Der Wanderfalke – nun auch wieder Brutvogel in Polen. *Greifvögel und Falknerei*, 1999, 48–56.

Upton, R.C. (1980) *A Bird in the Hand – Celebrated Falconers of the Past*. Debretts Peerage, London.

von Hohenstaufen, F. (1248) *De Arte Venandi Cum Avibus*. Manuscript.

Wallace, M. P. (2001) Recovery efforts for the California Condor. *Abstracts, 4th Eurasian Congress on Raptors.* Raptor Research Foundation, Seville, Spain.

Waller, R. (1982) *Der Wilde Falk ist Mein Gesell,* 4th edn. Neumann-Neudamm, Berlin, Germany.

Williams, N.P. & Evans, J. (2000) The application of DNA technology to enforce raptor conservation legislation within Great Britain. In *Raptors at Risk*, eds. R.D. Chancellor & B.-U. Meyburg, pp. 859–867. World Working Group on Birds of Prey and Owls, Berlin.

Wink, M. & Sauer-Gürth, H. (2004). Phylogenetic relationships in diurnal raptors based on nucleotide sequences of mitochondrial and nuclear marker genes. In *Raptors at Risk*, eds. R.D. Chancellor & B.-U. Meyburg, pp. 483–498. World Working Group on Birds of Prey and Owls, Berlin.

Xaodie, Y. (2005) Falconry in the minorities of China. Presentation at the Symposium 'Falconry: a world heritage', Abu Dhabi, 13–15 September 2005.

Yalden, D.W. (1987) The natural history of Domesday Cheshire. *Naturalist,* 112, 125–131.

Yazdani, A. (2005) Falconry: a cultural heritage in Iran. Presentation at the Symposium 'Falconry: a world heritage', Abu Dhabi, 13–15 September 2005.

12

Gamebird Science, Agricultural Policy and Biodiversity Conservation in Lowland Areas of the UK

Nicholas J. Aebischer

Game & Wildlife Conservation Trust, Fordingbridge, Hampshire, UK

Introduction

Gamebirds have been trapped and shot in Europe for many centuries. In lowland Britain, the grey partridge *Perdix perdix* and European quail *Coturnix coturnix* are native, whilst the pheasant *Phasianus colchicus* and red-legged partridge *Alectoris rufa* became established from introductions carried out probably in the 11th century and in the late 18th century respectively (Brown & Grice, 2005). These gamebirds are sedentary (apart from European quail, which winters in Africa and is not shot in Britain), so numbers available as quarry in the autumn and winter depend strongly on breeding densities and reproductive success. In late autumn, their numbers are bolstered by the arrival of overwintering ducks, geese and waders such as snipe *Gallinago gallinago* and woodcock *Scolopax rusticola* from countries to the north and east.

Recreational Hunting, Conservation and Rural Livelihoods: Science and Practice, 1st edition. Edited by B. Dickson, J. Hutton and W.M. Adams. © 2009 Blackwell Publishing, ISBN 978-1-4051-6785-7 (pb) and 978-1-4051-9142-5 (hb).

Historically, in Britain, the landowner holds the rights to the game on his land (Tapper, 1992). With improvements in firearm design, gamebird shooting increased in popularity. The sport developed from walked-up shooting, involving a handful of shooters on foot accompanied by dogs to locate and retrieve quarry, to driven shooting whereby beaters drove birds over a line of waiting guns. As high gamebird densities were needed for driven shooting, estate owners employed professional keepers to manage and enhance gamebird densities over much of the UK, providing habitats and ruthlessly controlling predators. The grey partridge was the main managed quarry species in lowland Britain during much of the 19th century and the first half of the 20th century. Income from leasing shooting rights or selling shoot days could be high, and the species was of considerable economic importance. In hill areas, bird shooting was concentrated on red grouse *Lagopus lagopus scoticus* driven over guns, and estates developed management strategies based on burning heather *Calluna vulgaris* and predator control (Hudson, 1992).

From the 1950s, the grey partridge underwent a prolonged decline (Potts, 1986), so many sporting estates turned from shoots based on managing wild grey partridges to shoots based on releasing artificially reared pheasants and, to a lesser extent, red-legged partridges. Others abandoned shooting altogether as a commercial activity and intensified arable farming operations. The deterioration of farm incomes during the late 1990s drove farms to diversify, leading to the intensification of shoot management based on released gamebirds (Aebischer & Ewald, 2004). The income can be important: a recent economic study found that in 2004 there were 26,000 providers of driven lowland game shooting in the UK offering 150,000 shooting days for a gross income of £375 million (PACEC, 2006).

The case of grey partridge in the UK demonstrates the importance of applied scientific research on quarry species both to the survival of a declining species and to the economic interests of rural landowners. Unlike in North America, where land managed for agriculture and for gamebirds is often separate, European lowland gamebirds are dependent on actively farmed arable landscapes, whose nature is closely bound to agricultural policy. British gamebird research has therefore included issues of agricultural policy and land management. Since the 1950s, the widespread decline of the grey partridge has directed research towards diagnosing the causes of decline, then at testing management options designed to counteract those causes. Before reviewing this research, however, it is useful first to provide as background the changes in agricultural policy that have taken place since the World War II.

Agricultural background

World War II emphasized to European policymakers the need for security of food production. Polices were developed that rewarded farmers for producing food. They were central to the Common Agricultural Policy (CAP), adopted in 1962 by the fledgling European Economic Community (now the European Union (EU)) and later by the UK when it joined in 1973 (Fennell, 1997). At the same time, technological advances in the agricultural industry paved the way for increasing yields through improved crop varieties, better machinery, the use of chemical fertilisers and the use of pesticides. The combination of policy and technology was so successful that many crops were over-produced, leading to wastage and rising CAP expenditure. This stimulated policy change (Fennell, 1997), starting with a quota on dairy production in 1984, then in 1988 a ceiling on agricultural support expenditure. During this period, in 1985, the CAP integrated environmental considerations by allowing special aid to farmers who agreed to conserve or improve the environment in areas 'of recognized importance from an ecological and landscape point of view', designated as Environmentally Sensitive Areas (ESAs).

In the late 1980s, pressure on agricultural subsidies from the Uruguay round of the General Agreement on Trade and Tariffs (GATT) talks led to the MacSharry reforms of 1992, which reduced levels of support by 29 per cent for cereals and 15 per cent for beef. They also created payments to withdraw land from production (known as 'set-aside'), payments to limit stocking levels, and measures to encourage early farmer retirement and afforestation. Further pressure, linked to the expansion of the European Union, the need to stabilise the CAP budget and the World Trade Organisation (WTO), led to a more fundamental reform of the CAP in 2003 based on decoupling subsidies from a particular crop. The new 'single farm payments' are linked to respect for environmental, food safety and animal welfare standards. Member States have some flexibility in the extent of decoupling, which the UK has embraced as an opportunity to introduce new environmental stewardship schemes on farms: the Entry Level Scheme open to all, and the competitive Higher Level Scheme. These reforms came into force in 2005.

Historical overview of gamebird science

The notion of wildlife management was formalised in the 1930s, under the influence of Aldo Leopold, the pioneering game conservationist (Leopold, 1933). In this new branch of science, most attention was focused on species with economic implications, in other words those that were hunted and those that were pests (e.g. Middleton, 1935). In the 1960s, public recognition of the impact of organochlorine pesticides (especially on birds of prey) (e.g. Prestt & Ratcliffe, 1972) raised the profile of ecological research, and was the first demonstration of the impacts of the intensive agricultural methods being actively pursued in lowland Europe on wildlife in the wider countryside. Soon after this, concern grew over declining bags of grey partridges on farms (Potts, 1986). This stimulated research into the causes of the decline in the 1970s, and the potential remedies (from the 1980s onwards). Evidence of declines in other farmland birds, whose numbers were much more difficult to monitor because they were not shot, came much later, starting with Marchant *et al.* (1990). The body of scientific research on the impacts of intensive agriculture on gamebirds matured considerably earlier than that on other farmland birds. It was therefore gamebird research that had a major influence on the reform of UK agricultural policy in ways that favoured the conservation of biodiversity, with beneficial effects on a much broader range of species in the wider countryside.

Scientific research into causes of decline

The decline of the grey partridge can be tracked back through time using bag records (numbers of birds shot). Falls in the bags of wild grey partridges began in the 1950s, and averaged around 5 per cent per annum over the ensuing 30 years (Potts, 1986). These concerns triggered a long-term research programme that started in the late 1960s.

The decline was quickly associated with a fall in productivity, specifically a reduction in chick survival during the first six weeks after hatching (Potts, 1980, 1986). Grey partridge chicks are nidifugous and chicks forage for themselves. Their diet during the first two weeks consists almost entirely of insects, a rich source of protein (Ford *et al.*, 1938; Potts, 1980, 1986). The survival of such young chicks was closely linked to the availability and consumption of certain preferred invertebrate groups, especially phytophagous insects (Potts, 1980; Aebischer, 1997a). The timing of this decline was coincident with the rapidly increasing use of herbicides in arable farming (Potts, 1980). By 1965, nearly all cereal fields were treated with herbicides, which greatly reduced the abundance of arable weeds that acted as host plants to insects. Southwood & Cross (1969) estimated that the abundance of invertebrates within cereals probably halved as a result. Whereas the action of herbicides on invertebrates was indirect, the action of insecticides, which became widespread during the 1970s, was direct. Vickerman & Sunderland (1977) showed that insecticide use during the summer had the potential to reduce chick-food invertebrates by over 90 per cent. In practice, grey partridge chick survival rates averaged a third lower on areas where insecticide was used extensively than on ones with little or no insecticide use (Aebischer & Potts, 1998).

The problem of low chick survival was exacerbated by a reduction of nesting cover. Grey partridges nest on the ground, and conceal their nest in rank herbaceous vegetation on grassy banks, uncut field margins, hedge bottoms or autumn-sown cereals. Field enlargement and the removal of field boundaries to increase agricultural efficiency probably reduced nesting cover in the UK by 24 per cent by 1978 (Potts, 1980), and inappropriate cutting or treatment with herbicide to prevent crop invasion crop by 'weeds' (Boatman, 1992) further reduced nesting cover quality.

In turn, the disappearance of cover increases vulnerability to predation. Adult partridges, incubating females and their eggs are vulnerable to mammalian and avian predators, and grey partridge declines have been linked to increased predation (Potts, 1980). Until the 1950s most large estates in the UK employed gamekeepers to kill predators. Since then, the killing of many predators has become illegal because of their rarity and the number of gamekeepers involved in active predator control has fallen. The numbers of foxes *Vulpes vulpes* and corvids (particularly carrion crows *Corvus corone*) have increased (Tapper, 1992) as have partridge nesting losses (Potts, 1986).

Scientific research into effective conservation measures

Proposed remedies are most effective when the proposals are based on an understanding of biological processes and tested experimentally. Taking the causes described above, I describe what is known about the biological processes and examine the applied research that has led to ways of mitigating the problems.

To identify ways of bolstering chick survival, the first step is to understand the foraging patterns of chicks. Green (1984) found that grey partridge adults and their chicks spent 97 per cent of their time inside cereal crops, mostly close to field margins. This led to experimental trials of 'conservation headlands', where herbicide use on the outer 6 metres of cereal crops was restricted to selective compounds only, which controlled pernicious weeds but allowed an understorey of insect-rich host plants to develop at the base of the crop; no insecticides were used after 15 March (Sotherton, 1991). Compared with conventional headlands, weed cover and diversity were three to four times as high, densities of the insect groups consumed by partridge chicks up to three times as high, and mean grey partridge brood size 1.5 times as high. Although yields in conservation headlands were reduced, the overall financial penalty was less than 1 per cent (Boatman & Sotherton, 1988). In terms of wider biodiversity, rare arable weeds like pheasant-eye *Adonis annua* and shepherd's needle *Scandix pecten-veneris*, that cannot withstand modern herbicide regimes, reappeared in crop headlands treated in this way, as did many pollen- and nectar-bearing plants that attract butterflies and hoverflies (Sotherton, 1991).

As regards crop husbandry generally, it follows from the research that, in order to maximise the availability of beneficial invertebrates to gamebird chicks, insecticide use should be sparing and applied in strict accordance with known numbers of insect pests at or above threshold levels. Broad-spectrum insecticides should be avoided where possible, and the aphid-specific compound pirimicarb used for summer aphid control (Potts, 1997). Ideally, the outer 12 metres of crop should be left unsprayed to avoid drift into adjacent margins or watercourses.

An alternative to modifying the management of the cereal crops was to grow crops explicitly for gamebirds. Fodder crops such as kale have long been recognised as providing gamebird-holding cover over winter for shooting purposes, and kale strips were often established specifically for that purpose. From the 1970s onwards, additional crops such as sunflower, maize, canary grass, millet and sorghum were advocated as cover for gamebirds (Coles, 1975). Research into the use of game crops by non-game bird species began in the 1990s, and experiments found that the most attractive crops to farmland songbirds were second-year (seeding) kale, quinoa, triticale and millet (Stoate *et al.*, 2004).

Restoration of nesting cover requires establishing areas of rank tussocky grass and other concealing vegetation. These can be placed around field margins, as grass strips along the boundary or hedgerow, provided that they are cut on a rotational basis every two to three years, so that every year tall dead grass is present early in the season (Aebischer, 1997a). To address the issue of field enlargement, large fields can be subdivided in a non-permanent fashion using raised grass strips or 'beetle banks' to provide additional cover. These strips are raised by ploughing and sown with tussock-forming grasses such as cocksfoot *Dactylis glomerata* or Yorkshire fog *Holcus lanatus*. As well as providing gamebird nesting cover, these strips have been found to be attractive to skylarks *Alauda arvensis* and harvest mice *Micromys minutus* (Stoate & Leake, 2002). Over winter, they hold huge numbers of hibernating predatory beetles that disperse into the surrounding crop in the spring and consume crop pests such as aphids (Collins *et al.*, 1996).

Reducing losses to predation has traditionally been approached by lethal means, i.e. controlling numbers of predators directly. There is no doubt that this is effective for lowland gamebirds: a cross-over experiment on two sites over seven years, implementing legal predator control first on one and then the other, demonstrated that after three years the breeding density of grey partridges averaged 2.6 times higher and autumn density 3.5 times higher with predator control than without it (Tapper *et al.*, 1996).

There has been little research into mitigation of predation through habitat management. Watson (2007a) found that grey partridges used cover more at sites where predation risk was high. The peak in their mortality occurred in late winter/early spring, a time of the year when the availability of cover was particularly low (Watson, 2007b). By implication, it would be helpful to provide types of cover that remain in place throughout this period. If food were made available in close proximity, then the birds would never have to venture far from cover.

Integration of gamebird science with agricultural policy

The first ESAs were designated in 1987, and others in 1993, encouraging farmers to safeguard areas of the UK countryside where the landscape, wildlife or historic interest was of national importance. The emphasis was on the conservation of traditional extensive grassland landscapes, and included financial incentives for the reversion of arable land to grass. The wildlife benefits were low (e.g. Kleijn *et al.*, 2001) or even negative for gamebirds (Aebischer & Potts,

1998), although in the light of these findings, revisions to the UK prescriptions in 1993 and 1997 included payments in some ESAs for arable options such as conservation headlands and elements of mixed farming (Aebischer, 1997b).

A five-year voluntary 'set-aside' scheme was originally introduced by the European Community in 1988 to reduce cereal surpluses. The land taken out of production was either planted to woodland or rapidly reverted to a thick tumbledown grassy sward that was of little interest to game. Following the MacSharry reforms, agricultural subsidies were made conditional on setting aside arable land on holdings producing more than 92 tonnes of cereals, initially 15 per cent later reduced to 10 per cent. The reforms introduced a new rotational set-aside scheme that required the vegetation to be cut before 1 July or cultivated between 1 May and 1 June to control weeds. This proved disastrous to wildlife because it was attractive nesting cover, but the cutting regime caused widespread destruction of partridge, pheasant and skylark nests, and exposed the survivors to intensive corvid predation (Poulsen & Sotherton, 1992).

In response to an outcry from game and wildlife organisations, from 1993 the regulations governing set-aside management were changed to allow non-residual herbicides instead of cutting for weed control on rotational set-aside. Non-rotational set-aside was introduced, allowing the creation of wildlife habitats by planting grass strips and unharvestable seed mixtures known in England as 'wild bird cover' (WBC).

EU regulations stipulated that set-aside areas must be at least 20 m wide and cover no less than 0.3 ha. This enabled land managers interested in game and wildlife to distribute the set-aside in strips around the farm, increasing landscape heterogeneity, rather than concentrating it in one large block (Sotherton, 1998). Optimal use of rotational set-aside is to leave it to regenerate naturally after harvest, providing overwinter stubbles where birds can feed on weed seeds and grain, and where chick-food insects in the soil can hibernate undisturbed by cultivation. The regenerated vegetation can provide insect-rich brood-rearing habitat in the following summer (Moreby & Aebischer, 1992). Equally good brood-rearing habitat combined with overwinter food and cover can be created under the WBC option by sowing an unharvestable mixture such as kale and quinoa or triticale and mustard. Kale as a component of the mixture is particularly valuable because it is biennial, so if left for two years it will provide cover at the critical late winter/early spring period. The grass option allows the creation of nesting habitat next to arable crops, and acts as an alternative to Beetle Banks. Up to 2 metres next to a hedge or wood may be left uncut each year, so rotational cutting every two to three years will prevent scrub formation while maintaining nesting cover.

The approaches that restore chick-food insects and nesting habitat can be combined for maximum benefit. For example, land managers can place conservation headlands next to a field margin with suitable nesting cover, or either side of a beetle bank. The different set-aside options can be combined within the same 20-m strip to achieve the same juxtaposition of nesting and brood-rearing habitats.

In parallel with mandatory set-aside, the MacSharry reforms introduced payments for agri-environmental measures, offering incentives to land managers to adopt sympathetic management practices. In the UK, this led to the introduction of the Countryside Stewardship Scheme (England), Tir Cymen (Wales) and Countryside Premium Scheme (Scotland) in 1993. Influenced by the gamebird research that had demonstrated wider benefits to plants and wildlife, and also by research from the Royal Society for the Protection of Birds on the severely declining stone curlew *Burhinus oedicnemus* and cirl bunting *Emberiza cirlus*, these schemes made grants available for many of the options described above, for instance conservation headlands, unsprayed headlands, wildflower strips, beetle banks, grass margins, hedgerow planting and traditional hedgerow management (Aebischer *et al.*, 2000). Acknowledgement of much wider declines than hitherto recognized among farmland birds (Fuller *et al.*, 1995) led the Game & Wildlife Conservation Trust, the Royal Society for the Protection of Birds and the government agency English Nature (now Natural England) to propose a package of wildlife measures targeted at the in-field and field-margin components of arable and mixed agriculture. In 1998, the British government approved a pilot Arable Stewardship Scheme that promoted mixed farming by supporting spring cropping and undersowing, provided equivalents of the wildlife-friendly options of set-aside in the form of overwinter stubbles, cover crop mixtures and summer fallow, and brought related measures like conservation headlands, beetle banks and grass margins together under one umbrella. The pilot scheme was found to benefit grey partridges and wintering passerines (Bradbury *et al.*, 2004).

Since 2000, the UK government has seized the opportunity offered by the mid-term review of the CAP to implement radical proposals on how to create a sustainable, competitive and diverse farming sector (Curry, 2002; Smallshire *et al.*, 2004). Among EU member states, it has gone the furthest in decoupling production from subsidies with the Single Farm Payment scheme, and tying subsidies instead to good agricultural practice and wildlife-sympathetic land management. In particular it introduced a new Environmental Stewardship scheme that replaces previous agri-environmental schemes while incorporating and augmenting all the wildlife-friendly elements described above. In England and Wales, these options come in the form of the Entry Level Scheme (Anon., 2005a), open to all farmers who apply, and the Higher Level Scheme (Anon., 2005b), which supports more intensive habitat management with a competitive, targeted approach. The Land Management Contracts offer similar opportunities in Scotland (Anon., 2005c).

Science speaks through demonstration

Much of the success in integrating gamebird science with agricultural policy can be down to the availability of demonstration farms, where policy makers can see with their own eyes the biodiversity benefits wrought by the types of management described above.

Loddington Farm in Leicestershire, managed since 1992 by The Game & Wildlife Conservation Trust under the auspices of the Allerton Research and Educational Trust, has played a pivotal role as a demonstration farm. Intended to be a platform to show how game management and farming could be integrated to the benefit of wildlife, the policy has been to modernise the farming, to adopt novel conservation measures and to employ a full-time gamekeeper to encourage wild game and other wildlife through the sympathetic habitat management techniques described earlier and predator control (Stoate & Leake, 2002). Within five years of starting management, autumn numbers of pheasants increased fourfold and those of brown hares over tenfold. Numbers of breeding songbirds doubled overall, with the group of nationally declining species outperforming that of nationally stable or increasing ones; the former group continued to decline on neighbouring farms. Loddington's approach also takes into account wider environmental objectives, using set-aside and hedgerows to promote water quality and prevent soil erosion. Throughout, its profitability has remained comparable with those of conventional farms.

Discussion and conclusion

The agricultural policies implemented in the UK after World War II were the main cause of the decline of the key lowland game bird species, the grey partridge. Research associated with grey partridges has identified and demonstrated effective solutions to this decline but which also address growing concerns about biodiversity. This research has played an important role in shaping new agricultural policies. Indeed, an independent review found that the grey partridge research provided the best evidence among farmland bird studies for the impact of indirect effects of pesticides (Campbell *et al.*, 1997). Gamebird research thus led the way in identifying the environmental problems associated with agricultural intensification, then in devising applied solutions compatible with modern agriculture and hence acceptable to land managers.

This is not, however, the same as having the solutions implemented. Farmers need to make a profit to survive, but implementing wildlife conservation measures implies a cost in terms of time, manpower and inputs. Traditionally, the costs were offset by the revenue from wild gamebird shooting but, post-decline, it takes at least five years of labour to restore a wild shoot (Stoate & Leake, 2002), which is why release-based shooting of pheasants and red-legged partridges has been widely adopted. To allow them to adapt to the wild, captive-reared gamebirds are released at least one month before shooting. Typically, pheasants are acclimatised in large woodland pens holding hundreds or even thousands of birds at once, whereas red-legged partridges are released in smaller groups from pens in cover crops. Release-based shooting provides managers with an incentive to carry out habitat management that will improve the game-holding capacity of land after release (e.g. woodland rides, cover crops), but not to provide nesting and brood-rearing habitats. This partial management can provide biodiversity benefits (Sage *et al.*, 2005b; Draycott *et al.*, in press), but there are also negative consequences particularly to woodland flora when densities of released pheasants exceed 1000 per hectare of release pen (Sage *et al.*, 2005a).

The introduction of 'Environmental Stewardship' in England and Wales has radically changed farmers' incentives, as it ties subsidies to good agricultural practice and wildlife-sympathetic land management. Many costs of management are covered, and admission to the Higher Level Scheme is specifically dependent on implementing options to benefit particular species of declining wildlife on farmland. These latest changes to the economics of farming and land management mean that the opportunities for restoring biodiversity to farmland are now better than they have been for many decades.

References

Aebischer, N.J. (1997a) Gamebirds: management of the grey partridge in Britain. In *Conservation and the Use of Wildlife Resources,* ed. M. Bolton, pp. 131–151. Chapman & Hall, London.

Aebischer, N.J. (1997b) Effects of cropping practices on declining farmland birds during the breeding season. In *Proceedings of the 1997 Brighton Crop Protection Conference – Weeds,* pp. 915–922. British Crop Protection Council, Farnham.

Aebischer, N.J. & Potts, G.R. (1998) Spatial changes in Grey Partridge (*Perdix perdix*) distribution in relation to 25 years of changing agriculture in Sussex, U.K. *Gibier Faune Sauvage,* 15, 293–308.

Aebischer, N.J., Green, R.E. & Evans, A.D. (2000) From science to recovery: four case studies of how research has been translated into conservation action in the UK. In *Ecology and Conservation of Lowland Farmland Birds,* eds. N.J. Aebischer, A.D. Evans, P.V. Grice & J.A. Vickery, pp. 43–54. British Ornithologists' Union, Tring.

Aebischer, N.J. & Ewald, J.A. (2004) Managing the UK Grey Partridge *Perdix perdix* recovery: population change, reproduction, habitat and shooting. *Ibis,* 146 (Suppl 2), 181–191.

Anon. (2005a) *Entry Level Stewardship Handbook: Terms and Conditions and How to Apply.* Rural Development Service, Department for Environment, Food and Rural Affairs, London.

Anon. (2005b) *Higher Level Stewardship Handbook: Terms and Conditions and How to Apply.* Rural Development Service, Department for Environment, Food and Rural Affairs, London.

Anon. (2005c) *Land Management Contract Menu Scheme 2005: Notes for Guidance.* Scottish Executive, Edinburgh.

Boatman, N.D. (1992) Herbicides and the management of field boundary vegetation. *Pesticide Outlook,* 3, 30–34.

Boatman, N.D. & Sotherton, N.W. (1988) The agronomic consequences and costs of managing field margins for game and wildlife conservation. *Aspects of Applied Biology* 17, 47–56.

Bradbury, R.B., Browne, S.J., Stevens, D.K. & Aebischer, N.J. (2004) Five-year evaluation of the impact of the Arable Stewardship Pilot Scheme on birds. *Ibis,* 146 (Suppl 2), 171–180.

Brown, A.F. & Grice, P.V. (2005) *Birds in England.* T. & A.D. Poyser, London.

Campbell, L.H., Avery, M.I., Donald, P.F., Evans, A.D., Green, R.E. & Wilson, J.D. (1997) *A Review of the Indirect Effects of Pesticides on Birds. JNCC Report No. 227.* Joint Nature Conservation Committee, Peterborough.

Coles, C.L. (1975) *The Complete Book of Game Conservation.* Stanley Paul, London.

Collins, K.L., Wilcox, A., Chaney, K. & Boatman, N.D. (1996) Relationships between polyphagous predator density and overwintering habitat within arable field margins and Beetle Banks. In *Proceedings of the 1996 Brighton Crop Protection Conference – Pests & Diseases*, Vol. 2, pp. 635–640. British Crop Protection Council, Farnham.

Curry, D. (2002) *Farming and Food: A Sustainable Future. Policy Commission on the Future of Farming and Food.* Cabinet Office, London.

Draycott, R.A.H., Hoodless, A.N. & Sage, R.B. (2008) Effects of pheasant management on vegetation and birds in lowland woodlands. *Journal of Applied Ecology*, 45, 334–341.

Fennell, R. (1997) *The Common Agricultural Policy: Continuity and Change.* Clarendon Press, Oxford.

Ford, J., Chitty, H. & Middleton, A.D. (1938) The food of partridge chicks (*Perdix perdix* L.) in Great Britain. *Journal of Animal Ecology*, 7, 251–265.

Fuller, R.J., Gregory, R.D., Gibbons, D.W. *et al.* (1995) Population declines and range contractions among lowland farmland birds in Britain. *Conservation Biology*, 9, 1425–1441.

Green, R.E. (1984) The feeding ecology and survival of partridge chicks (*Alectoris rufa* and *Perdix perdix*) on arable farmland in East Anglia. *Journal of Applied Ecology*, 21, 817–830.

Hudson, P.J. (1992) *Grouse in Space and Time: The Population Biology of a Managed Gamebird.* Game Conservancy Ltd, Fordingbridge, Hampshire.

Kleijn, D., Berendse, F., Smit, R. & Gilissen, N. (2001) Agri-environment schemes do not effectively protect biodiversity in Dutch agricultural landscapes. *Nature*, 413, 723–725.

Leopold, A. (1933) *Game Management.* Charles Scribner's Sons, New York.

Marchant, J.H., Hudson, R., Carter, S.P. & Whittington, P. (1990) *Population Trends in British Breeding Birds.* British Trust for Ornithology, Tring.

Middleton, A.D. (1935) Factors controlling the population of the partridge (*Perdix perdix*) in Great Britain. *Proceedings of the Zoological Society of London*, 106, 795–814.

Moreby, S.J. & Aebischer, N.J. (1992) Invertebrate abundance on cereal fields and set-aside land: implications for wild gamebird chicks. In *Set-aside, BCPC Monograph No. 58*, ed. J. Clarke, pp. 181–187. BCPC Publications, Farnham.

PACEC (2006) *The Economic and Environmental Impact of Sporting Shooting in the UK.* Public and Corporate Economic Consultants, London.

Potts, G.R. (1980) The effects of modern agriculture, nest predation and game management on the population ecology of partridges (*Perdix perdix* and *Alectoris rufa*). *Advances in Ecological Research*, 11, 1–79.

Potts, G.R. (1986) *The Partridge: Pesticides, Predation and Conservation.* Collins, London.

Potts, G.R. (1997) Cereal farming, pesticides and grey partridges. In *Farming and Birds in Europe*, eds. D.J. Pain & M.W. Pienkowski, pp. 150–177. Academic Press, London.

Poulsen, J.G. & Sotherton, N.W. (1992) Crow predation in recently cut set-aside land. *British Birds*, 85, 674–675.

Prestt, I. & Ratcliffe, D.A. (1972) Effects of organochlorine insecticides on European birdlife. In *Proceedings of the XV International Ornithological Congress*, ed. K.H. Voous, pp. 486–513. Brill, Leiden.

Sage, R.B., Ludolf, I.C. & Robertson, P.A. (2005a) The ground flora of ancient semi-natural woodlands in pheasant release pens in England. *Biological Conservation*, 122, 243–252.

Sage, R.B., Parish, D.M.B., Woodburn, M.I.A. & Thompson, P.G.L. (2005b) Songbirds using crops planted on farmland as cover for game birds. *European Journal of Wildlife Research*, 51, 248–253.

Smallshire, D., Robertson, P. & Thompson, P.G.L. (2004) Policy into practice: the development and delivery of agri-environment schemes and supporting advice in England. *Ibis*, 146 (Suppl 2), 250–258.

Sotherton, N.W. (1991) Conservation Headlands: a practical combination of intensive cereal farming and conservation. In *The Ecology of Temperate Cereal Fields*, eds. L.G. Firbank, N. Carter, J.F. Derbyshire & G.R. Potts, pp. 373–397. Blackwell Scientific Publications, Oxford.

Sotherton, N.W. (1998) Land use changes and the decline of farmland wildlife: an appraisal of the set-aside approach. *Biological Conservation*, 83, 259–268.

Southwood, T.R.E. & Cross, D.J. (1969) The ecology of the partridge III. Breeding success and the abundance of insects in natural habitats. *Journal of Animal Ecology*, 38, 497–509.

Stoate, C. & Leake, A. (2002) *Where the Birds Sing: 10 Years of Conservation on Farmland*. The Game Conservancy Trust & Allerton Research and Educational Trust, Fordingbridge.

Stoate, C., Henderson, I.G. & Parish, D.M.B. (2004) Development of an agri-environment scheme option: seed-bearing crops for farmland birds. *Ibis*, 146 (Suppl 2), 203–209.

Tapper, S.C. (1992) *Game Heritage: An Ecological Review from Shooting and Gamekeeping Records*. Game Conservancy Ltd, Fordingbridge.

Tapper, S.C., Potts, G.R. & Brockless, M.H. (1996) The effect of an experimental reduction in predation pressure on the breeding success and population density of grey partridges (*Perdix perdix*). *Journal of Applied Ecology*, 33, 965–978.

Vickerman, G.P. & Sunderland, K.D. (1977) Some effects of dimethoate on arthropods in winter wheat. *Journal of Applied Ecology*, 14, 767–777.

Watson, M., Aebischer, N.J. & Cresswell, W. (2007a) Vigilance and fitness in grey partridges *Perdix perdix*: The effects of group size and foraging-vigilance trade-offs on predation mortality. *Journal of Animal Ecology*, 76, 211–221.

Watson, M., Aebischer, N.J., Potts, G.R. & Ewald, J.A. (2007b) The relative effects of raptor predation and shooting on overwinter mortality of grey partridges in the United Kingdom. *Journal of Applied Ecology*, 44, 972–982.

13

The Re-Introduction of Recreational Hunting in Uganda

Richard H. Lamprey[1] and Arthur Mugisha[2]

[1]Fauna & Flora International, Nairobi, Kenya
[2]Fauna & Flora International, Kampala, Uganda

Historical background

Uganda is renowned for its diversity of landscapes and fauna, and during the 1950s and 1960s, many national parks, game reserves, controlled hunting areas and animal sanctuaries were established to protect important wildlife areas. Tourism became a major foreign exchange earner as visitors came from around the world to view the vast herds of elephant *Loxodonta africana*, Uganda kob *Kobus kob* and buffalo *Syncerus caffer* in Queen Elizabeth National Park, to take the launch trip to the base of Murchison Falls, or to climb the Rwenzori Mountains.

However, over the 1970s and 1980s, Uganda's wildlife populations underwent a catastrophic decline. Following the breakdown in law and order during the regime of Idi Amin in the 1970s, conservation areas were encroached upon, and the wildlife hunted on a massive scale (Douglas-Hamilton *et al.*, 1980; Eltringham & Malpas, 1980, 1983; Edroma, 1984; R.C.D. Olivier *et al.*, 1989; Aerial monitoring of large mammal populations in the Queen Elizabeth

Recreational Hunting, Conservation and Rural Livelihoods: Science and Practice, 1st edition.
Edited by B. Dickson, J. Hutton and W.M. Adams. © 2009 Blackwell Publishing,
ISBN 978-1-4051-6785-7 (pb) and 978-1-4051-9142-5 (hb).

National Park, Uganda. Unpublished Report to Uganda National Parks, Kampala, Uganda). Armies living in the bush poached most of Uganda's elephants and all of its rhinos (*Diceros bicornis* and *Ceratotherium simum*), and lodges were ransacked and parks infrastructure destroyed.

Since 1986, Uganda has been governed by the National Resistance Movement (NRM) and experienced greater peace and stability. The government has emphasised that tourism to the protected areas should be a key driver of the country's economic recovery. In 1995–1996, surveys were conducted to determine the status of national parks and reserves after the years of upheaval (Lamprey & Michelmore, 1996; Lamprey *et al.*, 2003). The surveys revealed that many protected areas were massively encroached and that wildlife populations had been reduced to critically low levels. Over 65,000 people lived illegally or with uncertain status within parks and reserves, and many controlled hunting areas (CHAs: see below) were completely settled. Several key wildlife species had become extinct. Oryx had been entirely extirpated from their range in Karamoja, Derby's eland *Taurotragus derbianus* from West Nile, the bongo *Tragelaphus euryceros* from Mt. Elgon, and both the black and the white rhino from their ranges in the north. Over the entire country, wildlife populations had been reduced by 95 per cent from their 1960s numbers.

In 1996, a new wildlife statute was passed in Uganda. This merged Uganda National Parks and the Game Department into a single agency, the Uganda Wildlife Authority (UWA), charged with managing wildlife both within and outside protected areas. With parks and reserves encroached and run down, and wildlife populations shattered by poaching, UWA has faced an unprecedented challenge in reversing the declines of the last 30 years. New policies and legislation for environmental protection and wildlife management have been formulated to guide this process.

Sport hunting and the hunting ban of 1979

The earliest legislation to provide for wildlife conservation and utilisation in Uganda was the Game Ordinance, enacted in its earliest form in 1926. Under the Ordinance, the Game Department (GD) established game reserves, regulated trophy hunting, and controlled problem animals. In 1963, the year after Uganda became independent from Britain, the Game Ordinance was replaced by the Game (Preservation and Control) Act, which refined hunting regulations and listed the species that could be hunted on different types of permit. The Act recognised the rights of local communities to utilise wild-life, and to have a voice in its management. It made provision for customary hunting by specifying that 'at the request of the members of any tribe or the inhabitants of any village, the district commissioner of the area, may, with the approval of the Chief Game Warden, authorise the tribesmen or villagers to hunt animals within such area subject to such conditions as to mode of hunting, number, species and sex of animals as may be specified'. In addition, the Act specified the establishment of 'local game committees' for the purpose of 'advising the minister on any matter concerning the conservation of game in that area'.

In respect of regulating hunting in specific areas, or of vulnerable species, the Act specified 'controlled hunting areas' (CHAs) where 'the Minister may prohibit the hunting of [certain] species [or] prescribe the maximum number of that species which may be hunted'. CHAs were created by the Ugandan Parliament, and formally gazetted in statutory instruments that provided a boundary description for the area, and listed the hunting quotas. This principle of legally gazetted hunting quotas was unusual in the East African context and provided a strong basis for the control of hunting; in neighbouring Tanzania and Kenya, the quotas were simply set by the game department annually. However, other than regulating the hunting of scheduled species, the Government had no control whatsoever of any form of land use in CHAs. At the time, the rural population of Uganda was still relatively low and pressure on land was much less than it is today; the threat to CHAs from human population expansion was not foreseen.

During the chaos of the Amin era, tourism revenues were reduced to virtually nothing. However, the GD was able to maintain some low-key activities, using revenues generated from sport hunting licences. Inevitably perhaps, this process became corrupted, as, in the face of massive inflation, many hungry townspeople viewed hunting – whether licensed or not – as a way to obtain cheap meat. In 1979, the year the Amin regime was overthrown sport hunting was banned by Ministerial decree, as

> the country's rich wildlife was fast disappearing, not only from poaching alone, but also because almost all hunters were exceeding their allowances of what they could hunt legally. In many instances the Game Guards … authorised to arrest any hunter exceeding his bag limit, were not strong enough to resist small gifts of meat or money, or both.   (Game Department Annual Report, 1979)

The hunting ban achieved little in stemming hunting for meat. Instead, the GD lost virtually all of its revenues, and by the late 1980s it was incapable of any regulatory activities whatsoever. By the 1990s, an older generation of wildlife managers had retired, and with them the concept that sport hunting could play a part in wildlife conservation. 'Recreational hunting' was incomprehensible to a new generation of wardens emerging from conservative wildlife training courses in Uganda's universities and colleges. In legal terms, the 'controlled hunting area' was considered defunct, as hunting was banned, and many CHAs were now settled. Nevertheless, CHAs continued to be perceived by local communities as 'protected areas'; the statute therefore provided an opportunity for the incorporation of CHAs into a redesigned protected area network if they still had conservation value. The CHA review was completed in 2002 with the approval by Parliament of the Protected Area System Plan for Uganda (Lamprey et al., 1999).

Re-introduction of sport hunting through wildlife use rights

Under the Wildlife Statute 1996, the Government continues to own Uganda's wildlife. However, the statute also recognises that wildlife may be better conserved if, under certain conditions, landowners and communities can manage wildlife for their own benefit. The statute defines 'wildlife use right' (WUR) as a right granted to a person, community or organization to make use of wildlife in accordance with a grant or 'licence'. To ensure that WURs are not abused, the law must be clearly interpreted, institutional linkages and fee structures established, and wildlife species placed in categories assigned to the different WURs.

The Statute specifies six categories of wildlife use rights, which the Minister may vary, revoke or add to on the advice of the Uganda Wildlife Authority (UWA). The Statute also provides measures for licensing professional trappers and hunters and specifies methods that may be pursued for hunting and taking and trading wildlife species and specimens. Sport hunting for meat and trophies may take place under the 'Class A WUR', which also covers game cropping and tribal hunting.

Implementation by UWA of the hunting WUR was initiated cautiously in 2001 on privately owned land neighbouring the Lake Mburo National Park. The process was closely scrutinised in all its stages by conservation NGOs and vociferous anti-hunting lobby groups. The first hunting programme was introduced as a 'pilot scheme', with quotas derived from rigorous wildlife counts, and with stringent regulations imposed on hunting practice; experience gained would be used in implementing WURs in other parts of the country. Below we examine and discuss the successes and challenges of this first case study.

Description of study area

The study area adjoins the northern and eastern boundaries of the Lake Mburo National Park, in Mbarara district in south-western Uganda (see Figure 13.1). The area, in what was originally termed Kaaro Karungi but now commonly known as Nshara Ranches, encompasses privately owned parcels of land that had originally been planned for a commercial ranching scheme in the 1960s (Mugisha, 1993).

The landscape comprises gently undulating hills with thicket and savanna, interspersed with rock outcrops. Originally, the Nshara landscape included the present Lake Mburo National Park and was the prime pastoral grazing grounds for the characteristic long-horned Ankole cows within the Ankole kingdom by the turn of the 19th century (Infield, 2002). The area remains the primary grazing land for the BaHima people; depending on season, some 60,000–100,000 cattle are grazed within the ecosystem.

This ecosystem is a biologically diverse area with thickets, *Combretum* and *Acacia* woodlands and wetlands, extending north from Lake Mburo into the dry rangelands of Ankole. The ecosystem harbors significant populations of savanna game species that include zebra *Equus burchelli*, topi *Damaliscus lunatus*, buffalo, eland *Taurotragus oryx* and waterbuck *Kobus ellipsiprymnus*, and is particularly important in harbouring Uganda's only population of impala *Aepyceros melampus*. Other wildlife species include hippos *Hippopotamus amphibius* and crocodiles *Crocodylus niloticus*; in the past there were also lions *Panthera leo*.

The area receives an annual rainfall of about 800 mm. During the long dry season between June and September most of the water sources in the area dry up except for the lakes and associated wetlands and rivers, most of which lie within and to the south of the Park. During the wet season on the other hand, both livestock and wildlife disperse northwards to seek grazing in the ranching areas and hills.

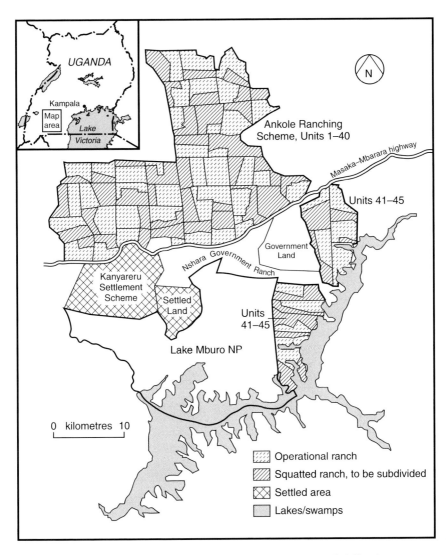

Figure 13.1 **Map of the Lake Mburo National Park (LMNP) following gazette-ment in 1983, and the adjacent ranchland area of the Ankole Ranching Scheme (ARS) (ranch units 1–50). Also shown is the Nshara Dairy Ranch and adjacent Government land. Inset top left is the location of the area in Uganda.**

Creation of Lake Mburo National Park

Over the past 50 years, the rich biodiversity in the Nshara rangelands has attracted the attention of government wildlife agencies, conservationists and sport hunters. In the 1930s, much of this landscape was gazetted as the Masha Animal Sanctuary (MAC), for game conservation. In 1964 the lake area and its peripheral rangelands south of the Mbarara–Masaka highway were declared the Lake Mburo Game Reserve that permitted human activities other than 'permanent' settlement.

In 1962, the newly independent Government established the 'Ankole Ranching Scheme' (ARS), in which rangeland areas north of the game reserve were divided into five square mile blocks, cleared of tsetse (Ford, 1971) and allocated to a few elite cattle owners. These ranchers were supported with loans and grants to undertake commercial beef ranching with exotic breeds (Mugerwa, 1992). Later in the 1970s, 10 more blocks were excised from the reserve and added on to the ranching scheme. This was in disregard of the fact that local pastoralists traditionally used the area, and that there was wildlife, including lions, that would conflict with the livestock scheme. A few years later, a dairy and stock-improvement project – Nshara – was established in the same area by excising another 30 square miles from the Reserve.

To mitigate the conflicts that arose out of this inconsistent land policy, the Government of the day instigated two radical measures. Firstly, lions and other predators were shot throughout the area, to protect livestock herds. Secondly, the government deliberately marginalised the local pastoralists, labeling them Rwandese who were not entitled to own land in Uganda (Mugisha, 1993). Some pastoralists left for other parts of the country where they could access pastures for their livestock in peace. However, many remained, because they had nowhere else to go, and they continued to graze their cattle on the land of absentee 'official' landowners or within the game reserve.

In 1983, the Government, without any consultations or social impact assessment, gazetted the whole area to the south of the Masaka–Mbarara highway, including the Nshara land and the 10 private ranches east of Lake Mburo, as Lake Mburo National Park. Private landowners with legal land titles, and local landless pastoralists, were evicted at short notice from the new park without compensation. This was locally perceived as a political move to punish communities that were not supportive of the then Government. To dispossessed pastoralists, the exclusion of cattle from the key area of the former Nshara landscape made the whole concept of the park meaningless and detestable in addition to the fact that they were denied access (Infield, 2002).

Government land use policies that drove land tenure from communal to private, and promoted both commercial livestock husbandry and wildlife management above local social needs, were highly contentious in the Lake Mburo area. In 1986, the Uganda Peoples' Congress (UPC) government was toppled by the National Resistance Movement (NRM) whose top leadership was largely comprised of the children of the dispossessed pastoral families from the Nshara area. Following this regime change, these families 'repossessed' the national park, looted park property and made concerted efforts to exterminate wildlife, so that the government would never again consider it for conservation.

In the late 1980s, to address the unrest in Lake Mburo, the National Resistance Movement (NRM) Government declared its intention to degazette 60 per cent of the national park, and to settle the genuinely disadvantaged pastoralists in this degazetted park land, and on other land made available by ranch subdivision. A Ranch Restructuring Board was established to implement this task, and by 1995 the board had reallocated the available land to commercial farmers and former pastoral families. Although the land was now more fairly distributed, the Lake Mburo ecosystem was fragmented into large and small parcels (Figure 13.1).

The problem of small plots was further compounded by the presence of wild animals that continued to move seasonally through the area. Livestock farmers viewed grazing wildlife as competitors for pastures and water, and as carriers for livestock diseases. Despite the concerted efforts of UWA to improve community relations (Hulme, 1997; Hulme & Infield, 2001; Infield, 2002), local people remained antagonistic to the park. They intensively hunted the wildlife for meat themselves, or encouraged local hunting groups to exterminate it. Since the ranchlands were the wet season dispersal area for Lake Mburo's wildlife, the intensity of the poaching critically jeopardised the viability of the park itself.

During the early 1990s, as economic conditions and road transportation links improved, poaching increased as bushmeat from Lake Mburo could be taken further afield to urban markets. Fraser Stewart (1992) estimated that a minimum of 600 impala and 400 bushbuck were killed annually, but other sources believe the actual totals to be at least double these figures.

Wildlife trends in Lake Mburo

Over the last 20 years, a number of wildlife censuses have been conducted in the Lake Mburo ecosystem. The first systematic aerial census ('systematic reconnaissance flight' – SRF), conducted according to the standardised aerial sample count methodology for East African savannahs (Norton-Griffiths 1978), was implemented in 1982 and covered only the former Lake Mburo Game Reserve, which lay to the south of the Masaka–Mbarara highway (Eltringham & Malpas, 1983); the northern ranches were not covered. The next, in 1992, covered the entire ecosystem including the park and the Ankole Ranching Scheme (R.C.D. Olivier, 1992; Aerial total counts in Uganda National Parks. Unpublished Report to Uganda National Parks, Kampala, Uganda). Beginning in 1995, UWA began a regular programme of SRFs, with a census zone that includes the park and the ranches, in total 1500 km^2 (Lamprey & Michelmore, 1996). These counts are conducted at approximately two-year intervals. Figure 13.2 shows the distribution of large wild mammal species in the ecosystem in 1998.

Analysis of the 1999 SRF confirmed an alarming trend since 1992; the impala population had declined by 90 per cent, from 15,000 to just 1600 animals. Declines were also recorded for topi and eland. Clearly, wildlife populations in the ranches were being severely impacted by heavy poaching, and if Lake Mburo NP was to remain a viable park, new management approaches were needed. In response, UWA examined ways in which local landowners might be persuaded to place a greater value on wildlife. These proposed mechanisms included cropping and sport hunting.

Impala cropping pilot project

In 1999, UWA considered the options for conservation in Lake Mburo, and concluded that the most effective measure would be to raise the value of wildlife for local people through wildlife utilisation schemes, particularly organised game cropping. At this time, the meat value of a single impala to local hunters was estimated at US$8, when sold in local markets (Averbeck, 2002). It was felt that a cropping scheme to supply certificated impala meat to selected restaurants in Kampala would yield significant returns to landowners (Hautzinger & Gafabusa, 1999; Report on utilization of impala carcasses. Uganda Industrial Research Institute, Uganda Meat Technology Centre, Uganda, unpublished).

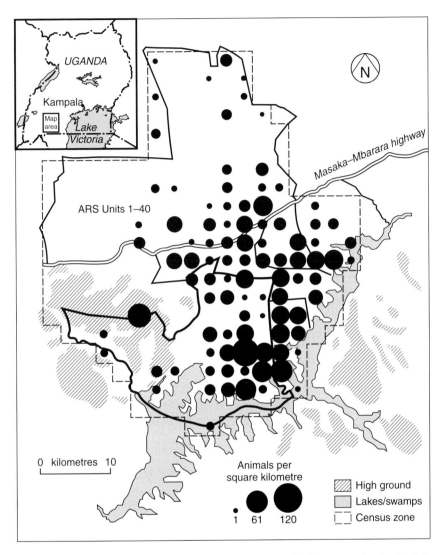

Figure 13.2   **Distribution and abundance of larger wild fauna species in the Lake Mburo ecosystem in 1996 (Lamprey & Michelmore, 1996).**

Table 13.1   **Revenue generated from the impala cropping pilot scheme, Lake Mburo.**

Item	Revenue/animal (Uganda Shillings)	US$ equivalent/ animal	Total revenue (US$)
Skin	40,000	36	3,636
Meat	20,000	18	1,818
Total	60,000	55	5,455

Note: 1 US$ = Uganda Shillings 1100 (in 1999).

Averbeck (2002) assessed potential off takes for both cropping and sport hunting, and estimated that at an off take of 10 per cent, some 200 impala could be utilised annually for meat and hides. Alternatively, at 3 per cent off take, 60 impala could be allocated on quota to sport hunters.

In the pilot cropping scheme, 100 impala were cropped and the meat was sold to selected outlets in Kampala. The skins were offered for sale to the public from the UWA headquarters, and funds generated were given to the respective landowners on whose land impala were harvested. Each impala yielded about US$50 in meat and skin value. Table 13.1 shows revenue earnings from the cropping scheme.

There were three important outcomes to this cropping scheme. Firstly, it broke the ice on the sensitive issue of implementing the new statutory provision for a 'hunting WUR' – this was the first time licensed off take had been approved since the ban of 1979. Secondly, the exercise served to build trust and confidence among the local landowners towards UWA. The landowners began to believe in the possibility of benefiting from wildlife on their land.

The third important outcome of this programme was that with such low cropping quotas, the scheme was simply not economically viable. Cropping costs far exceeded the revenue generated from the meat and hides. It had been calculated previously that about 20 people would have to be employed, ranging from hunters to skinners to processors, and that the scheme would require high capital and recurrent expenditures of slaughterhouses, generators and vehicles. Even if the quota was greatly expanded to include other species, such a scheme would only generate an annual profit of US$5000. Therefore, UWA considered sport hunting as another option to increase the value of impala and other species.

Sport hunting pilot programme

In 2001, at the end of the pilot cropping programme, UWA started a four-year community-based pilot sport hunting programme on private land neighbouring Lake Mburo National Park. The hunting 'game fee' would be substantially higher than the pure meat/skin value derived by cropping, and a high proportion of the game fees would accrue to local landowners. Guidelines for the exercise were formulated in participation with the identified landowners, who were mobilised to establish the Rurambira Wildlife Association (RWA), named after the local parish (the smallest administrative unit in Uganda). Meetings were organised at local level, targeting land owners to discuss possibilities of managing wildlife outside the park. The views and concerns of the people were analysed and draft guidelines to address those concerns were again presented to the RWA for their consensus, before UWA approved them. The sport hunting scheme is implemented over combined land holdings totalling about 100 km^2.

The hunting quota was determined based on a 2 to 3 per cent off take of the total population estimate (Averbeck, 2002). Other factors included the birth rate, behaviour and ranges of the individual species. The unit price for each animal was based largely on game fees charged in Tanzania. The revenue to different stakeholders is based on a proportion of the game fees, according to the scheme shown in Table 13.2.

Table 13.2 **Proportion of game fees accruing to stakeholders from the Lake Mburo pilot sport hunting scheme.**

Stakeholders	Proportion of game fees (%)
Rurambira Wildlife Association[*]	65
The Sub-county (Nyakashashara)	5
Community – Protected Area – Institution (CPI)[†]	5
Uganda Wildlife Authority (UWA)	25

[*]A community based organization (CBO) in Rurambira parish, which includes the ranches (Nos. 45–50) in which hunting is permitted.
[†]The body (committee) established under UWA policy to act as the link between the protected area and the sub-counties.

Source: UWA (2002).

The sport hunting pilot scheme has run since 2001, through a concession agreement between UWA, the RWA and the hunting outfitter Game Trails Uganda (GTU), a company founded and managed by a Ugandan national. Quotas and game fees are little changed since the start of the scheme, and for the period up to 2005 about US$34,000 was generated annually for distribution amongst the stakeholders according to the proportions shown in Table 13.2. Table 13.3 shows the quota, utilisation and game fee revenue from the scheme in 2001/2 and 2004/5. Over the period 2001–2005, the total revenue generated from game fees was US$137,020, with US$89,062 accruing to the RWA. The RWA has used its share of the revenue for community development projects such as schools, access roads and water improvement in the parish. Game fee revenues of US$6851 (5 per cent of the total) have accrued to each of the other local bodies, comprising the Nyakashashara sub-county and to the Community–Protected Area Institution (CPI) – the coordinating body between protected areas and neighbouring communities.

The continued aerial surveys indicated that wildlife populations were beginning to recover. In August 2004 and February 2006, aerial censuses revealed that wildlife populations were dispersing further into the ranches than recorded in the previous surveys, and that by 2004 the impala population had increased to 4500 (see Figure 13.3). Zebra are also increasing rapidly (see Figure 13.4), as are buffalo and warthog.

Evidence suggests that the sport hunting pilot scheme is achieving its objectives. In a UWA evaluation in 2003, local communities were interviewed to determine their attitudes, with the following broad findings:

- 81 per cent of the respondents were aware of the programme.
- 80 per cent of pastoralists and 100 per cent of the fishing community (using the lakes in the ecosystem) were willing to protect wildlife if they benefited from it.
- Asked if the pilot programme should continue, over 90 per cent of cattle keepers, about 60 per cent of crop farmers and 60 per cent of fisher respondents were in favour of continuing the pilot scheme.

Discussion and conclusion

The Lake Mburo sport hunting scheme is still in its early stages. As the programme is extended, its success or otherwise will become more apparent.

Table 13.3 Quota utilisation and game fee revenues (US$) for the Lake Mburo pilot sport hunting scheme in 2001/2002, and 2004/2005. For these two periods, revenues are assessed over the 12 months of the UWA calendar year, from 1 July to 30 June.

Common name	Scientific name	Quota	Unit price	2001/2		2004/5	
				Utilised	Game fees	Utilised	Game fees
Baboon	Papio cynocephalus	15	90	0	0	0	0
Buffalo	Syncerus caffer	10	600	10	6,000	10	6,000
Bushbuck	Tragelaphus scriptus	10	250	9	2,250	8	2,000
Bushpig	Potamochoerus porcus	15	150	1	150	0	0
Duiker	Cephalophus	4	130	1	130	3	390
Eland	Tragelaphus oryx	7	500	7	3,500	3	1,500
Hippopotamus	Hippopotamus amphibious	6	500	3	1,500	1	500
Impala	Aepyceros melampus	50	250	22	5,500	22	5,500
Oribi	Ourebia ourebi	6	150	6	900	6	900
Reedbuck	Redunca redunca	5	250	5	1,250	5	1,250
Topi	Damaliscus lunatus	4	350	4	1,400	5	1,750
Warthog	Phacochoerus aethiopicus	14	250	14	3,500	9	2,250
Waterbuck	Kobus ellipsiprymnus	10	500	10	5,000	8	4,000
Zebra	Equus burchelli	31	500	5	2,500	17	8,500
Totals		187		97	33,580	97	34,540

Source: UWA.

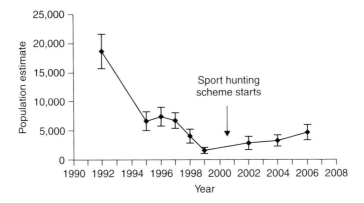

Figure 13.3   **Population estimates for impala in the Lake Mburo ecosystem, 1992–2006.**

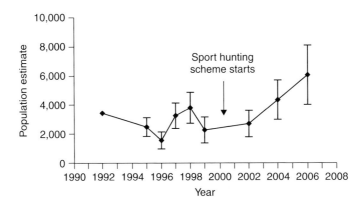

Figure 13.4   **Population estimates for zebra in the Lake Mburo ecosystem, 1992–2006.**

However, our early analysis suggests that the scheme is having an important impact in the area. Firstly, the aerial survey data do suggest that wildlife populations are increasing significantly in areas outside the park – this is confirmed by wardens and visitor reports on the ground.

Secondly, the communities are deriving significant revenues from wildlife, and are able to use these revenues for community projects. A high proportion of game fee revenues accrue to the community; at 65 per cent this is far higher than in Tanzania, where 25 per cent is apportioned only to the district, or in Mozambique where 20 per cent is set aside for communities. Clearly, the higher the benefit, and the closer it is targeted to community needs at local level, the greater the effectiveness of the programme. This is reflected in a marked improvement in the attitude of local people to the park and to UWA.

Another success is that landowners felt that they were closely involved in the design, implementation and evaluation of the scheme; the role of UWA and Game Trails Uganda was to provide technical backup in setting quotas and selling the hunting safaris. Given the complexity of the hunting issue, and the criteria for issuing a hunting WUR, it is unlikely that landowners themselves could have obtained hunting rights for their own land. This stakeholder linkage is seen as important in selling the concept of wildlife use rights and sport hunting elsewhere in Uganda; during 2007 a number of community groups were taken from northern Uganda to assess for themselves the impact of the scheme, as a prelude to discussions over the introduction of sport hunting in their own areas.

Other lessons learned from the exercise include the following:

- To remain in business, GTU must ensure that wildlife populations are not hunted illegally by others. Therefore GTU have introduced anti-poaching patrols in the ranch areas with support from the landowners and these are proving to be effective in curbing poaching.
- Wildlife managers had anticipated that the sport hunting exercise would also assist in reducing 'vermin' animals, such as bush pigs and baboons. However sport hunters are not interested in shooting these species.
- The operational area of the scheme is small, being confined to ranches immediately adjacent to Lake Mburo National Park; many of landowners further from the park feel left out.
- Mechanisms for directly sharing revenues among landowners are not yet in place. The RWA has instead used revenues for general community projects (e.g. schools).

The Lake Mburo hunting scheme is intended to be a pilot project. As such, the costs of the exercise have been high, particularly in the aerial monitoring, where each survey will cost US$6000–8000. For continued monitoring of the programme, UWA will need to maintain the aerial survey programme; this will determine whether the apparent upward trend in wildlife numbers is real, rather than just 'data noise' from a census system based on sampling. This is turn will have cost implications; the survey is conducted at high precision and intensity, and is therefore expensive to implement. Thus the programme as it currently stands is barely sustainable, but neither is it intended to be; once policymakers are convinced that sport hunting is working as a viable conservation tool – and for this the intensive monitoring data are needed – the lessons learned can be applied more cheaply to the implementation of WURs across the country. If sport hunting is to be implemented in other areas, the inter-survey interval may be extended from two years to five years, and hunting outfitters might be required to contribute to the costs.

Empirically, if the hunting success in Lake Mburo is to be sustained, it must compete with other forms of land use. Table 13.4 gives estimates of revenues per hectare accruing to the community from different form of land use in the Lake Mburo ecosystem. Clearly, in areas with sufficient rainfall to support agriculture, hunting is not a competitive land use, and in the long term cultivation is likely to prevail. However, in the drier northern parts of the Lake Mburo ecosystem where livestock production is the primary land use, hunting may increase revenues per hectare by about 30 per cent. Hunting will also significantly increase revenues in larger ranches where landowners wish to maintain pastoralism for cultural reasons, whatever the rainfall of the area.

Table 13.4　**Revenues from different forms of land use, Lake Mburo.**

	Uaganda Shillings/ ha/yr	US$ equivalent/ ha/yr[§]
Mixed agriculture[*]	506,500	281.4
Livestock production[*]	10,500	5.8
Illegal hunting (unsustainable off take)[†]	6,500	3.6
Sport hunting as currently practiced[‡]	3,000	1.7

[*]From Emerton (1999).

[†]Extrapolated from Emerton (1999), on assumption that this is 'lost opportunity' within LMNP, that the 'huntable area' is 130 km², and that similar wildlife densities exist outside in the ranches.

[‡]Based on current sport hunting concession, data as in Table 13.4, on ranches 46–50 only (75 km²).

[§]Based on 1 US$ = Uganda Shillings 1800 (2005 rate).

Of great significance to the future of the sport hunting scheme is the decision of the Ranch Restructuring Board to fragment ranches into smaller landholdings (see Figure 13.1). The effect of this subdivision remains to be seen, but already the rapid increase in the number of sheep and goats, from 14,000 in 2004 to 27,000 in 2006, may be attributed to changes in pastoral practices, as people diversify their livelihoods on small plots. As these land units are of insufficient size for large scale cattle keeping, it is probable that they will eventually be converted to agricultural fields and fenced. This will have a detrimental effect on wildlife movements within the area, which in turn will threaten the viability of the sport hunting scheme.

The game fee returns per km^2 in Lake Mburo far exceeds those for a typical Tanzanian hunting block. Comparing revenue from game fees only (the 'revenue to the landowner'), Lake Mburo earns US $ 170 per km^2, compared to US $29 per km^2 in Ugalla East (Lamprey, 1993), and US $21 per km^2 for the 41 hunting blocks of Selous Game Reserve (Lamprey, 1995). Lake Mburo revenues per km^2 are much higher because the 'hunting block' is smaller, being just 75 km^2, and because the quotas, expressed as a proportion of the population, are higher than in Tanzania and used more intensively. The Lake Mburo impala quota, for example, is set at 2 per cent of the estimated population, compared to 0.7 per cent in Tanzania's Ugalla GR and 0.6 per cent in Selous GR.

In their early assessment, Okua *et al.* (1997) discussed the potential of sport hunting in Uganda. The study concluded that with low wildlife populations outside protected areas, 'there is limited potential for trophy hunting in which hunters pay premium prices to hunt a wide variety of animals including the "big five". However, there is some scope for hunters who want to collect otherwise unobtainable species such as kob'. There were two possible scenarios to improve revenue earnings from hunting. First, the government may allocate hunting blocks with a greater species mix than the small blocks in this Lake Mburo case study – in effect this would mean hunting in much higher potential protected areas. Second, hunters might be provided with two to three blocks to spread overheads. It was thought that this would limit the number of operators to one or two only, and would probably attract only those who have an existing hunting business outside Uganda (e.g. in Tanzania). It was estimated that an outfitter in Uganda has to generate over US$70,000 per block in client fees for the business to be viable.

On the basis of the study by Okua *et al.* (1997) and the Lake Mburo case study, it may be concluded that if UWA wishes to promote sport hunting as a viable tool for conservation, it will need to enlarge the estate for hunting. This will involve establishing hunting blocks within protected areas, especially wildlife reserves (WRs). Possible sport hunting areas might, for example, include Bugungu WR, Karuma WR and East Madi WR.

In Uganda, the economics of wildlife utilisation on community land outside protected areas can only be analysed in the context of land tenure and use. Most rangeland areas in Uganda are now fragmented into large and small landholdings, some with titles, some without, and many with squatters. The management of sport hunting in such areas, divided up by new or derelict fences, is challenging, if not impossible. Against this background, we conclude that sport hunting can be a successful conservation tool in Uganda, but it may only be viable in larger open ecosystems. Where such areas remain, it is critical that both central and local governments recognise that sport hunting has the potential to sustainably benefit those who often bear the greatest costs of wildlife conservation, namely the local landowners and communities.

Acknowledgements

The authors would like to thank the Executive Director of the Uganda Wildlife Authority (UWA), Mr Moses Mapesa, for his encouragement in our presentation of the Lake Mburo sport hunting scheme. Mr John Makombo, Deputy Director Field Operations of UWA, kindly made data on hunting quotas and revenues available.

References

Averbeck, C. (2002) *Population Ecology of Impala (Aepyceros melampus) and Community-based Wildlife Conservation in Uganda.* PhD Thesis, Technical University of Munich, Germany.

Douglas-Hamilton, I., Malpas, R., Edroma, E., Holt, P. Laker-Ajok, G. & Weyerhaeuser, R. (1980) *Uganda Elephant and Wildlife Survey.* Uganda Institute of Ecology, Report to IUCN.

Edroma, E.L. (1984) Drastic decline in the numbers of animals in Uganda National Parks. In *Rangelands: A Resource Under Seige,* eds. P.J. Joss, P.W. Lynch & O.B. Williams. Australian Academy of Science, Canberra.

Eltringham, S.K. & Malpas, R.C. (1980) The decline in elephant numbers in Rwenzori and Kabalega Falls National Parks, Uganda. *African Journal of Ecology* 18, 73–86.

Eltringham, S.K. & Malpas, R.C. (1983) *The Conservation Status of Uganda's Game and Forest Reserves in 1982 and 1983.* Uganda Game Department, Entebbe, Uganda.

Emerton, L. (1999) *Balancing the Opportunity Costs of Wildlife Conservation for Communities around Lake Mburo National Park, Uganda.* African Wildlife Foundation, Nairobi, Kenya.

Ford, J. (1971) *The Role of the Trypanosomiases in African Ecology.* Clarendon Press, Oxford.

Fraser Stewart, J.W. (1992) Integrating local communities with protected area and wildlife management in Uganda. Field Document No. 6 for FAO Project DP:FO UGA/86/010, for Ministry of Tourism, Wildlife and Antiquities, Kampala.

Game Department (1925–1981) Annual Reports of the Game Department. Government Printer, Entebbe. Archived at Wildlife Division, Ministry of Tourism, Wildlife and Antiquities, Kampala.

Hulme, D. (1997) Community conservation in practice: a case study of Lake Mburo National Park. Community Conservation Research in Africa. Working Paper No. 3, University of Manchester.

Hulme, D. & Infield, M. (2001) Community conservation: reciprocity and park–people relationship; Lake Mburo National Park, Uganda. In *African Wildlife and Livelihoods*, eds. D. Hulme & M. Murphree. James Currey Ltd., Oxford.

Infield, M. (2002) *The Culture of Conservation: Exclusive Landscapes, Beautiful Cows and Conflict Over Lake Mburo National Park, Uganda.* PhD Dissertation, University of East Anglia.

Lamprey, R.H. (1993) *Management Plan for the Ugalla Game Reserve, Western Tanzania.* Wildlife Division, Dares Salaam, Tanzania.

Lamprey, R.H. (1995) *General Management Plan for the Selous Game Reserve*, Tanzania. Wildlife Division/GTZ, Dares Salaam, Tanzania.

Lamprey, R.H. & Michelmore, F. (1996) *Surveys of Uganda's Protected Areas, Phase I and Phase II.* Ministry of Tourism, Wildlife and Antiquities, Kampala, Uganda.

Lamprey, R.H., Buhanga, E., Omoding, J. e*t al.* (1999) *A Wildlife Protected Area System Plan for Uganda.* Ministry of Tourism, Trade and Industry, Kampala.

Lamprey, R.H., Buhanga, E. & Omoding, J. (2003) *A Study of Wildlife Distributions, Wildlife Management Systems, and Options for Wildlife-based Livelihoods in Uganda.* International Food Policy Research Institute and USAID, Kampala, Uganda.

Mugerwa, W.K. (1992) *Rangeland Tenure and Resource Management: an Overview of Pastoralism in Uganda.* Makerere Institute for Social Research, Kampala.

Mugisha, A.R. (1993) *A Case Study of Nshara Grazing Area, Mbarara, Uganda.* MSc, Rural Resources and Environmental Policy, Wye College, London University.

Norton-Griffiths, M. (1978) *Counting Animals.* Handbook No. 1. African Wildlife Foundation, Nairobi.

Okua, J.M., Buhanga, E., Opolot, F.H.O., Anywar, J., Karugaba, K. & Heath, B.R. (1997) *Guidelines for the Implementation of Wildlife Use Rights.* Uganda Wildlife Authority, African Wildlife Foundation.

Uganda Wildlife Authority (UWA) (2002) Management of wildlife outside protected areas. *Proceedings of a workshop*, Jinja, 30 July–1 August 2002. Uganda Wildlife Authority.

Does Recreational Hunting Conflict with Photo-Tourism?

Richard Davies[1], Kas Hamman[2] and Hector Magome[3]

[1]Wildlife Business Consultant
[2]Director for Biodiversity Conservation, CapeNature, South Africa
[3]South African National Parks, Pretoria, South Africa

Introduction

Some people hold the view that hunting and photo-tourism are mutually exclusive forms of land use. In this chapter, the compatibility of hunting and photo-tourism is assessed in two different protected areas in South Africa. They are shown to be compatible in one, and partially compatible in another. The two areas have similarities, but differences in landscapes and the nature of photo-tourism products create different outcomes.

The two study areas are the South African protected areas (PAs) of Pilanesberg and Madikwe (Figure 14.1). Both are state-owned and administered by the North West Parks and Tourism Board, a statutory provincial board charged with biodiversity conservation and tourism marketing. The first of these reserves was established in the late 1970s and the latter in the early 1990s. They were previously farmed through extensive cattle ranching. The reason for their establishment was primarily to optimise socioeconomic benefits for surrounding communities, within a context of being financially sustainable for their annual operational requirements. Although Pilanesberg

Recreational Hunting, Conservation and Rural Livelihoods: Science and Practice, 1st edition.
Edited by B. Dickson, J. Hutton and W.M. Adams. © 2009 Blackwell Publishing,
ISBN 978-1-4051-6785-7 (pb) and 978-1-4051-9142-5 (hb).

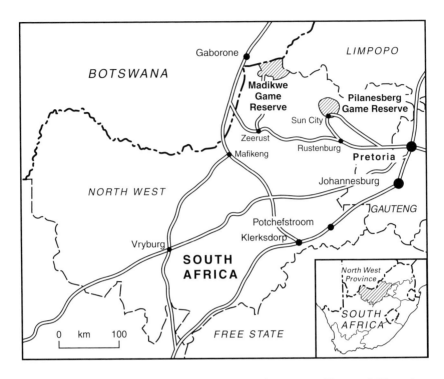

Figure 14.1 **North West Province, South Africa, showing Madikwe and Pilanesburg protected areas.**

is called a 'national park', it is actually proclaimed under provincial rather than the national legislation that would give it national park status. State-owned conservation land in South Africa may be administered by the province where it is located or by the National Parks Board (SANParks). The designation and hence administration is more a consequence of history than directly related to biodiversity or other attributes.

The North West Parks and Tourism Board and its predecessor, Bop Parks (Bophuthatswana National Parks Board, marketed as 'Bop Parks'), have always subscribed to the concept of sustainable utilisation (Management Plan, 2000). Natural resources are viewed as important in providing benefits to people to improve their quality of life, especially the rural poor and those living closest to the Parks.

Pilanesberg National Park covers approximately 500 km². Madikwe is somewhat larger at just over 600 km² but there is an additional 200 km² of private land which has been fenced into and included in the overall conservation management of the reserve. The topography and geology of the two areas are very different. Pilanesberg is an ancient 1300-million-year-old volcano which has left a rough and rugged topography. Madikwe in contrast is bordered in the south by the Dwarsberg Mountains; this topography then gives way to a large flat dolomitic plateau before once again descending over a ridge (Rant Van Tweedepoort) running approximately across the middle of the reserve before again flattening out into the vast northern plains. These differences contribute to the type of tourism products and the ability to carry out hunting.

Tourism in Pilanesberg mostly involves self-drive tourists who use the 188 km of park roads to undertake game viewing drives in their own vehicles. There are also some game drive operators who use the same roads for guided open vehicle safaris. The guests can overnight in one of several lodges located in and around the reserve. The tourism product in Madikwe is fundamentally different to that in Pilanesberg. It is focused at the mid to upper market and is therefore more expensive and exclusive. The largest difference is that there is no self-drive option. All guests must report to one of the many lodges, from where they are taken on guided drives throughout the park. The lodges have game viewing protocols which all must adhere to; this optimises the visitor's experience. There is also close interaction between the park staff and the concessionaires regarding park management issues.

In both reserves almost all the game species (i.e. large mammals) were reintroduced, having formerly occurred in the area before becoming locally extinct to make way for agriculture. This included the 'big five' (lion, elephant, rhino, buffalo and leopard).[1] In Pilanesberg over 6000 head of game (Boonzaaier & Collinson, 2000) were reintroduced, and in Madikwe over 8000 head. Some species still occurred in the area and their numbers were either supplemented if their initial population was deemed too low or were allowed to increase naturally. At Madikwe it was planned to include the 'big five' from the outset. At Pilanesberg, lions were only introduced in 1993, because of the financial and economic benefits which they would generate for the North West Parks and Tourism Board, their concessionaires and the wider economy (Vorhies & Vorhies, 1993).

Both reserves are fenced with a game fence 2.1 m high, which comprises a game proof wire mesh, five electrified wire strands and three cables to ensure that no animals leave the reserve. The fence also includes a 1 m wide wire mesh apron running along the ground. This is attached to the main fence on the inside and is packed with rocks to prevent burrowing or digging animals from breaching the integrity of the fence. This is required as the reserves are surrounded by high population densities and Pilanesberg in particular has three large towns in excess of 5000 inhabitants on the reserve boundary. The fence prevents any movement of large mammals out of the reserves, and also prevents livestock from entering. This implies that most of the larger mammals need to be actively managed to maintain population numbers within specified limits. For the more prolific species this requires removal of individuals or groups where limits are exceeded.

To achieve this level of management, intensive monitoring is required. For some species this is at the individual level. Each year the game in both reserves is counted and a census report produced. The census technique varies according to the species of game and there is a policy which regulates this. For species that are valuable (either in biological or financial terms), and for some potentially dangerous species, a separate and specific monitoring system is applied. For some species such as black and white rhino, lion and wild dog *Lycaon pictus*, an individual register is maintained for each animal. For lion and wild dog in particular, it is possible to identify the parents of any offspring.

This level of monitoring is required because the relatively small area of the parks implies that these animals need to be closely managed to ensure the long-term survival of the population and to avoid the dangers of inbreeding. Those animals which occur naturally at low densities are managed as one 'metapopulation' between the two reserves, thus forming part of a larger breeding gene pool. The rhino monitoring (especially for the black rhino) also provides very specific information regarding the age structure and territory of individuals, especially breeding adults.

On the basis of an annual census and other monitoring information, a game removal quota is determined. This quota is then allocated to hunting, capture and live removal. This allocation is done to optimise financial-return, practical and humane considerations. The hunting quota is set to ensure that the revenue is optimised for the different income-generating options: hunting cannot occur while tourists are in an area, primarily for safety reasons. This will then be balanced against what is practical within these constraints. Usually the hunting is conducted in sections of the park where general photo-tourism is less important, or areas that are not accessible to photo-tourists, mostly due to a lack of roads and tracks. In Pilanesberg these are the more remote hilly areas of the park where access and roads accessible to normal sedan-type vehicles would be costly to maintain and develop. However, certain species may be hunted in the photo-tourism areas, especially those species which occupy territories which occur almost exclusively in these areas. If good trophy individuals of either lion, rhino, buffalo or leopard do occur in a photo-tourism areas, that area may then be closed to photo-tourists while the hunters are operating. These closures would generally, but not always, occur in the less busy periods especially during mid-week when occupancies in the tourism accommodation are significantly lower than at weekends.

Safety is an important consideration in choosing hunting areas. This is to ensure that no stray bullets or ricochets will pose a danger to those not involved with the hunting operations. In hilly areas these zone are usually limited to a specific valley where the hillsides creates a barrier. Pilanesberg, with its many deep valleys provides this safety zone within a relatively small area of the reserve. In Madikwe the terrain is much flatter and this results in large areas, up to one quarter of the reserve, being closed to all other activities to maintain this safety zone. This is more disruptive to other users and here the benefits of hunting have to be significant to compensate for the losses incurred by other income-generating activities.

Pilanesberg National Park

The habitat in Pilanesberg is described by Acocks (1975) as sour bushveld. It is situated approximately 150 km north-west of the large cities of Johannesburg and Pretoria. It also borders on the well known Sun City Resort Complex, which can accommodate over 2000 overnight guests. This is privately owned and operates a recreational complex which provides a range of activities including two world-class 18-hole golf courses, many other sporting facilities, water sports such as water slides, para-sailing, artificial wave pool, casino, etc. Although not all the guests from Sun City visit Pilanesberg there are guided open-vehicle game drives which operate from the resort; these can transport up to 350 people at any one time on safari vehicles into the reserve, with a further 200 seats available if required.

Pilanesberg is a very popular destination and caters to a wide range of market users from self catered campsites (costing US$20[2] per site per night accommodating up to six people) to luxury overnight accommodation costing in excess of US$600 per person per night. Guests stay in overnight formal accommodation facilities and campsites. No 'off-road' driving is permitted by any visitors.

Pilanesberg is undoubtedly one of the most visited parks containing the 'big five' in Africa when expressed as visitors per square kilometre. A situational analysis which was undertaken in Pilanesberg in 1999 indicated there were 1345 visitor beds and 432 caravan/camping sites within the park (Contour, 1999a). In 2004 an additional lodge accommodating 180 overnight guests was added to the existing number.

A survey which was conducted in Pilanesberg over the very busy Easter long weekend in 1999 (Contour, 1999b) and it showed that there were over 9000 visitors in 2426 vehicles in the Park. There are 188 km of road network for use by the tourists and this therefore translates into a visitor use of three vehicles per kilometre, although in the more popular central basin the vehicle density was up to 10 vehicles per kilometre. The same survey by Contour (1999b) indicated that Pilanesberg had significantly higher usage per square kilometre of park than the Kruger National Park and Hluhluwe/Imfolozi Game Reserves, when compared against actual infrastructure and visitor usage. Both these reserves contain the 'big five' and cater to a similar market. The Kruger National Park is in excess of 2000 km^2 and is therefore perhaps a poor comparison because of its size. For this reason Hlhluwe/Imfolozi was also compared. Situated in KwaZulu-Natal, it is 960 km^2, or almost twice the size of Pilanesberg. A summary of the salient points from the Contour (1999a) comparison is presented in Table 14.1.

It is apparent from Table 14.1 that Pilanesberg has very high visitor numbers and usage. Neither the Kruger National Park nor Hluhluwe/Imfolozi Game Reserves operate any sport hunting activities. This is not permitted under their policy framework; both these protected areas however remove live game for relocation to both state owned and private land.

Madikwe Game Reserve

Tourism in Madikwe is focused on guided safaris which are operated by a number of concessionaires who operate lodges in the reserve. These are long-term leases and they all share common traversing rights over the whole reserve. There has been a gradual development of lodges over the last 10 years. The first

Table 14.1 **A comparison of the density of tourist roads, tourist beds and visitors per square kilometre of park, the actual number of tourists per square kilometre of park and the number of vehicles per day per kilometre of road over the busy Easter long weekend 1999 for Kruger National Park, Hluhluwe/Imfolozi Game Reserves and Pilanesberg National Park.**

Comparative item	Kruger National Park	Pilanesberg National Park	Hluhluwe-/ Imfolozi Game Reserve
No. of square kilometres of park per km of tourist road	0.741	0.291	0.42
Number of km^2 of park per visitor bed	0.472	0.024	0.25
Average no. of visitors per day per km^2 of park	1.34	17.67	4.40
No. of vehicles per year per km^2 of park	128	1 203	429
Average number of vehicles per day per kilometre of road	0.26	0.83	0.5
Average number of vehicles per day per km of road, over Easter weekend	30	213	NA
No. of tourist beds per km^2 of park	2.12	39.67	4.02

lodge was completed in 1994 and was developed in the extreme north-east corner. A second lodge also in the extreme north but closer to the western corner was completed in 1996 and few smaller satellite lodge adjacent to it over the next few years. For the next four years there were very few new lodges developed until once again from 2000 to 2004 a further 25 were completed, mostly small 10-bed facilities. During these early years there were only 140 tourist beds and these were all in the extreme north of the park. The product was still new and the occupancies and achieved rates were relatively low.

By 2005 the lodge development plan was far advanced and over 500 lodge beds had been developed. These were now scattered over most of the reserve and not confined to the northern areas only.

Hunting in Madikwe has mostly occurred on the plains in the south of the park. These areas were far away for game drives originating in the north, and hence in the early years hunters and tourists rarely overlapped. As lodge developments increased in number and distribution the photo-tourism use of the whole reserve increased. This brought the two groups into conflict, since large areas, up to a quarter of the reserve, needed to be closed to photo-tourism use to provide a safety area. With photo-tourists now at higher densities than the early years, and now using all the reserve for their game-viewing activities, concessionaires objected when large game-viewing areas were closed. They put pressure on the reserve management to reduce or stop hunting. The relatively high contribution which the lodges now make in concession fees (see Table 14.2) means this is now more productive than the hunting, which was more competitive when photo-tourism use of the reserve was lower and more restricted in the areas used.

Hunting and sustainable use

Hunting has been used as a mechanism to generate income and other socio-economic benefits and as an integral part of the sustainable use strategy of the North West Parks and Tourism Board (hereafter Board). Although the financial benefits are more obvious, hunting also generates significant employment. The actual hunting operation employs at least six staff per hunting safari. In addition to this there are the other jobs such as taxidermy, more than 80 per cent of the hunters having this done locally. Hunters also spend additional time in the county as non-hunting tourists and this also creates subsequent employment. Importantly, hunting is an export industry and for a country like South Africa this contributes significantly to foreign exchange earnings.

Hunting is a widely used income generation strategy in most of the parks under the control and management of the Board. It has proved to be financially very rewarding for the Board and for the Pilanesberg National Park and Madikwe Game Reserve in particular. By generating this valuable net income the hunting operations help contribute to the biodiversity conservation strategy of the North West Parks and Tourism Board.

Hunting in South Africa takes two forms, 'biltong' hunting and trophy hunting. The latter is often split into the local trophy market and the foreign trophy market. 'Biltong' hunting is essentially hunting for meat and is a very popular activity especially in the winter months, accounting for a large percentage of the animals hunted in South Africa each year. It usually focuses on the more abundant and cheaper animals. It has not been used as a game removal strategy

Table 14.2   The number of animals hunted and the income generated there from (US$) with the average income earned per animal, the number of 'big five' animals hunted and the total number of all game within Madikwe Game Reserve for the period 1997–2004. The percentage of animals hunted as a proportion of the total number in the park is also presented.

Year	1997	1998	1999	2000	2001	2002	2003	2004
No. of animals hunted	124	97	128	137	131	91	109	64
Total income	$89,248	$73,532	$81,837	$124,657	$111,871	$202,160	$415,782	$320,201
Income per animal	$720	$758	$639	$910	$854	$2,222	$3,815	$5,003
No of 'big five' animals	3	2	2	7	6	11	13	15
Total game numbers	13,454	16,011	11,826	16,232	12,791	8,318	5,019	6,594
Hunted as % of total	0.64%	0.45%	0.67%	0.62%	0.66%	0.72%	1.55%	0.83%
US$:ZAR rate	R4.52	R5.89	R6.04	R6.85	R8.05	R10.37	R7.47	R6.27

in either Pilanesberg or Madikwe because the relative net financial returns are low and the number of animals hunted needs to be high to be cost effective. This high level of disturbance makes the game wary of vehicles and human activity and this impacts negatively on the photo-tourism product.

Both reserves have generally and especially in the last decade catered almost exclusively for the foreign trophy market. Thus providing the best return per animal and at the same time removes relatively few animals.

Financial and economic implications

Records were obtained for Madikwe and Pilanesberg for the number of animals hunted, the income generated for all activities and the income per animal for the period 1997 to 2004 and the number of 'big five' animals hunted (Tables 14.2 and 14.3).

Total game numbers are counted in the annual census reports from the North West Parks and Tourism Board. Not all the animals on the hunting quota are included in the census. All large abundant herbivores are counted primarily for the impact they have on the environment; but the smaller and less abundant herbivore species (such as warthog, common duiker, steenbok, etc.) are not. Also the smaller predators (caracal, African wild cat, honey badger etc.) and primates (chacma baboon and vervet monkeys) are not counted for logistical and practical reasons but they are included in the hunting packages. These smaller species make up a very small portion of the hunting quota, both in numbers and financial value. The number of animals hunted, as a percentage of the total number in the park, refers only to those animals included in the census. In this section, the relative financial contributions of hunting, live capture and photo-tourism to operating costs will be explored. The economic contribution is also mentioned, but has not been quantified in as much detail.

Income generated (in US dollars) is the amount the operator pays to the Board for the right to hunt and for trophy fees. The operator would sell these on with a mark-up to clients. Although prices are paid to the Board in South African rand (ZAR), these have been converted to US dollars as most of the animals are ultimately sold in this currency. The interbank rate used for this conversion is that quoted on www.oanda.com for 30 June each year.

Madikwe relied on both live capture and hunting to generate income in the early years (Figure 14.2). Hunting provided in excess of 47 per cent of the

Table 14.3 The number of animals hunted in Pilanesberg National Park and the net income generated by the Board from these animals with the average net income earned per animal hunted (US$) and the number of 'big five' animals hunted, the total number of animals in the park and the number of these hunted, expressed as a percentage of the total number for the years 1997–2004, the exchange rate per US dollar to the South African rand is also presented for each year.

Year	1997	1998	1999	2000	2001	2002	2003
No. of animals hunted	143	172	186	201	172	172	123
Total income	$137,046	$129,332	$133,808	$339,803	$211,600	$360,266	$651,484
Income per animal	$958	$752	$719	$1,691	$1,230	$2,095	$5,297
No. of 'big five' animals	7	5	5	12	8	16	23
Total game numbers	9,306	7,563	8,074	7,817	5,408	4,795	4,340
Hunted as % of total	0.97%	1.40%	1.40%	1.62%	2.05%	2.57%	2.30%
US$:ZAR rate	R4.52	R5.89	R6.04	R6.85	R8.05	R10.37	R7.47

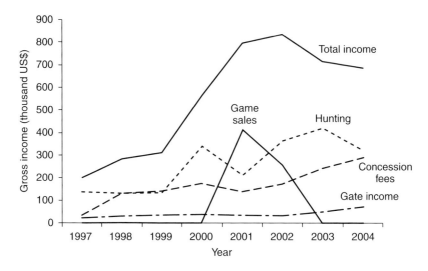

Figure 14.2   **Income by categories for Madikwe 1997 to 2004.**

income generated in Madikwe for the first seven years. Figure 14.3 provides a breakdown of the income earned by the different categories from 1997 to 2004. No live game was removed from 1994 to 2000, during this period numbers where increasing and could not sustain live harvesting. In 2001 game numbers had increased to the point were breeding animals could be removed and there was a significant increase in live game income. The hunting was possible in the early years as usually only males are removed from the population. This has virtually no impact on population growth if done properly. Initially hunting accounted for a large proportion of the income. The concession income and gate income (entry fees paid by photo-tourism) has increased and will continue to increase as the lodges are completed and occupancies increase. These have increased to the point where they now collectively generate more income than hunting. Live game income could not be obtained for 2003 and 2004.

The income generated from hunting has been analysed in more detail and a breakdown of this with game numbers is presented in Table 14.2. Important points to notice in these figures are that the actual number of animals hunted is very small, but the net income per animal is high. The average income per animal has increased over time as the number of more valuable 'big five' animals has increased, while the total number of animals hunted is actually decreasing.

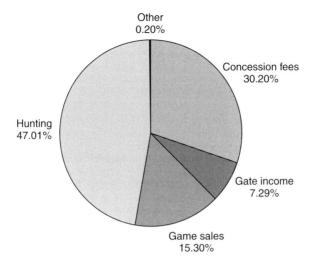

Figure 14.3 **Cumulative income from different categories for Madikwe 1997 to 2004.**

Data on hunting income at Pilanesberg are presented in Table 14.3. The total number of animals hunted in Pilanesberg, i.e. successfully killed by hunting, from 1997 to 2003 is 1169; of these the 'big five' comprised 76 animals or 6.5 per cent of the hunting quota. The number of animals hunted annually has varied between 201 and 123. The total number of game in the park and the number hunted as a percentage of the total population for the periods 1997 to 2003 varies between 2.57 per cent and as low as 0.97 per cent. The net income paid by hunters to the Board has increased from US$129,332 in 1998 to US$651,484 in 2003.

The net income from hunting presented in Table 14.3 has been further broken down into categories (Figure 14.4). Trophy fees account for the largest percentage of the income earned, while camp rental and the income from the sale of hunted meat are very similar. There is some fluctuation in the net income per animal from US$719 in 1999 to US$5297 in 2003. One of the major reasons for this is the number of 'big five' animals which are offered, but also the changes in exchange rate between the rand and the dollar. The 2003, 2002 and 2000 seasons, in that order, earned the most but also hunted the most 'big five' animals. It is a strategy of Kgama Wildlife Operations (Pty) Ltd (who supplied the statistics used here and manage the tenders on behalf of the Board)

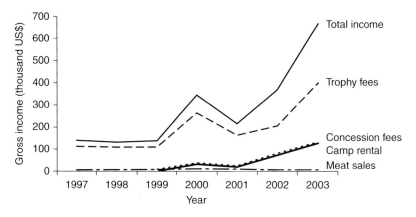

Figure 14.4 **Income by category, Pilanesburg 1997 to 2003.**

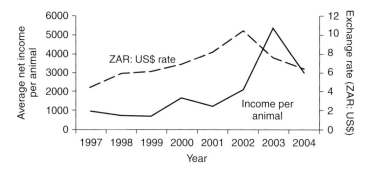

Figure 14.5 **Average net income per animal and exchange rate.**

to build hunting packages around valuable 'big five' animals. This adds value to the normally lower-priced species and extends the duration of the hunting safari. This obviously increases net income and the economic benefit from the hunting operations.

There is, however, a close correlation between the average income earned per animal and the South African rand–US dollar exchange rate, as indicated in Figure 14.5. Although income per animal has increased, at least some of this can be attributed to increasing strength of the dollar. The peak in 2003 may be due to game which was sold under stronger dollar rates, as game is often sold up to nine months in advance. Figure 14.2 also supports this by indicating that camp fees have increased and meat sales have declined, indicating longer stays at the camp and less meat being sold.

Income from hunting comprises four components, namely trophy fees, concession fees, camp fees and meat sales. The trophy fee is the fee the operators pay for the rights to hunt the animal, while the concession fee is the right to operate a concession in the park for a defined period each season. The camp fees are for hiring facilities which have been erected in the park exclusively for hunting. Meat sales are the income earned from meat and other products which are sold after the animals are hunted. Only the trophy (horns, skull, etc.) and the skin may be kept by the hunters and some meat for the pot. The contribution of each of these to the total income earned by the hunting operations is presented in Figure 14.4.

Trophy fees comprise the largest component, ranging from 84 to 56 per cent, while the concession fees, camp rental and meat sales make up the remainder. The meat sales are an additional by-product. The proximity of Pilanesberg to large urban centres with disposable income adds to the benefits the Board earns from the hunting operations, although this has declined over time.

For the 1999 and 2000 financial years, hunting comprised between 15 and 18 per cent of the total income earned by the parks. These were the years for which good quality financial information for the other operations could be obtained. Hunting income is also earned in 75–90 days per year during the short hunting season in winter, while the other operations are mostly year-round. Recent reports for 2005 and 2006 indicated that black rhino were hunted in both years. In 2005 the fees paid to the Parks Board for one male black rhino was US$210,000 and in 2006 the figure was US$195,000. Neither of these were special 'trophy' quality. There is support for this from rhino specialists as there are excess male black rhinos available at present and these removals will have no effect on population growth.

The net financial margin to the Board from the different operations is difficult to determine, as not all costs are allocated to cost centres against income. However, the costs of the hunting operation rely primarily on good habitat management and good security. The tourism and live sales components also rely on these issues for their products; these costs could therefore be shared between these income-generating enterprises. There is a significant difference in costs between the enterprises in terms of the infrastructure, control, maintenance and management components.

In Pilanesberg many of the tourism operations are outsourced to private companies and hence the costs of running these are not carried by the Board.

As with the hunting operations, the marketing, equipment, capital requirements and actual management of the hunters are in the hands of private operators, whose turnover is not reflected in these figures.

The Board however does carry out some tourism functions itself and these
need to be funded. These include the control and charging of entry fees at the
entrance gates, along with associated administration including auditing and
financial procedures. In addition the road and track network must be developed and maintained. It is often argued that this would be required for general
park management; this is partly true but the quantity and quality of the roads
and other infrastructure could be significantly lower than for photo-tourism,
accounting for a significant saving.

There are no detailed figures as to what these costs are, but the gravel roads
in Pilanesberg have a replacement cost of approximately US$1.6 million and
an estimated maintenance cost of approximately US$200,000 per annum
(G. Youldon, North West Parks and Tourism Board technical officer, personal
communication). The administration associated with entry permits for visitors has not been estimated. The net return to the Board for the photo-tourism
is therefore lower than for the hunting and live removal operations.

Summary and conclusions

Both Madikwe and Pilanesberg have so far been able to accommodate photo-
tourism and hunting. Photo-tourism use, expressed as the number of visitors
to Pilanesberg per unit area of park or per kilometre of road, is the highest to
be found in any 'big five' reserve in South Africa and probably in Africa. In spite
of this, hunting can still be undertaken through careful zoning and management. The hunting however only uses a very small percentage of the total game
population and is undertaken during a very short season of 70–90 days. The
financial contribution to the Board is significant and the economic impact is
also probably high because of the services required by hunters such as car and
aircraft hire, taxidermy and, most importantly, the hunting outfitters. Most of
these are local and the benefit is therefore usually in the immediate area.

Pilanesberg is a much older and more mature tourism and hunting product
than Madikwe, whose tourism is still growing. The topography of Pilanesberg
implies that some areas cannot be accessed effectively by photo-tourists and
it is in these areas that professional hunters can maintain a presence without
impacting on other users. Hunters can also operate in valleys with little or no

danger to other users from stray bullets or ricochets. This means that smaller areas can be closed to other uses for limited periods.

The situation in Madikwe is somewhat different. The area is much flatter, and to maintain safety of guests it is often necessary to close larger areas temporarily while hunting operations are under way. These areas are mostly accessible to concessionaires and their guests, unlike Pilanesberg where they are generally not. The game lodges which operate in Madikwe and know and understand the area are often aggrieved when areas of the park are closed to their guests. They know where game can be found and when more popular areas are 'out of bounds' this creates tensions.

In the early years, Madikwe was very lightly populated with lodges, and the visitor densities were low. The hunting operations were and still are confined to the southern half of the park, between the Rant Van Tweedepoort and the Dwarsberg Mountains. As more lodges have been built, especially closer to the 'hunting' area, this has increased the conflict. This conflict is expressed within the Park Communication Forums where all the users interact from time to time. The income from tourism is likely to grow significantly in the future as the lodges mature (Davies & Leitner, 2003). The growing number of tourism concessions and resulting discontent showed with temporary closure of large areas for hunting has led to suspension of hunting operations, probably permanently.

Species such as rhino and lion which are very closely monitored can be very selectively hunted. This not only adds financial and economic value to the Parks Board but also contributes to ensuring the long-term survival of some species through the monitoring and management through removing old and/ or individuals which may inbreed. Individuals from within the population are identified and these are offered as and when appropriate to hunters.

Madikwe and Pilanesberg are completely enclosed by game fences and are therefore closed systems, at least for the larger mammals; they are also relatively small and cannot maintain genetically viable populations of some species without management intervention. In both reserves the high-intensity monitoring of certain species for other management purposes has allowed very specific targeting of individuals for removal. These are often old males who have bred and are possibly past their 'biological prime'. In many instances these are sought after by hunters and can make a final contribution to the management of the park. A black rhino was identified for removal in the mid-1990s. A campaign was launched with conservation interest groups on the biological merits of hunting this individual. The feedback from these groups was positive, but while the process was under way the animal died from natural causes. Black rhino are very tightly monitored and controlled in Southern Africa and a specialist working group has been established to ensure the various populations grow at the optimum rate possible to improve the conservation status of the species. This could imply that more old animals, especially males could be targeted for hunting in the future. With the income made available for management of these protected areas.

Through appropriate management and zoning, photo-tourism and hunting can exist side by side. This is possible in Pilanesberg due to the topography and some areas being inaccessible to photo-tourists. In Madikwe the growing number of photo-tourism concessions and the topography which necessitates closing large areas during hunting operations has shown that these two activities cannot be combined effectively.

Acknowledgements

CapeNature is thanked for providing the funds for the writing of this paper. The efforts of the staff of the North West Parks and Tourism Board and Kgama Safaris Pty Ltd are greatly appreciated for the information which they supplied to make this possible as well as contributing to the earlier drafts of this paper.

Notes

1 African elephant (*Loxodonta africana*); African buffalo *Syncerus caffer*; Black rhinoceros (*Diceros bicornis*); white rhinoceros (*Ceratotherium simum*); lion (*Panthera leo*); leopard (*Panthera pardus*).
2 Based on 1US$:ZAR6.70.

References

Acocks, J.P.H. (1975) Veld types of South Africa. *Memoirs of the Botanical Survey of South Africa.* No. 28.
Boonzaaier, W.B. & Collinson, R.F.H. (2000) *Pilanesberg National Park Management Plan 2000.* Contour Managers and Collinson Consulting, Rustenburg, South Africa.

Contour (1999a) *A Situational Analysis Report Volume II. Pilanesberg National Park Management Series.* Contour Project Managers, Rustenburg, South Africa.

Contour (1999b) *Easter 1999 Visitor Survey.* Contour Project Managers, Rustenburg, South Africa.

Davies, R.J. & Leitner, P. (2003) Financial and economic implications. In *Madikwe Game Reserve – A Decade of Progress*, eds. R.J. Davies & M. Brett. North West Parks and Tourism Board, Mmabatho, South Africa.

Vorhies, D.& Vorhies, F. (1993) *Introducing Lions into Pilanesberg: An Economic Assessment.* Unpublished report prepared for the Bophuthatswana National Park & Wildlife Management Board, July 1993.

Governance

When Does Hunting Contribute to Conservation and Rural Development?

Bill Wall[1] and Brian Child[2]

[1]Independent Consultant, Alaska, USA
[2]Center for African Studies & Department of Geography,
University of Florida, Gainesville, FL, USA

Introduction

Around the world, hunting has proved to be a powerful tool to promote conservation when conducted in ways that are biologically sound within appropriate governance and institutional settings (Wall & Kernohan, 2003; Wall, 2005; Mahoney, this volume, Chapter 16; Booth & Cumming, this volume, Chapter 17). This chapter aims to define the characteristics of a hunting programme that promotes conservation and rural development, and which therefore earns the name 'conservation hunting'. The idea that the term 'conservation hunting' serves a useful distinction has several roots, but it has been furthest developed in Canada as a prototype tool to encourage the better design of hunting programmes (Wall, 2005). We believe it will eventually become a generic label for hunting programmes that share a range of characteristics that encourage conservation and contribute to rural economies.

The primary focus of this work is recreational (or 'sport') hunting which is practised largely in the developing world by tourists from industrialised

Recreational Hunting, Conservation and Rural Livelihoods: Science and Practice, 1st edition.
Edited by B. Dickson, J. Hutton and W.M. Adams. © 2009 Blackwell Publishing,
ISBN 978-1-4051-6785-7 (pb) and 978-1-4051-9142-5 (hb).

countries. This form of recreational hunting, in which the hunting tourist is very likely to return home with a trophy of his or her quarry (and is therefore often called 'trophy hunting'), has the potential to see very significant sums of money transferred from rich countries (of Europe and North America especially) to developing economies – and even more importantly, to remote rural areas where there are few alternative economic opportunities (Bond, 1999; Virk, 1999). Conservation hunting programmes in the context of tourist hunting aim to integrate the social, economic and ecological drivers of sustainability and to use a diverse spectrum of quarry species to make wild and remote lands more economically productive, thus creating opportunities for sustainable *in situ* conservation.

Ultimately, the ability to use hunting to create sustainable conservation benefits and support rural livelihoods hinges as much on issues of governance as on any biological characteristic of the target species, a fact that is often ignored by wildlife managers. In view of this, we identify and contrast two major systems for the governance of hunting and by combining this with information on national governance characteristics establish a typology which explains, amongst other things, why the North American Model is unlikely to be successful in many developing countries.

Different approaches to hunting and the impact of governance

To understand the effectiveness of different strategies to achieve conservation hunting we need not only to understand what the broad strategies are that are being deployed around the world, but also to better understand the political construction of the society in which hunting is occurring.

Personalised and impersonal governance systems

North (2006) describes, convincingly, a fundamental dichotomy in governance which has a significant impact on conservation hunting. At one end we have relatively effective economies characterised by persistent competition in the political and economic market-places, where a level competitive field is kept in place by strong judicial systems and well-defined property rights. These systems protect the conditions for impersonal transaction by providing judicial recourse where people are cheated, and generate considerable economic participation and wealth. They are called 'impersonal economies'.

At the other extreme we find weak or inequitable economies characterised by highly interdependent political and economic elites, where individuals seldom trust political, judicial or economic institutions (including property rights) so that economic transactions depend heavily on personalised exchanges and reciprocity, i.e. who you know. In these economies, technical bureaucrats often play a highly political role (North, 1990), elites are disproportionately powerful and individual property rights are weak so that the elites are the gatekeepers to many resources including wildlife. These systems are called 'personalised economies' and within them the potential for many ordinary people to generate value-added products or services is reduced by a lack of access to resources and a dependence on personalised exchange. This political and economic process significantly increases the risks and basic transaction costs of conducting business. It typically engenders processes and transactions that are not transparent and increases opportunity for corruption to occur throughout any government-regulated programme.

Public and personal hunting systems

There are two rather different generic models for governing hunting, and it is instructive to match these to the political contexts just described. The first is generically known as 'public hunting' where most aspects of the hunting system are generally organised and managed by the State (or a sub-unit of the State, such as a province) as a public service. This includes the management of the quarry species, quota setting, licences and other practices to restrict and allocate hunting opportunities. Licence fees and taxes are set and taken by the State and in most situations are reinvested in management of the resource. Public hunting underpins the North American Model of hunting described by Mahoney (this volume, Chapter 16).

'Personal hunting' differs from public hunting in that the principal programme manager is not the State, but the landholder (landholders may range from individuals to rural communities occupying land that is legally owned by the State) who directly benefits from the revenues that are generated by the economic activity that takes place on their land, with 'their' wildlife resources. 'Personal hunting' is probably the best name to describe this system because the key characteristic is that the benefits are focused on those who manage the land and consequently decide the future of wildlife.

However, it is *essential* to recognize that in most parts of the world the authority for hunting is actually vested in the State. In some cases the State has chosen to devolve it to create an organizational structure that approaches a personal hunting model.

In some cases it is difficult to determine the precise boundary between the State and private sectors, especially in the newly emerging states in Central Asia (Subbotin, personal communication). Furthermore, those who benefit are often not individual landowners, but the members of poor rural communities managing their resources through common-property regimes (Child, 1995). Increasingly, this situation is being promoted as a mechanism to create tangible financial incentives for conservation in places as far apart as Asia (see Frisina & Tareen, this volume, Chapter 9) and Africa (Jones, this volume, Chapter 10).

The dichotomy established here intentionally avoids any consideration of 'who should or does own wildlife', which is a complex field and which, in any case, is not central to the success of conservation hunting. As noted, even if the State claims the ownership of wild animals it can still (and often does) construct a hunting system that provides material rewards to landholders for good conservation management.

Which systems deliver conservation hunting?

Specific well-evolved political conditions are a prerequisite for the success of public hunting models and these have been well demonstrated through presentation of the North American Model (see Mahoney, this volume). The conversion from a market hunting approach to a restorative approach, to recreational hunting in the United States required a strong economy, an enlightened public demanding conservation, an army of biological scientists and enforcement officers, good governance with transparency in laws, regulations and enforcement, as well as a citizenry with a high degree of respect for the law. In most cases, because of the effectiveness of regulations and enforcement, public hunting does not negatively affect the species or ecosystems involved. Specifically, where the revenues it generates are invested in the management of the resource on public lands (and sometimes private lands too), one can conclude that public hunting contributes to conservation. On the other hand, public hunting at times struggles to provide landholders with strong incentives to encourage wild animals on their land. Many sectors of society do not perceive this to be a severe problem in a country like the US, where other legislation like the Endangered Species Act can be used to influence the way that landholders deal with wild species if they become sufficiently threatened or endangered, but the situation is very different in many developing countries where the capacity and effective reach of the State is restricted. In recent times it is notable that the public hunting model in the US has evolved in some states to embrace more of the personal hunting strategy, where economic benefits and regulatory power and responsibility have been devolved to the private landowner or operator. This has created in essence a hybrid system which has embraced conservation benefits from both systems (Wall & Bourland, 1992).

Public hunting invariably fails in personalised economies because the very factors that underpin the North American Model are missing. The public hunting model has not worked effectively for conservation in Africa and other developing regions such as Central Asia and even Mexico (Wall & Kernohan, 2003; Manterola & Valdés, 2003; Virk, 1999). Currently there are many difficulties arising in the Russian Federation with a significant change in governance approach and an economy in transition (Subbotin & Wall, 2003). In personalised economies wealthy people with vehicles and firearms tend to get preferential access to hunting opportunities, alienating landholders who protest at the inequity of the system (M.W. Murphree, 2000, unpublished presentation). In many developing countries, governance, accountability mechanisms and state agencies are unable to capture and channel licence or tax revenues to manage and support conservation programmes as they do in the North American Model. Moreover, the spending power of the citizenry is low, and civic hunting organisations are neither able to raise charitable funds for conservation nor exert effective public scrutiny. Finally, there is rarely the opportunity to raise tax revenues from the sale of hunting equipment for reinvestment in the wild resource base.

Attempts to secure conservation hunting based on public systems in personalised economies may be doomed to failure, but conservation hunting does exist in these economies, even in so-called 'failed states', where it has been based on the personal hunting model. This is partly because the State plays much less of an active role in personal hunting systems, which shorten and simplify economic and accountability feedback loops. Importantly, it is also because personal hunting systems can return the financial and other benefits of hunting to the individuals who directly decide the future of the hunted resource. This is particularly significant in biodiversity-rich developing countries where key wildlife species commonly live outside state protected areas, on communally-held or private land. These are generally areas where wildlife has to compete or fit in with other land uses, especially farming, which is a particular problem for large carnivores and herbivores (Ahmed *et al.*, 2001; Taylor & Clark, 2005).

The simple logic of conservation under these conditions is that landholders might tolerate, or even encourage, wild animals on their land if they can gain a significant and reliable income from them. If this is not the case they will inevitably discourage or even eradicate species that impact negatively on their livelihoods. The conflict between African farmers and elephants (*Loxodonta africana*) is well known (Martin & Conybeare, 1992). Conversely, how often is a large North American herbivore seen in the soybean fields of the Mississippi River valley? This logic lies at the heart of southern Africa's conservation policies for private and communally held land. The underlying assumption is that wildlife has to compete and can do so if the institutional framework internalizes costs and benefits at the level of the landholder. This sort of approach is sometimes termed 'incentive-driven conservation' (Hutton & Leader-Williams, 2003). It is not intended to replace protected area conservation, but to use the synergistic resources proactively to create a much larger economic and ecological landscape (Wallington *et al.*, 2005). Although any type of hunting can result in conservation benefits, most if not all conservation hunting programmes in the developing world are based on sustainable tourist hunting for trophies with profit as a key incentive.

Principles underlying successful conservation hunting

Five principles for sustainable resource use at the level of the community were framed in 1993 (Murphree, 2000) and apply equally well to conservation hunting – and not only in developing countries:

1 Effective management of natural resources is best achieved by giving it focused value for those who live with them.
2 Differential inputs must result in differential benefits.
3 There must be a positive correlation between quality of management and the magnitude of benefit.

4 The unit of proprietorship should be the unit of production, management and benefit.

5 The unit of proprietorship should be as small as practicable, within ecological and sociopolitical constraints.

The first three principles encapsulate the economically instrumentalist approach that lies at the heart of sustainable hunting. The latter two follow 'common property theory' and relate to organisational effectiveness and governance. There are two common threads in these principles: the concept of individual discretionary choice, and 'subsidiarity' – the natural principle that regulatory authority should rightly arise by upward delegation to higher levels of organisation and not from centralisation of power by elites (see Murphree, 2000, for an excellent explanation). These are important because many wild resources are threatened by a political ecology where benefits and authority accumulate towards the top of the hierarchy (globally or nationally), but costs and responsibilities are imposed on landholders and people least able to afford them. This governance structure is highly inequitable, and long feedback loops mean it is unresponsive and unaccountable. Recent progress in private and community conservation has usually resulted from the creation of new economic institutions that redress these dysfunctional benefit and power relationships (Murphree, 2000). The importance of correctly aligned rights and responsibilities to sustainability is increasingly recognised in the literature (e.g. Fabricius *et al.*, 2004). A simple way to conceptualise these rights is to imagine the three legs on which the system needs to balance: the right to benefit, the authority to manage, and the right to use resources in the highest valued uses and with the best partners possible (the right of disposal). For example, wildlife communities in Zimbabwe, Namibia and Botswana put up with crop-raiding elephants because they can sell a trophy elephant for ten thousand dollars, eat the resulting meat, set their own quotas and decide how to allocate these quotas (i.e. they have managerial authority and the right to benefit) and choose who to sell them to (i.e. they have the right of disposal) (Metcalfe, 1993; Child *et al.*, 1997). The relationship between different hunting and governance systems is summarised in Table 15.1.

Conclusions

A conservation-hunting programme is one that contributes to the viability of wildlife species, ecosystems and biodiversity by using sustainable harvests

Table 15.1 **The relative effectiveness of public hunting and incentive-driven hunting systems in different governance environments.**

Type of hunting	Characterization of governance regime	
	Impersonal economy Competitive political and economic transactions are governed by property rights; rule of law is respected; reliable justice and effective civil society in place.	*Personalized economy* Highly interdependent political and economic elites, weak or corrupt bureaucracy, weak property rights and legal systems, rule of law not respected and ineffective civil society.
Public hunting	Usually effective with popular public benefits, transparency and civic oversight. However, many do not deliver conservation incentives at the landholder level.	Commonly results in corruption, the capture of benefits by elites and unsustainable harvesting.
Personal hunting	Can be effective, delivering benefits to the landholder, but sometimes seen as elitist and of limited value to society.	Highly effective, particularly where Murphree's principles are in place so that rights, benefits and authority are genuinely devolved to the landholder creating conspicuous economic incentives for conservation.

to generate net economic benefit that in turn creates culturally relevant conservation incentives. Such programmes are managed through legitimate and effective governance and regulatory systems. In the context of foreign trophy hunting, experience suggests that key elements of success are:

- A socioeconomic process which shares appropriate benefits, power and responsibility between tenured landholders, be these private landholders, communities or the State;
- The creation of an economic incentive for land use practices that supports the maintenance of wild game species, habitats, native ecosystems and other biodiversity;

- Adaptive management and simple monitoring systems that track a few key indicators of success such as data on trophy quality;
- Well designed and legitimised regulatory frameworks that encompass economic and biological parameters, while minimising the cost of bureaucracy on profits and commercial flexibility; and
- A relationship between hunting and other intensive uses of habitats, such as grazing of domestic livestock, that is compatible with the maintenance of productive wildlife populations.

A successful conservation hunting outcome hinges on governance and in the developing world, where privatised economies are common, it depends on the correct alignment of economic institutions to maximise benefits and to ensure that those who bear the costs of wild animals on their land also acquire both the benefits and responsibilities that accrue from hunting them. Even in countries where the public model fails, incentive-led conservation has proved highly successful on state, private and communal land under certain conditions, with the profits from trophy hunting often being the commercial engine to drive these programmes. The key characteristic of successful conservation hunting based on personal hunting systems is the devolution of benefits, management authority and responsibility, and the right of the tenured landholder to sell hunting in the market. Where these rights are strong enough, the profitability of trophy hunting is such that the programmes usually work, at least at a basic level. Where they fail, it tends to be in situations where the proprietary rights of landholders are weak and the benefits they get are captured by other stakeholders through excessive bureaucracy, government fees and charges and outright corruption (Leader-Williams *et al.*, this volume, Chapter 18).

The challenge to all those involved in the tourist hunting industry is to ensure that these situations do not arise, and that they are soon corrected where they do. To help achieve this it would be helpful if the individual hunter was better informed about the governance of hunting in the country he or she is visiting, as well as the contribution his or her individual safari will make to improve conservation and rural livelihoods. Most people who can afford to travel overseas to hunt want the money they spend to contribute to the conservation of game species, and, increasingly, to support local livelihoods (Wall, 2005). The key question is: does it, and can we monitor and promote this link? Only when the hunter is armed with this knowledge will he or she be able to ensure their money flows into effective conservation and rural development.

References

Ahmed, J., Tareen, N. & Khan, P. (2001) Conservation of Sulaiman markhor and Afghan urial by local tribesmen in Torghar, Pakistan. In *Lessons Learned: Case Studies in Sustainable Use*, eds. J. Ahmed *et al.* IUCN, Gland.

Bond, I. (1999) *CAMPFIRE as a Vehicle for Sustainable Rural Development in the Semi-arid Communal Lands of Zimbabwe: Incentives for Institutional Change.* PhD Thesis, Faculty of Agriculture, University of Zimbabwe, Harare.

Child, B., Ward, S. & Tavengwa, T. (1997) *Zimbabwe's CAMPFIRE Programmeme: Natural Resource Management by the People.* IUCN-ROSA Environmental Issue Series No. 2. IUCN-ROSA, Harare.

Child, G. (1995) *Wildlife and People: The Zimbabwean Success.* WISDOM Foundation, Harare.

Fabricius, C., Kock, E., Magome, H. & Turner, S. (2004) *Rights, Resources & Rural Development: Community-based Natural Resource Management in Southern Africa.* Earthscan, London.

Hutton, J.M. & Leader-Williams, N. (2003) Sustainable use and incentive-driven conservation: realigning human and conservation interests. *Oryx,* 37, 215–226.

Manterola, C. & Valdés, M. (2003) *The UMA System Wildlife Management and Hunting in Mexico (The Ownership of Wildlife: PUBLIC VS. PRIVATE).* Paper presented at the 3rd International Wildlife Management Congress, 1–5 December 2003, Christchurch, New Zealand.

Martin, R.B. & Conybeare, A.M.G. (1992) *Elephant Management in Zimbabwe,* 2nd edn. Department of National Parks and Wildlife Management, Harare.

Metcalfe, S. (1993) Zimbabwe's communal area management programme for indigenous resources: CAMPFIRE. In *Natural Connections: Perspectives in Community-based Conservation,* eds. D. Western & M. Wright. Island Press, Washington, DC.

Murphree, M.W. (2000) Boundaries and borders: the question of scale in the theory and practice of common property management. Unpublished paper presented at the Eighth Biennial Conference of the International Association of Common Property (IASCP), 31 May 2000, Bloomington, Indiana.

North, D. (1990) *Institutions, Institutional Change and Economic Performance.* Cambridge University Press, Cambridge.

North, D. (2006) The natural state: or why economic development is so difficult to achieve. Paper presented at Tenth Annual Conference, 'Institutions: Economic, Political and Social Behavior', 21–24 September 2006, International Society for New Institutional Economics, Boulder, Colorado.

Subbotin, A. & Wall, W.A. (2003) An ecosystem approach to the development of con-servation-hunting programs for mountain sheep in Central Asia. Paper presented at the 3rd International Wildlife Management Congress, 1–5 December 2003, Christchurch, New Zealand.

Taylor, D. & Clark, T.W. (2005) Management context: people, animals and institutions. In *Coexisting with Large Carnivores: Lessons from Greater Yellowstone,* eds. T.W. Clark *et al.,* pp. 26–68. Island Press, Washington, DC.

Virk, A.T. (1999) *Integrating Wildlife Conservation with Community-based Development in Northern Areas, Pakistan.* Ph.D. Dissertation, University of Montana, Missioula.

Wall, W.A. (2005) A framework proposal for conservation-hunting best practices. In *Conservation Hunting: People and Wildlife in Canada's North,* eds. M.M.R. Freeman, R.J. Hudson & L. Foote. Occasional Publication No. 56. Canadian Circumpolar Institute, Edmonton, Alberta.

Wall, W.A. & Bourland, T. (1992) Timber, wildlife management and conservation: a story of multiple-use and conservation through economic utilization. Paper presented at the 3rd International Wildlife Ranching Symposium, November 1992, Pretoria.

Wall, W.A. & Kernohan, B. (2003) Key components of conservation-hunting programmes and their relationship to populations, ecosystems, and people. Paper presented at the 3rd International Wildlife Management Congress, 1–5 December 2003, Christchurch, New Zealand.

Wallington, T., Hobbs, R. & Moore, S. (2005) Implications of current ecological thinking for biodiversity conservation: a review of the salient issues. *Ecology and Society,* 10(1), 15.

Recreational Hunting and Sustainable Wildlife Use in North America

Shane Patrick Mahoney

Sustainable Development and Strategic Science Branch,
Department of Environment and Conservation,
Government of Newfoundland and Labrador, Canada

Introduction

As with any long-distant event the timing and means of humanity's first appearance in North America remains a subject for debate, but it probably occurred no more than 40,000 years ago, and the weight of evidence suggests considerably less. What is more certain is that these first North Americans were sophisticated hunters who are believed to have assisted in the great wave of large-mammal extinction that so markedly affected the North American landscape by about 11,000 years ago (Martin, 1967). If true, this hunting-assisted decline of wildlife was to repeat itself upon the arrival of modern Europeans some 10,000 years later.

In the intervening time American Indian and Inuit peoples relied continuously on the wild animals around them, developing sophisticated religious and mythological cultures that emphasised their relationship with the natural world and their dependence on the animals they hunted. These traditions placed some restriction on the taking of wild animals and, in combination with low human numbers and experienced prey, may have helped prevent a

Recreational Hunting, Conservation and Rural Livelihoods: Science and Practice, 1st edition.
Edited by B. Dickson, J. Hutton and W.M. Adams. © 2009 Blackwell Publishing,
ISBN 978-1-4051-6785-7 (pb) and 978-1-4051-9142-5 (hb).

repeated large scale extinction of animal populations (Krech, 1999). A more systematic and deliberate approach to wildlife depletion would eventually emerge, however, as markets and 'progress' laid a heavy hand on the wildlife of North America.

Wildlife slaughter and rescue in North America

The first historically evidenced European contact with North America was the failed attempt at Norse settlement at L'Anse aux Meadows, Newfoundland, in the year 1000. Five hundred years later western European nations were sending flotillas of ships to prosecute the whale and cod fisheries off North America's great eastern island, a frenzied engagement that spawned the slaughter of coastal seabirds such as the Labrador duck *Camptorhynchus labradorius* and great auk *Pinguinus impennis* for fish bait, oil and food.

This perverse wave crashed inland with the fur trade. The French, Dutch and English all entered the game early, collectively laying claim to a landmass already occupied by millions of native residents (Wilson, 1999). Eager for the metal trade goods the foreigners offered, aboriginals turned their talents to the task of providing furs. But this traffic also included a virulent brew of diseases, together with cultural and religious elements, that devastated the traditional societies, killing millions and destabilising long-developed relationships with nature and between tribes (Wilson, 1999). Increasingly, for those who survived, a new and competitive race developed to acquire as many furs and as much trade goods as possible. Firearms and alcohol became coveted possessions.

The weakening of native populations helped open up the continent's interior, providing access to increasing numbers of explorers, trappers, traders and, finally, settlers. Moving ever westward, these conquering citizens cleared the land and altered wildlife abundance, killing for food, profit and protection. Predators were always targeted, but invariably all wildlife was taken without limits. Descendants of Europeans in Canada and the US thought wildlife theirs for the taking, to use for personal consumption or to supply any developing market. Few laws stood in their way and those that did were largely ignored (Trefethen, 1975).

As cities prospered along the New World's eastern seaboard a great flow of wildlife moved from out of the west to ready markets, its transport greatly facilitated by railroad expansion in the mid-19th century. Systematically this market system extirpated game species from each newly accessible region, a pattern best exemplified by the demise of the once teeming herds of bison *Bison bison.* Within a matter of decades 30 million or more of these animals were reduced to a few hundred in Canada and the US (Lott, 2002). Along with the fur trade's decimation of the beaver (*Castor canadensis*), the bison's demise proved how rapidly unrestricted hunting could reduce wildlife abundance. Its slaughter was well publicised by journalists, however, and helped engender a vague notion of non-wasteful use of wildlife that would eventually mature into a profound conservation movement.

This slow awakening was energised in the late 19th century by an evident general decline in wildlife abundance in both Canada and the US and was spurred by a rising class of hunters committed to sustainable wildlife use, the elimination of market hunting, and a fair chase ethic borrowed largely from European hunting enthusiasts. This citizens' movement eventually matured into what we recognise today as the 'North American Model' of wildlife conservation (Geist, 1985). Its first origins may be traced to a complex intersection of intellectual and cultural developments that included an elevated notion of civic responsibility in North American society, and the rise of hundreds of local sportsman's clubs that fomented debate and public discourse on the wild resources of North America.

Remarkably, despite Canada's position within the British Empire (and later British Commonwealth), it opted for congruency with the conservation policies and laws of the US, enabling a continental constitution that rescued wildlife in North America, and has coordinated its use and protection, for over one hundred years. As such it helped to form the only continental, as well as the longest surviving, science-based conservation movement in the world.

Principles of the North American Model

Given the breadth and diversity of North American landscapes and regional cultures, it was inevitable that some nuances in wildlife conservation approaches would arise geographically. Nevertheless, across the continent several basic tenets are now generally applied. The Model is guided by seven pragmatic principles (Geist *et al.*, 2001), which have become colloquially referenced as the 'seven sisters for conservation' (Mahoney, 2004). Each of these is seminal to an array of policies, management prescriptions, legislative instructions and organisational diversity that have collectively supported and maintained open citizen access to a common property resource. The Model is not a product of a single time or intellect, and it has withstood many debates and trials; and continues to face challenges today. But understanding its structure and history offers the world one approach for conservation and sustainable use that has surely succeeded. Regulated hunting has been at the movement's heart since its inception.

Maintain wildlife as a public trust resource

Since Roman times the question of who owns wildlife has been the subject of legal debate, and while colonial North Americans considered individual access to wildlife a sacred right, the specific issue of who could own wild creatures was not addressed under law. This changed with a US Supreme Court decision in 1842 which clearly established wildlife as public property, held in trust for all citizens by the state.

The implications of this 'public trust doctrine' are many. Not only does it place great restriction on the privatisation of wildlife, it also engenders a system of paid professionals who manage, for the public good and in a coordinated manner, the wildlife resources of Canada and the US. Most importantly, however, it ensures that the public has a right to comment and actively work on behalf of the resource, as and when they choose. The wildlife is, after all, theirs. In North America, ownership of wildlife is not devolved to individuals, landowners, or occupiers of land.

Prohibit commerce in dead wildlife and its products

The scale of commercial exploitation throughout the 19th century was such that all North American wildlife of any economic value was threatened. Even after game laws were developed the financial incentives for marketing wildlife remained, the trade being vast and highly lucrative. Thus the decision to end commercial traffic in dead wildlife was neither easy to conceive nor simple to enforce. This radical assault on the free wheeling slaughter of wildlife came too late for species such as the passenger pigeon *Ectopistes migratorius*, heath hen *Tympanuchus cupido cupido* and Carolina parakeet *Conuropsis carolinensis*, but there can be no doubt that it pulled numerous others from the brink of extinction.

In terms of colonial history and long-held frontier attitudes towards wildlife, the suffocation of markets categorically changed the relationship between people and wildlife in the New World and engendered policies and laws that, until recent decades, have gone almost unchallenged. While its origins lay in the exhortations of concerned citizens, confident now that the wildlife indeed belonged to them, hunters and anglers were especially engaged, gaining their voice through publications and hunting club forums (Trefethen, 1975).

Allocate wildlife democratically and by law

Despite the drive to eliminate wildlife markets there was no intention to eliminate wildlife use. This was a profound subtlety which inspired calibration, but did not lead to preservationist approaches. The question was *how* to allocate wildlife and ensure its reasonable use. The answer lay in democratically structured legislation that would safeguard against the rise of elites and give every citizen equal say and access to the wildlife resource.

In combination with the public trust doctrine, the network of legislation that emerged to safeguard citizen rights was crucial to developing an ethic for conservation and sustainable use of wildlife in North America. North Americans have developed an extraordinary array of foundations, societies, clubs and conservancies to advocate for wildlife, conserve habitat and safeguard a myriad of wildlife experiences, including hunting and angling. Collectively these non-governmental organisations have raised billions of dollars in support of wildlife and sustainable use programmes. Some have been the mechanisms for wildlife reintroductions and recovery on a continental scale.

At the individual-citizen level the North American legal framework for wildlife use and access has inspired financial and political support that has no equal. It cannot be overemphasised that the equitability of wildlife allocation lies as the cornerstone of the entire North American Model. The citizenry, en masse, truly understand that wildlife is theirs, *de facto* and *de jure*. They act as though they own it.

Ensure wildlife use is for legitimate purpose

Although laws could govern access to wildlife there also had to be guidelines as to its appropriate use, especially given the origins of North American conservation. Eventually, the legitimate taking of animals came to mean killing for food and fur, self-defence and property protection.

Not surprisingly, the wastage of animals is directly addressed in North American legislation and policy, and is particularly stringent where hunting is involved. The image of thousands of bison carcasses rotting in the sun has cast a long shadow over the conscience of North America's citizenry. Today, perhaps no breach of ethics in hunting is more despised than the wanton killing of an animal and the wastage of its meat.

This principle of the North American Model has also eliminated other wasteful practices such as the killing of shore and wading birds for their feathers (to supply the haberdashery industry) and elk *Cervus elaphus*: in Europe, red deer) for their canine teeth (ivory for the jewellery market). It has led to the emphasis on harvesting wild animals as a source of high quality meat that is a characteristic, though not exclusive, motivation of hunting in North America.

Preserve hunting opportunity for all

In addition to the general application of law to wildlife use and disposition, the Model specifically addresses and heavily emphasises the issue of hunting. This is in no way surprising, as hunters and anglers were the leading spokesmen for conservation reform in the seminal late 19th century period. While the North American hunting movement borrowed from European experience in matters of 'fair chase' and the 'sport' terminology, its codification of every individual's access to and responsibility for wildlife was distinctly New World and revolutionary, certainly for its time. It has resulted in a relentless participation by hunters from all walks of life in wildlife management and conservation issues that not only affect their cherished activities, but the sustainability of wildlife for its own sake. The result is that hunters and anglers wield great muscle in the political arena of North America, and especially in the US. The North American Model has ensured that hunters are a force to be reckoned with, despite representing only about 6 per cent of the North American population.

Furthermore, hunting still enjoys widespread support from the North American public. The latest survey in the US, for example, shows that 77.6 per cent of citizens support legal hunting (Responsive Management, 2006), an actual increase of 4 percentage points from 1995.

Recognise and manage wildlife as an international resource

The special status of migratory wildlife was recognised early in North America. The international approaches required for its conservation and use, such as the 1911 Fur Seal Treaty and the famous and effective Migratory Bird Protection

Act of 1916, provided the legal frameworks to ensure consistency in Canadian and American approaches to the conservation of these species. Russia and Japan were also signatories to the Fur Seal Treaty.

The inclusion of wildlife in treaty law was one of the clearest signals that North American conservation had matured well beyond ensuring equal access to wildlife and the prescription of hunting privileges and practices. It signified that conservation of wild animals was the core ethic of the movement and that this responsibility was of national significance and a matter of national pride. This notion of transboundary responsibility and coordinated management of wildlife has become deeply ingrained in the North American Conservation Model, and is responsible for a dizzying array of ancillary policies, committees, colloquia, strategic approaches, and publications. This active networking helps ensure that governments at all levels remain cognizant of their treaty responsibilities.

The level of cooperation engendered under these agreements is also indicative of the sinewy strength of the Model. In recent years, under the North American Waterfowl Management Plan, millions of dollars derived from hunting activity in the US are actually diverted to conservation programmes for waterfowl enhancement in Canada. Thus provincial and state waterfowl managers not only share their expertise and talents, and coordinate their policies; funds are actually transferred between the Model's founding nations based upon where they are needed, regardless of where they are generated. Such arrangements are only possible where mutual trust and commitment are manifest and acknowledged. The long tenure of the North American Model has made such conditions secure.

Ensure science is the basis for conservation management and policy

During the 19th century interest in natural history and science flourished in North America, as it did in many European nations. Thus it was not surprising that early in the Model's formulation science was identified as crucial to safeguarding wildlife and ensuring against the overkill that was so evident at that time. The process was helped measurably by the focus given to this issue by President Theodore Roosevelt, an icon of North American conservation and a keen observer of wildlife throughout his life. Indeed the premise that science should form the basis of wildlife management is known as the Roosevelt Doctrine.

Although Roosevelt and others articulated this vision in the late 19th century, it was not until the 1930s and the writings of Aldo Leopold that this principle of the Model was aggressively exercised (Leopold, 1933). Quickly thereafter wildlife science became foundational to the North American Model, not only in theory but in practice as well. Its application is seen as one of the few safeguards against the politicisation of wildlife policy, and has generated an immense knowledge base upon which to evaluate best practices in conservation. Acquiring such knowledge has fallen to a diverse consortium state, provincial and federal officials, together with academics, all engaged in professional approaches to science, and doing so as their primary mandate.

Securing this knowledge has required massive financial commitment to wildlife research. Such funding is another success of the Model but its appropriation requires constant attention and encouragement. Furthermore, the marketability of scientific knowledge remains an ongoing challenge and seeking ways to ensure its application to decision-making processes often meets with obstacles and challenges. Nevertheless, wildlife science remains an integral and vibrant part of the North American Model.

Allocating and monitoring wildlife harvests in North America

The role of wildlife agencies

In North America provinces and states have primary responsibility and authority over the hunting of wildlife residing within their boundaries, with Federal Government responsibilities largely pertaining to international treaties and migratory species. In both Canada and the US Government wildlife agencies establish hunting zones and seasons, determine quotas and bag limits, prescribe rules for the age and qualifications of hunter applicants, administer hunting licence sales (usually with discriminatory prices for non-resident hunters), conduct wildlife research, enforce hunting laws and regulations, deliver various kinds of wildlife education programmes, and manage various kinds of wildlife refuges and sanctuaries where hunting may be prohibited or more tightly controlled.

In the US many wildlife agencies are responsible to a commission, board or council typically made up of governor-appointed members.

The commission has final authority to promulgate regulations, but in reality routinely ratifies most recommendations of the agency (Lueck, 2000). Similar structures do not exist in Canada, where final executive authority rests with the (appointed) deputy minister and (elected) Minister responsible for the provincial or territorial agency. In both countries agencies are staffed by professional wildlife managers and scientists, all of whom are Government employees.

Funding for wildlife management programmes

The funding for Government wildlife agencies in the US derives principally (approximately 65 per cent of all programme dollars) from hunters and anglers, either directly through within-state licence sales (35 per cent) or indirectly through federal taxes on hunting and fishing equipment (30 per cent). Additional funds are derived from a wide variety of sources including income tax check-off programmes, motorists who purchase wildlife licence plates, and even dedicated within-state taxes on commodities such as cigarettes. General tax revenues from the treasury are also provided to some state agencies, although the proportions vary considerably from state to state and approximately 21 agencies receive no funding at all from general tax sources (Lueck, 2000). While hunter-generated dollars are largely dispersed in programmes for game (hunted) species, hunter revenue also supports biodiversity programming in the broad sense.

In Canada the funding situation is quite different. In all provinces and territories funding to wildlife agencies is largely derived from the general treasury. Hunter licence fees are usually not directed towards wildlife management generally or hunting programmes specifically, but rolled into provincial coffers in combination with other Government revenues. Allocations to the wildlife agency are then determined in competition with the other demands on Government finances, such as transportation, health and education. In recent years exceptions to this general rule have emerged in a few Canadian jurisdictions.

The US agencies were in a funding situation comparable to Canada's, prior to 1934. In that year the head of the US bureau of Biological Survey, J.N. 'Ding' Darling, originated the Migratory Bird and Conservation Stamp (duck stamp), essentially a federal licence required annually by waterfowl hunters in that country. Since that time more than half a billion dollars derived from this stamp have been used to acquire and preserve millions of hectares of waterfowl habitat.

Only three years after the duck stamp initiative Carl Shoemaker, conservation director of the American National Wildlife Federation, drafted legislation (PittmanRobertson Act) authorising a tax on sporting arms and ammunition to assist states in wildlife management and restoration. The legislation stipulated that monies raised by state wildlife agencies from hunter licence fees could not be diverted to any other purpose but support of wildlife (principally game species), if Pittman–Robertson monies were to be made available to the state in question. In so doing, this legislation ensured some federal influence on American state wildlife programmes, and has certainly helped maintain wildlife agency loyalty to their hunter clientele. Nothing comparable to this legislation exists in Canada.

Quota setting techniques

Until early in the 20th century wildlife conservation in North America focused on game laws and protection, but by the 1920s it was clear that better information on the abundance of wildlife and the factors controlling animal (especially game) populations was required. Crucial in this regard were the investigations of Aldo Leopold (1928–1930) in the agricultural Midwest of the US, which strongly influenced the American Game Policy in 1930 and resulted in his book *Game Management* in 1933. Both these initiatives emphasised the need for improved biological information and for trained professionals to conduct game management programmes. Similar sentiments were developing cooperatively in Canada (Hewitt, 1921).

From these beginnings spread a continent-wide programme of wildlife research and monitoring that has massively increased our understanding of wildlife population dynamics, and greatly improved our abilities to manage both hunting efforts and wildlife abundance. Presently in North America virtually every state and provincial agency, supported by wildlife researchers at universities, establishes wildlife quotas utilising accepted scientific procedures appropriate to the species in question.

Wildlife quotas are usually established through population inventories using census information as well as hunter-trend statistics, and then allocations provided to resident and non-resident hunters based on some ratio that varies by jurisdiction. Depending upon the abundance of the species in question and the demand by hunters for licences, either an open access or draw system is employed to ensure fair opportunity for every qualified hunter to obtain a permit. This process is generally followed for all public lands in Canada and the US, with the major exceptions of national parks and certain wildlife refuges where hunting is prohibited. It is also generally true of much private land within registered hunting zones, except that access to private property is at the discretion of the landholder.

Challenges to the North American Model

Privatising wildlife ownership and developing wildlife markets

While wildlife is considered a public trust resource in North America, legally enshrined trespass rights result in *de facto* control (but not ownership) by landowners of wildlife on their property. This indirect authority is of major significance when we consider that more than 70 per cent of land in the US is privately owned (Butler *et al.*, 2005), as is a growing percentage in Canada. Obviously the attitudes of land owners will have major implications for the future of the North American Model and there is already growing demand to make wildlife a private resource, allowing landowners to manage it in accordance with market demands.

These social tendencies are encouraged by established industries such as agriculture and forestry, which have witnessed failures in traditional markets and see the 'natural' advantage of having a varied income from the lands they lease, own or manage. Quite predictably, some of the Government bureaucracies developed in support of these industries also view the private commercial use of wildlife as reasonable, and see its administrative guidance on such lands as an extension of their normal responsibilities. Seen from these perspectives agricultural or forest lands can produce many crops for the market, one of which is wild animals. The products derived from wildlife might vary from entertainment or wildlife-viewing to hides, meat and other body parts sold for consumption, to paid hunting.

Not surprisingly, the idea of managing wildlife for the two latter markets (especially) is highly controversial in North America, colliding as it does with some crucial underlying principles of the Model (Geist, 1988). Regarding fee hunting the principle concern is that hunting will become a more elitist activity with concomitant loss in broad participation and the social and political support this generates. There are also ethical concerns about the 'fair chase' standards that would apply to hunting animals on private lands and behind fences, an issue exacerbated by the increased probability of modifying wildlife behaviour under such circumstances (Butler *et al.*, 2005).

It is also felt that marketing pressures will encourage genetic manipulation of fenced wildlife to increase trophy hunting opportunities, distancing further this kind of hunting experience from the traditional hunting culture that has long supported wildlife recovery and management in North America. Meanwhile the creation of a commercial market for dead animal products is considered tantamount to a return of 19th century conditions that fostered wildlife's historical demise (Geist, 1989).

Despite the controversy, some means of landowner compensation is necessary, as tolerating wildlife on their property often has costs for owners. Furthermore, by rightfully denying hunter access, landholders can effect declines in hunting participation, something that would in itself surely undermine the North American Model as we know it. As a result, various mechanisms have developed to accommodate landowner demands and appeals from commercial interests.

Fee hunting and leasing of land for hunting are the most common forms of compensation and both have been ongoing for some time. In the mid-to-late 1980s Oregon, Colorado, California and Texas all introduced programmes whereby landowners could, with approval of the wildlife authority in their state, not only sell hunting permits and tags to hunters but also, by following state habitat management guidelines, increase harvest quotas on their land (Freese & Trauger, 2000). While programmes vary, at least seven states are now involved and corporations such as International Paper have also developed business models that include fee hunting and leasing programmes on their forest properties (Rasker *et al.*, 1992). Significant fee hunting also occurs on some American Indian lands (Czech, 1999).

Not all such initiatives are of recent origin. Leasing of private land for hunting access in Texas dates back to the 1920s (Teer & Forrest, 1968) and is a well established and accepted practice in that state. It is important to understand that, to date, these various programmes do not confer *de jure* ownership of wildlife to the land owner or corporation, but they come close to *de facto* privatisation, and as such are strongly opposed by supporters who fear the Model's public trust doctrine will be undermined (Ernst, 1987; Geist, 1989).

In addition to the fears that such programmes can disenfranchise the principle of full democratic access to hunting opportunity, the private ownership and market development for wildlife has raised additional concerns over exotic game species being introduced for hunting or game farming purposes. In Texas over seventy exotic ungulate species have been introduced to increase the fee hunting market, and at least six of these have established free-ranging populations (Freese & Trauger, 2000). While the biological perils of these exotics running free in the North American wild are largely unstudied, experience suggests that the consequences for native wildlife could include disease, genetic pollution and competition.

Regardless of these concerns, fee hunting and leasing of private lands for hunting purposes are supported by many who see market forces and secure private property rights as the future to wildlife habitat conservation in large areas of North America (Burger & Teer, 1981). This issue is complex and threatens to strike a potential fault-line between upholding the principles of a highly successful conservation programme (the North American Model) and preserving, in the face of social and economic change, the very wildlife the original programme was designed to protect.

Declining hunter numbers

Simultaneous with the pressures for wildlife privatisation is a second forceful challenge to the North American Model; namely, the decline in hunter numbers. Latest statistics indicate that hunter numbers continue to fall gradually and that currently in the US only 6 per cent of the population participates in this activity (Responsive Management, 2006). Numbers are similar in Canada. Many social, political and economic factors impinge on this trend, but the impact of a sharp decline in hunting activity on conservation programmes would be catastrophic for many state wildlife conservation programmes, some 65 per cent of which are directly funded by hunter expenditures.

The financial implications for conservation are not the only concern however. In both Canada and the US hunters represent one of the most consistently supportive and vocal constituencies in wildlife conservation efforts and a decrease in their numbers will seriously affect the social and political debates around such issues. While there is no question that a broad public constituency considers wildlife conservation important, there remains the crucial question of whether the non-hunting community could ever replace both the financial and political investments of wildlife's consumptive users in the long term. What is clear is that presently hunters barge a huge proportion of the conservation freight in North America, and declines in their numbers pose a direct and substantive challenge to conservation on this continent.

North American and southern African approaches compared

While land tenure and wildlife ownership are the focus of an escalating debate in North America, they have always been central to policies of wildlife use and conservation in southern Africa (Booth & Cumming, this volume). Thus, while it is true that responsibility for wildlife rests generally with the Government, landowners in southern Africa may exert considerable control over its use depending upon the land tenure system in place. In Zimbabwe, for example, landowners and rural district councils are largely free to utilise wildlife on their land, while in Botswana fewer ownership rights over wildlife have been devolved from Government to individuals or collectivities. Varying policies apply in Zambia and Malawi. In South Africa, landowners with certified enclosed lands have wide authority to commercially utilise wildlife, not just for hunting but also for sale of carcasses or captive breeding.

Southern African countries generally agree that where considerable private or communal access to wildlife occurs, wildlife is better served than where strong state controls are exercised. The latter circumstance serves, in their experience, to remove incentives to conserve the resource, an approach at odds with North American policy and experience. As in North America, southern African countries recognise the importance of hunting for wildlife conservation; however, they also support game ranching and farming, and the sale of meat and other products from wildlife culling operations, practices that in manner and extent differ appreciably from North American approaches.

One of the great challenges to southern African wildlife policy has been how to best establish quotas and ensure appropriate harvest controls for wildlife that is managed by private or communal landholders. In general, few of these operators are in a position to undertake rigorous scientific assessments of the wildlife they market, and best guesses are frequently used to establish off take. Where aerial surveys or statistical models are used, they are usually provided by state-operated agencies. In North America scientific application to wildlife harvest is extensive, highly developed, and in many cases, legislated as a prerequisite to harvest. The provision of such scientific expertise is considered a

Government responsibility, and vast monies are spent annually by Canada and the US in this regard.

Conclusion

One of the reasons conservation history is so complex is that it is born out of the political and social evolution of nations, while at the same time reflecting a deeply founded human fascination with nature. Conservation approaches, or models, must be viewed in this light. The vastly different histories of North American and southern African nations have inevitably led to mark-edly different conservation strategies, and neither should be viewed as better than or a suitable replacement for the other. Each has its advantages and its problems.

Nevertheless, there is great merit in the mutual exploration of both. Opportunities will stem from our realisation that while all nations are in a constant state of flux, providing for nature's conservation is humankind's ongoing, shared and daunting challenge. What has worked in the past to maintain our cherished hunting traditions, and those creatures which inspired them, may not work in the future. What we once set aside may need to be rediscovered, modified and reapplied. We have too little time to learn from just our own successes, or defeats. People and nature have always been in a race with themselves.

References

Burger, G.V. & Teer, J.G. (1981) Economic and socioeconomic issues influencing wild-life management on private land. In *Proceedings of Wildlife Management on Private Lands Symposium,* eds. R.T. Dumke, G.V. Burger & J.R. March, pp. 252–278. Wisconsin Chapter, The Wildlife Society, Madison.

Butler, M.J., Teaschner, A.P., Ballard, W.B. & McGee, B.K. (2005) Commentary: Wildlife ranching in North America – arguments, issues, and perspectives. *Wildlife Society Bulletin,* 33(1), 381–389.

Czech, B. (1999) Big game management on tribal lands. In *Ecology and Management of Large Mammals in North America,* eds. S. Demarais & P.R. Krausman, pp. 277–291. Prentice Hall, Upper Saddle River, New Jersey.

Ernst, J.P. (1987) *Privatization of Wildlife Resources: A Question of Public Access.* Paper presented at the 42nd annual meeting of the South Dakota Wildlife Federation, Pierre, South Dakota.

Freese, C.H. & Trauger, D.L. (2000) Wildlife markets and biodiversity conservation in North America. *Wildlife Society Bulletin*, 28(1), 42–51.

Geist, V. (1985) Game ranching: Threat to wildlife conservation in North America. *Wildlife Society Bulletin*, 13(4), 594–598.

Geist, V. (1988) How markets in wildlife meat and parts, and the sale of hunting privileges, jeopardize wildlife conservation. *Conservation Biology*, 2(1), 15–26.

Geist, V. (1989) Legal trafficking and paid hunting threaten conservation. *Transactions of the North American Wildlife and Natural Resources Conference*, 54, 171–178.

Geist, V., Mahoney, S.P. and Organ, J.F. (2001) Why hunting has defined the North American model of wildlife conservation. *Transactions of the 66th North American Wildlife and Natural Resources Conference.*

Hewitt, C.G. (1921) *The Conservation of the Wildlife of Canada.* Charles Scribner and Sons, New York.

Krech, Shepard III. (1999) *The Ecological Indian: Myth and History.* W.W. Norton & Company, New York.

Leopold, A. (1933) *Game Management.* University of Wisconsin Press, Madison.

Lott, D.F. (2002) *American Bison: A Natural History.* University of California Press, Berkeley.

Lueck, D.L. (2000) An economic guide to state wildlife management. *PERC Research Study*, RS 00–2.

Mahoney, S.P. (2004) The seven sisters: pillars of the North American wildlife conservation model. *Bugle Magazine*, September/October Issue.

Martin, P.S. (1967) Prehistoric overkill. In *Pleistocene Extinctions: The Search for a Cause*, eds. P.S. Martin & H.E. Wright Jr., pp. 75–120. Yale University Press, New Haven.

Rasker, R., Martin, M.W. & Johnson, R.L. (1992) Economics: theory versus practice in wildlife management. *Conservation Biology*, 6(3), 338–349.

Responsive Management/National Shooting Sports Foundation (2008) The future of hunting and the shooting sports: Research based recruitment and retention strategies. Produced for the United States Fish and Wildlife Service under Grant Agreement CT-M-6-0. Harrisonberg, VA, 216pp.

Teer, J.G. & Forrest, N.K. (1968) Bionomic and ethical implications of commercial game harvest programs. *Transactions of the North American Wildlife and Natural Resources Conference*, 33, 192–202.

Trefethen, J.B. (1975) *An American Crusade for Wildlife.* Winchester Press, New York.

Wilson, J. (1999) *The Earth Shall Weep.* Grove Press, New York.

The Development of a Recreational Hunting Industry and its Relationship with Conservation in Southern Africa

Vernon R. Booth[1] and David H.M. Cumming[2]

[1]Freelance Wildlife Management Consultant, Harare, Zimbabwe
[2]Percy FitzPatrick Institute, University of Cape Town,
Cape Town, South Africa

Introduction

Modern humans originated in Africa about 200,000 years ago. However, the interaction between predatory hominids, such as *Homo erectus*, and large mammals in Africa extends back some two million years or more. There is ample evidence of stone-age cultures existing across southern Africa for tens of thousands of years, followed by iron-age cultures and livestock herding in the region during the last 2,000 years. The rock art of the region extends back to the end of the last ice-age some 12,000 years ago and clearly depicts the hunting of elephant (*Loxodonta africana*) and lesser quarry. There is some evidence of ivory trade with the east coast from Mapungubwe in the Limpopo valley in about 1,100 AD (Campbell, 1990) and earlier evidence of ivory trade from the east coast of Africa to the Arabian peninsula and India, as part of the age-old dhow trade (Martin & Martin, 1978). Clearly, hunting in Africa has had a long history with the result that the extent of large mammal extinctions in recent times (i.e. the last 12,000 years) were less marked than they were in

Recreational Hunting, Conservation and Rural Livelihoods: Science and Practice, 1st edition.
Edited by B. Dickson, J. Hutton and W.M. Adams. © 2009 Blackwell Publishing,
ISBN 978-1-4051-6785-7 (pb) and 978-1-4051-9142-5 (hb).

North and South America where large mammals were exposed, for the first time, to human predation (Lyons *et al.*, 2004).

European exploration, influence and penetration of the sub-continent began with the rounding of the Cape of Good Hope by Vasco da Gama in 1497 and the establishment of Portuguese trading posts along the East Coast of Africa. The Dutch and British East India Companies established bases in the Cape during the 17th century, which led eventually to the establishment of the Cape Colony under British rule, followed by the emergence of the Boer republics of Natal, Orange Free State and the Transvaal by 1840 (Pakenham, 1991). European exploration of the southern African interior was closely associated with hunting large game, particularly elephant for sport and trade, and the sale of ivory served to finance many of these early expeditions (MacKenzie, 1988). Exploration later expanded into commercial hunting and trading expeditions for ivory and hides to support growing export markets. In addition, large numbers of local hunters were enlisted and supplied with firearms, resulting in the rapid depletion of wildlife across the sub-continent and the near extinction of elephant south of the Zambezi. Animal diseases introduced with imported livestock took a further toll and the rinderpest pandemic of the 1890s decimated wildlife and livestock alike across the region. However, before the advent of firearms and European colonisation, hunting for food and for trade in meat and artefacts (e.g. those made from hides, horns and tusks) was an integral part of indigenous cultures throughout the region. Hunting techniques included large ceremonial drives of game into pitfall traps or nets, the use of poisoned arrows, spears and traps, and endurance hunting by the San to run down large antelope such as eland (*Taurotragus oryx*) and kudu (*Tragelahpus strepsiceros*).

The European colonies in southern Africa all became independent nation states during the second half of the 20th century.[1] The development of recreational hunting legislation and practice was thus strongly influenced by colonial history and differed accordingly across the region (Spinage, 1991). Additional important influences included: (a) the collapse of wildlife populations in the late 19th century; (b) emerging conservation ideals that resulted in protective legislation and the formation of game reserves; (c) colonial legislation governing land tenure and access rights to wildlife; (d) meat markets and beef production; (e) rapid rural human population growth and agriculture; (f) visions of meeting Africa's protein needs by farming and cropping wildlife; and (g) a growing demand after World War II for recreational hunting in Africa.

This chapter outlines the dominant influences on the development of wildlife management policy, including hunting, in southern Africa, before describing the major features of present day recreational hunting in the region which largely revolves around high-fee paying visitors and is locally known as sport, trophy or most commonly, safari hunting. In conclusion we consider the major underlying reasons why sport hunting in southern Africa has evolved along different lines from that in North America.

Dominant early influences on recreational hunting policy and practice

Protecting the game

Although hunting was widely practiced and vigorously pursued by indigenous pre-colonial societies it was also bound by cultural norms and custom (Mackenzie, 1988). These were completely ignored by colonial administrators and the first attempts to regulate hunting were established in the Cape in 1657 by van Reibeeck (Brynard, 1977 in Child, 2004; Spinage, 1991). Hunting regulations were introduced in the colonies and Boer Republics during the 19th century but they had little influence on the continuing decline of wildlife populations. The early part of the 20th century saw the establishment of several major game reserves such as Hluhlue-Umfolosi in 1895, Sabi in 1898 (later to become part of the Kruger National Park in 1926), Etosha in 1907, Gorongosa in 1921, Wankie (now Hwange) in 1928, and Addo Elephant National Park in 1931. It also saw the introduction and enforcement of stricter controls on firearms and the selective issuance of hunting licenses. The laws effectively barred indigenous Africans from hunting legally and their access to firearms was greatly curtailed and strictly regulated.

Land tenure, legislation and resource access rights

Colonisation in southern Africa resulted in the new rulers effectively laying claim to all land which was subsequently either alienated to freehold title, retained under trust for traditional communities (Tribal Trust Lands), or allocated to various state land categories such as leasehold land, forest reserves, nature reserves and national parks. The end result was the creation of dual agricultural systems with large-scale commercial agriculture on freehold land held by white minorities, and small-scale agro-pastoralism on communal land occupied by the black majority. Hunting was controlled on all land by the state through the establishment of hunting seasons and the issue of hunting licenses for various categories of game. Once in possession of a hunting license the hunter still required the permission of the land owner or landholder to hunt. Recreational hunting was almost exclusively conducted by resident nationals who combined the hunting experience with a predilection for game meat. The skin, horns or other animal parts were seldom kept by hunters as trophies, and generally recreational hunters in southern Africa adopted the 'fair chase' ethical code and abided by the regulations governing hunting in their respective countries. Hunting was then very inexpensive.

In its essentials the sport hunting system that persisted until the 1950s in most of southern Africa was dominated by European influences and was limited to the white minority. Major elements of the prevailing system were: (a) wildlife was owned and controlled by the state on public and private land, (b) hunting licenses of various categories were issued by the governments but the landholders' permission was required to enter and hunt on all categories of land, (c) hunting seasons were provided for in the legislation, (d) access to sporting weapons was strictly controlled over most of the region and effectively limited to whites. The majority of the population lived in designated 'communal lands' and was deprived of the right to hunt or to use traditional hunting methods. For example, in 1911 the administration in Southern Rhodesia outlawed annual traditional hunting ceremonies that used game drives and nets. Access to firearms was also strictly controlled so those living in communal lands were barred from using legal hunting methods.

Beef industry, fences and disease control

The first half of the 20th century was characterised by a major expansion of settler-farmers onto newly claimed colonial state land, with major investments in cattle ranching to provide beef to the Witwatersrand gold mines in South Africa and to meet growing demand for beef from Europe (Phimister, 1978; Milton, 1997). This resulted in the depletion of wildlife on commercial farms where game animals were seen as competitors for scarce grazing resources. This second onslaught on recovering wildlife populations was exacerbated by veterinary regulations predicated on the assumption that wildlife was a major source of livestock diseases. The view that one could not 'farm in a zoo' persisted through the 1960s. A prime example of conflicting interests between wildlife conservation and the commercial livestock industry was the use of game elimination as a tsetse control measure in Zimbabwe where it accounted for the extermination of more than a million animals between 1920 and the mid-1970s (Child & Riney, 1987). Game elimination programmes to control tsetse were also implemented in Zululand (South Africa), Zambia, Botswana and Mozambique (Nash, 1969).

The control of Foot and Mouth Disease in Botswana, Namibia, South Africa and Zimbabwe led to the erection of fences across vast areas to control movement of game and cattle and had major impacts on wildlife populations (Taylor & Martin, 1987; Williamson *et al.*, 1988). Fencing became mandatory in South Africa and Namibia for wildlife farms and ranches. Conservancies in Zimbabwe that reintroduced buffalo (*Syncerus caffer*) could only do so once a double perimeter game fence had been installed. The overall impact of fencing in much of the region was to reduce wildlife movements, isolate populations genetically, and, in some instances, markedly reduce sustainable densities as in the case of wildebeest (*Connochaetes taurinus*) in Timbavati Game Reserve (Ben-Shahar, 1993).

Game farming and cropping schemes

Increasing investment and development in Africa following World War II gave rise to growing interest in the conservation and sustainable use of large mammals to meet the protein needs of rapidly expanding human populations in Africa. Publication of 'The Wild Riches of Africa' (Huxley, 1961) and a number of other initiatives in the 1960s greatly influenced the development of game ranching and sustainable use of wildlife, particularly on private land, in much of southern Africa (Child, 2004). However, game cropping schemes mostly failed, for both technical and economic reasons (e.g. Parker, 1984), and gave way to more manageable and profitable safari hunting enterprises (e.g. Johnstone, 1975; Cumming, 1989).

The development of the safari hunting industry

East African influence

Bartell Bull (1992) captures the history of the 'safari industry' from its inception in the 1600s when the first European travellers arrived in the Cape, to the sophisticated commercial operations of the 1980s. In the early stages 'sport' hunting was undertaken by individuals who trekked across southern Africa, often collecting and describing the large mammals that they encountered. By 1900 the day of the lonely eccentric who lived for the lifestyle and earned a living from hunting elephant for ivory was over. A new breed of hunter began to emerge. These were well-heeled people, essentially on sporting holidays, who needed local expertise to plan and lead their expeditions to the field (Bull, 1992). Perhaps the most famous of these is the Teddy Roosevelt safari of 1910 to East Africa, in which Frederick Selous and Edward Buxton were used as principal advisors (Adams, 2004).

These epic safaris marked the beginning of the commercialisation of recreational hunting in East Africa. The years 1908–1914 were a time when 'princes, peers and American magnates' poured across in a continuous stream to hunt all over East Africa. Pioneer farmers in East Africa supplemented their income by organising and conducting these 'safaris', which in turn gave birth to the professional 'white hunter'. After World War I professional safari hunting companies or 'outfitters' emerged to conduct well-organised and equipped safaris within Kenya, Uganda and Tanzania.

After World War II safari hunting became more organised, with Kenya leading the way. A system of recreational hunting blocks was identified across the country, and laws and regulations governing the conduct of hunting safaris were developed. License fees and hunting permits were introduced to restrict hunting operations to designated areas. Professional hunters were registered after undergoing an apprenticeship and being approved by the strict rules that governed the East African Professional Hunters Association. This Association was held in high regard and later became the benchmark for the development of the sport hunting industry in Southern Africa in the 1970s.

The growth of safari hunting in southern Africa

The growing demand in southern Africa for hunting after World War II from both resident and international recreational hunters resulted in several countries establishing designated hunting areas – usually in lightly or unsettled areas that still held reasonable game populations. In Mozambique these were called Coutadas, in Zambia Game Management Areas, in Botswana Wildlife Management Areas. In Zimbabwe they were first known as Controlled Hunting Areas but they were later incorporated into the Parks & Wildlife Estate as Safari Areas. Between 1961 and 1973 the number of hunters using the limited number of Zimbabwe's Hunting Areas grew from 98 to 285 and by the end of this period 47 per cent were non-residents (Cumming, 1989). The demand for hunting camps exceeded supply and hunts were initially drawn on lottery but later auctioned, which remains the current practice to allocate Safari Areas to 'outfitters' and to sell individual hunts to hunters in the Zambezi Valley.

With the closure of all hunting in Kenya in 1974 and the demise of the classic East African Safari hosted by White Hunters (Dyer, 1979), the demand shifted to southern Africa and particularly to Botswana and Zimbabwe where the 'Big Five' were available on state land and where political and economic conditions were favourable. At much the same time there was a convergence of think-ing about the management of wildlife outside of protected areas in Namibia, South Africa and Zimbabwe which led to the development of innovative leg-islation that devolved responsibility for wildlife management to landowners. The legislation also allowed landowners to benefit financially from wildlife on their properties, with the result that wildlife-based tourism and the area of land devoted to wildlife conservation and sustainable use expanded rapidly on freehold land, and is still expanding rapidly in South Africa and Namibia. In Zimbabwe both the wildlife industry and wildlife populations expanded on private land through to the late 1990s, but declined following the onset of the land reform programme in 2000 (Bond & Cumming, 2006).

The impact of these changes is well illustrated by examining the wildlife sector in Zimbabwe during the 1960s and 1970s. Following the introduction of the Wild Life Conservation Act of 1960 landholders were allowed greater and greater discretion over the management of wildlife on their land, and could market its products with increasing freedom. The result was that the rapid decline in wildlife on private land over the previous 30 years was halted and even reversed (Bond & Cumming, 2006). With the new institutional framework in place wildlife producers were, in the 1970s, able to diversify into wildlife production to ameliorate the impact of declining terms of trade for ubiquitous agricultural commodities such as beef and cereals. Farmers who had previously complained that they could not 'farm in a zoo' now accepted that they could only farm sustainably and profitably in agriculturally marginal areas with low rainfall by switching to wildlife ranching, alone or in combination with livestock.

The lessons learned during the initial institutional reform process were consolidated and extended by the Parks and Wild Life Act of 1975. This Act made provision for National Parks and five other classes of protected areas; it also devolved responsibility for wildlife management to landowners/occupiers of alienated land. Landholders were allowed considerable freedom in the management and marketing of wildlife and its derivatives from their land. Discriminatory implicit taxation, such as through the imposition of government hunting license fees, unreasonable requirements for government surveys to determine quotas and permits, and other high transaction costs, were effectively abolished. Instead, landholders were encouraged to maximise the benefits earned by the resource within the limits of the land use policy for each property, an approach that was quickly adopted by the State itself. Finally, during the 1980s the legal provisions of the 1975 Act were extended to Rural District Councils, and thus partially to rural communities in communal lands, to manage and benefit from wildlife resources through the emerging Communal Areas Management Programme for Indigenous Resources (CAMPFIRE).

Zimbabwe's CAMPFIRE provided an early example of a model that was rapidly to be adopted through most of southern Africa. Programmes of Community Based Natural Resource Management (CBNRM), most of which were based on income from commercial safari hunting (Bond, 2001), rapidly developed in Botswana, Namibia and Zambia (Hulme & Murphree, 2001). Encompassing more than 75,000 km², Namibia's development of 'conservancies' within communal lands has been particularly successful in conservation terms with populations of many large mammal species within them having made spectacular recoveries (Jones, this volume; Jones & Murphree, 2001; Weaver & Skyer, 2005).

The key principles underpinning the legislative changes seen across much of the region towards the close of the 20th century were (a) that wildlife conservation would thrive where individuals had secure access rights to the land and the wildlife resources on it, and (b) the benefits from wildlife management should be consistent with the effort invested in managing the land and its wildlife resources. Excessive and inappropriate government controls and regulations that weakened or subverted these principles were considered likely to act as a disincentive to conservation, causing land owners and communities to adopt alternative land use practices and resource management strategies that would reduce, if not eliminate, wildlife populations.

Comparison with the North American experience

In southern Africa during the first half of the 20th century wildlife management, including recreational hunting, was based largely on European legislative frameworks. Other than in some well-policed protected areas, these policies and approaches failed to stem an ongoing decline in both wild species and the wild areas they lived in. Several factors contributed to the demise of both wild land and wildlife in southern Africa during this period. The more important of these were:

Rapidly expanding human rural populations: In most southern African countries more than 60 per cent of the population lives in rural areas compared with less than 10 per cent in North America and Europe. Overall, human populations in southern Africa expanded 20-fold during the 20th century. The majority of rural people are poor, small-scale, subsistence farmers who lack the resources, land, or inclination to support populations of large game animals on land under their care.

Competing land uses: Large-scale commercial ranches under free-hold title engaged in cattle ranching and large mammals were seen as competitors for grazing, as reservoirs of disease or, in the case of predators, as being responsible for direct losses of valuable stock. There were no compelling incentives, other than perhaps an individual farmer's personal whim, to maintain populations of wild animals that effectively belonged to the state.

Protectionist policies: Protected areas on state land were framed around the 1933 International Convention for the Protection of the Fauna and Flora of Africa under which the extraction or use of plants and animals from protected areas was prohibited, except for scientific purposes. The result severely limited the area of state land on which citizens could hunt. This situation contrasted markedly with that in North America where vast areas of state forest and grazing lands were available and managed for recreational hunting.

Inability to enforce: Within the state communal lands of southern Africa where the use of game species by local people was legally prohibited the state was unable to provide effective enforcement. The result was that informal and illegal subsistence hunting continued, and in some places it developed into a large-scale bushmeat trade that threatened the viability of many species that are sensitive to over-hunting (Jachmann, 1998; Barnett, 2000).

Disparities in wealth: There are major differences in wealth and disposable incomes between Southern African and North American citizens. Most people living in southern Africa are poor and live close to, or below, the 'poverty line' – a factor that greatly influences the potential viability, development and adoption of developed-world models for sport hunting.

Few recreational hunters and a tiny hunting economy: The resident population of recreational hunters was small and almost entirely confined to the minority colonial population. Their political influence was consequently small compared with that of the cattle ranching and agricultural industry. Revenues generated from hunting were small and were not re-invested in managing the wildlife resource.

Resource-poor wildlife management and supporting research agencies: In southern Africa no biologists were employed in agencies responsible for wild-life management and conservation in the region until the 1950s and early 1960s, depending on the country. The result was an almost complete lack of scientifically based wildlife management strategies, monitoring of wildlife population trends or levels of hunting. Even today the investment in wildlife research outside of South Africa is extremely low with a density of generally less than one biologist to 3,000 km^2 of wildlife land.

Biogeographical differences: The diversity of huntable large mammals in southern Africa greatly exceeds that of North America, and includes many less abundant and specialised antelope (e.g. roan (*Hippotragus equinus*), sable (*Hippotragus niger*), and tsessebe (*Damaliscus lunatus*) that, while attractive to hunters and capable of sustaining an harvest, are easily over-exploited. Given the limited resources of government agencies for survey and monitoring, not to mention law enforcement, a large-scale public hunting system would be difficult, if not impossible, to regulate under African conditions.

The cornerstone of the new policies was the devolution of secure resource access rights and the responsibility for the management and conservation of wildlife resources to the legal occupier of freehold or leasehold land, and the devolution of hunting rights from the state to local authorities on the huge tracts of state land that were set aside for traditional, communal small-holder farmers. While wildlife remained *res nullius* (i.e. belonged to no one until captured or confined) the new laws gave landowners and local authorities secure rights to use and benefit from 'their' wildlife. Safeguards, in the form of provision for hunting bans, were included in the legislation to cope with cases where the privileges and responsibilities associated with the new systems were abused. The result of the new legislation in the three countries (Zimbabwe, Botswana, Namibia) was a rapid proliferation of tourism based on recreational hunting and the creation of a vibrant commercial safari hunting industry. The changes in policy and law also resulted in marked increases in the numbers and distribution of a range of large game animals in the countries concerned. Importantly these changes were driven largely by the private sector, with little cost to government or the tax-payer.

Conclusions

It is hard to argue that a single southern African model for recreational hunting has evolved – the region is biogeographically, culturally and politically diverse – but despite its diversity the legislation, policy and practice governing recreational hunting across southern African countries seems to be converging on devolved, partially privatised models of high-fee paying safari hunting that generate revenues for those that live with wild animals, thus creating conservation incentives on state, communal and private lands alike. Generally speaking the principal drivers of this trend have been, firstly, the realisation by under-resourced governments that this is likely to result in both better conservation outcomes and greater financial returns (to landowners and the state itself) than earlier centrally controlled systems. Secondly, neither governments, nor the vast majority of the population in the region, have the resources to administer on the one hand, or the wherewithal to participate on the other hand, in the sort of government controlled recreational hunting models that have developed in North America. Thirdly, and perhaps equally important, is the consideration that rural communities are prepared to forgo individual hunting rights and opportunities in return for high financial returns from commercial sport hunting. There are also major differences in governance systems and economies that militate against successful government regulated sport hunting systems in southern Africa (see Child & Wall, Chapter 20).

A potential downside to current and emerging developments in conservation and sustainable use is that hunting may increasingly be limited to the wealthy and to international, high fee paying, tourists. Opportunities for affordable sport hunting by local residents, although still available in some countries, are becoming increasingly scarce, which contrasts markedly with the existing North American model. Furthermore, as Hulme & Murphree (2001) have noted,

> 'For generations, conservation policy in Africa has been socially illegitimate in the eyes of the continent's rural people. Community conservation has created opportunities for conservation to begin to develop a local constituency but the task of creating a conservation policy that is embedded in African society rather than imposed from above will be the work of generations'.

The currently predominant model of commercial recreational hunting in southern Africa is little more than 30 years old. Further evolution and change in the policy and practice of recreational hunting and conservation in the sub-region is clearly inevitable and the alternative directions in which it might plausibly develop need to be explored and evaluated.

Note

1 The following name changes occurred in southern African states on attaining independence: Bechuanaland became Botswana, Nyasaland became Malawi, Northern Rhodesia became Zambia and Southern Rhodesia became Zimbabwe and South West Africa became Namibia. Names of the remaining states of Angola, Mozambique and South Africa did not change.

References

Adams, W.A. (2004) *Against Extinction: The Story of Conservation*. Earthscan, London.

Barnett, R. (ed.) (2000) *Food for Thought: the Utilization of Wild Meat in Eastern and Southern Africa*. TRAFFIC East/Southern Africa, Nairobi.

Ben-Shahar, R. (1993) Does fencing reduce the carrying capacity for populations of large herbivores? *Journal of Tropical Ecology*, 9, 249–253.

Bond, I. (2001) Campfire and the incentives for institutional change. In *African Wildlife and Livelihoods: the Promise and Performance of Community Conservation*, eds. D. Hulme & M. Murphree, pp. 226–243. Weaver Press, Harare and James Currey, Oxford.

Bond, I. & Cumming, D.H.M. (2006) Wildlife research and development. In *Zimbabwe's Agricultural Revolution Revisited*, eds. M. Rukuni, P. Tawonezi, C. Eicher, M. Munyuki-Hungwe & P. Matondi, pp. 465–496. University of Zimbabwe Publications, Harare.

Brynard, A.M. (1977) Die Nasionale Parke van die Republiek van Suid-Afrika. *Koedoe Supplement*, 24–37.

Bull, B. (1992) *Safari: A Chronicle of Adventure*. Penguin Books, Harmondsworth, UK.

Campbell, A.C. (1990) History of elephants in Botswana. In *The Future of Botswana's Elephants*, ed. P. Hankock, pp. 5–15. Kalahari Conservation Society, Gaborone.

Child, G. (2004) Growth of modern nature conservation in southern Africa. In *Parks in Transition: Biodiversity, Rural Development and the Bottom Line*, ed. B. Child, pp. 7–27. Earthscan, London.

Child, G.F.T. & Riney, T. (1987) Tsetse control hunting in Zimbabwe, 1919–1958. *Zambezia*, 14, 11–71.

Cumming, D.H.M. (1989) Commercial and safari hunting in Zimbabwe. In *Wildlife Production Systems: Economic Utilisation of Wild Ungulates,* eds. R.J. Hudson, K.R. Drew & L.M. Baskin, pp. 147–169. Cambridge University Press, Cambridge.

Dyer, A. (1979) *The East African Hunters: The History of the East African Professional Hunters' Association.* Amwell Press, Clinton, New Jersey.

Hulme, D. & Murphree, M. (ed.) (2001) *African Wildlife and Livelihoods: the Promise and Performance of Community Conservation.* Weaver Press, Harare and James Currey, Oxford.

Huxley, J. (1961) The Wild Riches of Africa. Published by the *Observer* for the Fauna Preservation Society, London.

Jachmann, H. (1998*) Monitoring Illegal Wildlife Use and Law Enforcement in African Savanna Rangelands.* Wildlife Resource Monitoring Unit, ECZ, Lusaka.

Johnstone, P.A. (1975) Evaluation of a Rhodesian game ranch. *Journal of the Southern African Wildlife Management Association*, 5, 43–51.

Jones, B. & Murphree, M. (2001) The evolution of policy on community conservation in Namibia and Zimbabwe. In *African Wildlife and Livelihoods: the Promise and Performance of Community Conservation*, eds. D. Hulme & M. Murphree, pp. 38–58. Weaver Press, Harare and James Currey, Oxford.

Lyons, S.K., Smith, F.A. & Brown, J.H. (2004) Of mice, mastodons and men: human mediated extinctions on four continents. *Evolutionary Ecology Research*, 6, 339–358.

MacKenzie, J.M. (1988) *The Empire of Nature: Hunting, Conservation and British Imperialism.* Manchester University Press, Manchester.

Martin, E.B. & Martin, C.P. (1978) *Cargoes of the East: the Ports, Trade and Cultures of the Arabian States and Western Indian Ocean.* Elm Tree Books, London.

Milton, S. (1997) The Transvaal beef frontier: environment, market and the ideology of development, 1902–1942. In *Ecology and Empire: Environmental History of Settler Societies,* eds T. Griffiths & L. Robbin, pp. 199–212. University of Natal Press, Pietermaritzburg.

Nash, T.A.M. (1969) *Africa's Bane: The Tsetse Fly.* Collins, London.

Pakenham, T. (1991) *The Scramble for Africa,* Reprinted by Abacus, 1995. Little, Brown and Co., London.

Parker, I.S.C. (1984) Perspectives on wildlife cropping or culling. In *Conservation and Wildlife Management in Africa,* eds. R.H.V. Bell & E. McShane-Caluzi, pp. 233–254. Office of Training & Program Support, Forestry & Natural Resources Sector, U.S. Peace Corps, Washington, DC.

Phimister, I.R. (1978) Meat and monopolies: beef cattle in southern Rhodesia, 1890–1938. *Journal of African History*, 29, 391–414.

Spinage, C. (1991) *History and evolution of the fauna conservation laws of Botswana.* Occasional Paper No. 3. The Botswana Society, Gaborone.

Taylor, R.D. & Martin, R.B. (1987) Effects of veterinary fences on wildlife conservation in Zimbabwe. *Environmental Management*, 11, 327–334.

Weaver, L.C. & Skyer, P. (2005) Conservancies: integrating wildlife land-use options into the livelihood, development and conservation strategies of Namibian Communities. In *Conservation and Development Interventions at the Wildlife/ Livestock Interface: Implications for Wildlife, Livestock and Human Health*, eds. S.A. Osofsky, S. Cleveland, W.B. Karesh *et al.*, pp. 89–104. IUCN, Gland and Cambridge.

Williamson, D., Williamson, J. & Ngwamotsoko, K.T. (1988) Wildebeest migration in the Kalahari. *African Journal of Ecology*, 26, 269–280.

18

The Influence of Corruption on the Conduct of Recreational Hunting

Nigel Leader-Williams[1], Rolf D. Baldus[2] and R.J. Smith[1]

[1]Durrell Institute of Conservation and Ecology,
University of Kent, Canterbury, UK
[2]Tropical Game Commission, International Council for
Game and Wildlife Management

Introduction

Set in the more general context of how corruption can affect both wildlife conservation and wider societal interests, this chapter discusses how corruption might allegedly influence the conduct of recreational hunting. First, the chapter seeks a brief understanding of the influence of corruption on wider society. Second, it outlines what is known of the ways that corruption may influence conservation outcomes. Third, it reviews alleged cases of corruption in recreational hunting from around the world. Fourth, it discusses a detailed case of the conduct of recreational hunting in one country, and the measures needed for its reform. Finally, it discusses some key challenges for proponents of recreational hunting who wish to reform its governance practices. Our overview suggests that various developing, transformation or developed countries could have equally served as case studies, but our focus on Tanzania reflects 18 years of combined experience for two authors as senior advisors within the Government of Tanzania's Wildlife Department.

Recreational Hunting, Conservation and Rural Livelihoods: Science and Practice, 1st edition.
Edited by B. Dickson, J. Hutton and B. Adams. © 2009 Blackwell Publishing,
ISBN 978-1-4051-6785-7 (pb) and 978-1-4051-9142-5 (hb).

Corruption and its role in wider society

Corruption has a long history in the conduct of human affairs (Azfar *et al.*, 2001). As the old adage runs, 'everyone has their price'. Modern definitions of corruption range from general *Oxford English Dictionary* formulations like 'rendering morally unsound' and 'acting dishonestly or unfaithfully', to the much more specific 'unlawful use of public office for private gain' (Transparency International, 2007). Transparency International's definition accepts that only holders of public office practise corruption, and this chapter follows that definition. Nevertheless, instigating or agreeing to corrupt deals equally involves members of the public or the private sector, who act dishonestly by offering bribes or seeking an inequitable distribution of public services to their personal advantage (Transparency International, 2007).

Corruption can manifest itself among wider society in many ways, including: embezzling public funds; demanding bribes to overlook illegal activities; and, offering patronage, nepotism and political influence (Kaufmann, 1997). These forms of corrupt practice have consequences for wider society, by adding to transaction costs, impacting on investor and donor confidence, and limiting economic growth and productivity (Azfar *et al.*, 2001). Corruption is especially prevalent in countries with weak institutions or transitional governments (Barrett *et al.*, 2001). While the impact of corruption on social and economic development is widely recognised (Bardhan, 1997), conservation scientists have only recently begun to analyse its impacts (Smith, Muir *et al.*, 2003; Smith & Walpole, 2005; Wright *et al.*, 2007).

Corruption and its impacts on conservation outcomes

How might corruption influence conservation outcomes? Many areas of high conservation priority occur in developing countries (Smith, Muir *et al.*, 2003). Where corruption limits inward investment to, and suppresses development in, biodiversity-rich countries, it may actually have a positive outcome for biodiversity (Laurence, 2004). In contrast, biodiversity-rich areas may be threatened if corruption impacts on the effectiveness of conservation: for example by reducing the availability of funds, encouraging poor law enforcement, reducing political support for conservation, and/or increasing incentives to over-exploit resources. Consequently, corruption may influence conservation outcomes in complex ways (Barrett *et al.*, 2006), as illustrated by case studies of: corruptly managed commercial logging on traditionally owned, but untitled, land in Indonesia (Smith, Obidzinski *et al.*, 2003); the widespread failure to hand down appropriate sentences for illegal hunting of black rhinoceros *Diceros bicornis* and African elephant *Loxodonta africana* populations (Leader-Williams *et al.*, 1990); and the embezzlement of tourism revenue by local district councils in Masai Mara (Thompson & Homewood, 2002).

The first broader-scale study of the possible impacts of corruption on wider conservation outcomes was only published in 2003. Using national governance scores derived independently by Transparency International, multiple regression analyses sought to separate the possible influence of various socioeconomic parameters, including governance, on conservation outcomes. Poor governance scores, defined as being inversely related to levels of corruption nationally, were the most strongly correlated with the loss of black rhinos and elephants across a range of African countries (Smith, Muir *et al.*, 2003). Nevertheless, caution is needed when using such broad-scale data sets (Smith & Walpole, 2005; Barrett *et al.*, 2006), and further analyses are needed to increase understanding of how corruption might influence conservation outcomes (Ferraro, 2005).

Corruption and its impacts on recreational hunting

The conduct of recreational hunting is often linked with corrupt practices, particularly in poor countries that attract foreign tourist hunters willing to spend large sums of foreign exchange to hunt prime trophies. In turn, tourist hunting can attract outfitters who seek to circumvent legal controls over biological, ethical and financial aspects of the hunting industry through: exceeding or misusing quotas; poor hunting practices; and flouting of foreign exchange regulations. One solution to such management shortcomings is to ban tourist hunting. Tourists have hunted in at least 23 sub-Saharan African countries, but among the prime destinations, hunting has been banned in Kenya from 1977 to the present, in Uganda from 1967 until 2001 when hunting restarted on a trial basis, and in Tanzania from 1973 to 1977 (Price Waterhouse, 1996; Barnett & Patterson, 2006; Lamprey & Mugisha, this volume).

Despite assumed links between corruption, the conduct of recreational hunting and the consequent loss opportunities to hunt, no systematic study, to our knowledge, has previously examined the possible impacts of such linkages. Consequently, we undertook a web-based search using the two terms *corruption* and *hunting*, which identified several relevant web pages that featured alleged links between corrupt practices and the conduct of recreational hunting in different countries. While none was verified independently, these cases illustrate the alleged involvement of different levels of public office holder, ranging from field staff to senior public servants and political leaders (Table 18.1). However, these cases cannot generalise any formal relationships between corruption and the conduct of recreational hunting. In this sense, recreational hunting differs little from other aspects of conservation, where the influence of corruption remains poorly documented or researched. Hence more detailed case studies are needed, so we next consider Tanzania, a prime destination for classic African hunting safaris since the early days of the East African Protectorate (Hurt & Ravn, 2000), and a source of concern over the alleged influence of corruption on the conduct of its tourist hunting industry since the early 1970s (Planning and Assessment for Wildlife Management (PAWM) 1996a).

Governance and tourist hunting in Tanzania

This section draws on official and other reports, published literature and frequent allegations of practices parallel to those alleged for recreational hunting elsewhere (Table 18.1), that may also occur in Tanzania (see Table 18.2).

Governance in Tanzania

Tanzania gained independence from British colonial rule in 1964, and soon adopted a policy of socialist self-reliance. Since independence, Tanzania has suffered from a very low GDP, low per-capita incomes of ~US$200 per annum, a well-developed parallel economy and poor governance (Maliyamkono & Bagachawa, 1990; Transparency International, 2007). Indeed, the Commission Report on the State of Corruption in the Country (United Republic of Tanzania (URT), 1996) noted that corruption had grown since the 1970s through a combination of circumstances, including the poor economy, low salaries, lack of essential commodities, and restrictions on public servants earning extra income. When the economy was liberalised during the 1990s, the Commission noted the emergence of endemic and systemic corruption through a combination of factors, including businessmen developing close

Table 18.1 Web pages that feature alleged links between corrupt practices and the conduct of recreational hunting in different countries.

Issue (1), level (2) and country (3)	Alleged incident
(1) Misappropriation of fines for illegal hunting (2) Field officer (3) USA	State conservation officer in Kentucky allegedly directed two convicted poachers of deer and turkey to send fines to his postbox, and took further payments totaling US$9500 from poachers' mother.[*]
(1) Inappropriate involvement of officials in hunting endangered species (2) 'High level' government official (3) Vietnam	Alleged hunting of endangered gaur *Bos gaurus* by high-ranking government official and lawyer with whom he had close ties.[†]
(1) Inappropriate involvement of officials in offering hunting opportunities (2) 'High level' government official (3) Russia	Case of graft brought against head of Altai region's hunting control inspectorate, when discovered he also allegedly ran agency organising hunting trips.[‡]
(1) Misallocation of concessions (2) Director of National Parks and Wildlife Management (3) Zimbabwe	Charge of corruptly granting hunting rights to outfitter who had not paid required bid before beginning to unlawfully hunt, but charge fell because State failed to prove *prima facie* case against Director.[§]
(1) Lost opportunities for local communities (2) Members of Parliament (3) Croatia	Alleged attempt by Croatian Hunting Association to hijack share of concession fees formerly due to landowners by proposed changes to law on hunting, to vary distribution of funds collected through hunting ground concessions, and to reduce area covered by hunting grounds.[¶]
(1) Inequitable allocation of hunting opportunities to rich foreigners (2) Senior politicians (3) Pakistan	High level political support allegedly offered to hunting parties from neighbouring Arab countries to hunt large numbers of houbara bustards *Chlamydotis undulata*, although the species continues to decline.[**]

[*]www.biggamehunt.net/forums/viewtopic.php?t=13854&sid=d%E2%680%A6; [†]www.cpj.org/protests/03trs/Vietnam30apr03pl.html; [‡]www.guardian.co.uk/russia/article/0,1941743,00.html; [§]www.zimconservation.com/archives7–76.htm; [¶]see.oneworldsee.org/article/view/111868/1/; [**] www.american.edu/TED/pakistan-hunting.htm

ties with leaders, erosion of integrity among leaders, and lack of transparency in the economy. Consequently, the Commission noted two groups of corrupt officials in Tanzania (URT, 1996):

- those accepting bribes to supplement meagre incomes and make ends meet: this type of corruption is rampant in all economic sectors, including natural resources and tourism, for example, where wildlife officers take bribes to free poachers;
- those high-level officials, public servants and elected representatives whose earnings, property and savings portfolios exceed basic needs: this type of corruption feeds greed among the leadership, comprising elected politicians and chief executives in the public sector. Again, it includes the natural resources and tourism sectors, where, for example, interference occurs in executive decisions to allocate hunting blocks.

History of tourist hunting in Tanzania

The British colonial administration encouraged the development of safari hunting, while simultaneously restricting opportunities for traditional African hunting (Leader-Williams, 2000). Just before independence, Tanzania introduced Africa's first system to lease hunting blocks to outfitters in a network of game reserves that then covered 8 per cent of Tanzania's surface area (PAWM, 1996a). After independence, several game reserves including Selous were declared 'national projects', enabling retention of hunting revenues to fund wildlife management and infrastructural development, which soon proved successful when commercial poaching ceased and wildlife populations thrived (Nicholson, 1970, 2001).

Once the ban on tourist hunting was lifted in 1978, the parastatal Tanzania Wildlife Corporation (TAWICO) was given a monopoly on managing hunting blocks, but in practice sublet many blocks to private, and mainly European, outfitters. Because of ongoing corruption, TAWICO's monopoly was officially relaxed in 1984 when nine private outfitters were allocated hunting blocks for periods of up to four years (PAWM, 1996a; Nshala, 2001).

The Wildlife Department took over management of the increasingly lucrative tourist hunting industry in 1988, while TAWICO continued to offer hunting opportunities like other outfitters. To increase hunting opportunities, the Wildlife Department close to doubled the numbers of hunting blocks to c.130.

These covered ~180,000 km^2 or ~25 per cent of Tanzania's land surface, and were evenly distributed between unoccupied game reserves and areas occupied by people, thereby offering tourist hunting the opportunity to contribute to community-based conservation (PAWM, 1996a; Nshala, 2001).

Constraints in the management of tourist hunting

With assistance from USAID, the Government of Tanzania established the Planning and Assessment for Wildlife Management (PAWM) project in 1990. PAWM was mandated *inter alia* to advise the Director of Wildlife on constraints faced by the tourist hunting industry and to propose policies to maximise its potential. PAWM ran workshops to discuss tourist hunting in 1993 and community-based conservation in 1994 (Leader-Williams *et al.*, 1996a, 1996b). Here we use PAWMs work, together with work funded through German development assistance, to compare these constraints with alleged cases of corruption in recreational hunting elsewhere (Table 18.1).

Three broad levels of public office holder, ranging from field staff to senior public servants and political leaders, may be responsible for imposing constraints on the management of tourist hunting in Tanzania (cf. URT, 1996). In terms of level of public office holder, alleged infringements of hunting regulations in the field (Table 18.2) are most understandable (cf. URT, 1996), given that wildlife scouts earn an annual salary of <US$500, while a tourist hunter may pay US$100,000 for a three-week safari. Of much greater concern are constraints arising from decisions of mid- to senior-level staff in the Wildlife Department's headquarters who set quotas, issue licences, collect fees and allocate most hunting blocks (Table 18.2). Their decisions potentially impact many aspects of hunting, from ensuring its biological sustainability to maximising its financial returns and subsequent reinvestment in the resource, and to involving local communities fully in hunting conducted outside game reserves. Of equal concern is patronage and nepotism involving senior politicians, while perhaps the least tractable is the most senior politicians offering exclusive rights to foreigners presumably considered to be of financial or other strategic importance (Table 18.2).

In terms of issue, greatest concern perhaps centres on the manner of allocating hunting blocks (Table 18.2), as noted by the Commission Report on the State of Corruption in the Country in its evidence on the growth of systemic corruption in Tanzania (URT, 1996). Because outfitters do not compete to

Table 18.2   Constraints to effective management of tourist hunting in Tanzania, showing the key issue, the level of public office holder involved and the management concern each constraint raises.

Issue (1), level (2) and management concern (3)	Constraints to effective management	Source
(1) Infringements of hunting regulations (2) Wildlife scouts (3) Ethical and biological	Use of baits too close to national park boundaries; hunting from a vehicle; hunting at night with a spotlight; hunting within 500 m of a watering hole or at a salt lick; overshooting quotas; shooting second animal with better trophy; killing females.	Leader-Williams *et al.* (1996a); Spong *et al.* (2000)
(1) Managing quotas and issuing licences (2) Wildlife Department HQ staff (3) Biological	Quotas mainly set on basis of informed guesswork, with little reference to adaptive management; when blocks subdivided, each remains with previous quota; hunting licences issued to outfitters and filed and checked manually, making effective management of quotas difficult.	Severre (1996); Planning and Assessment for Wildlife Management (PAWM) (1996b); Nshala (2001); Baldus & Cauldwell (2004)
(1) Retention of increasingly complex range of pay-as-use fees by revolving fund (2) Wildlife Department HQ staff (3) Financial	Tanzania Wildlife Protection Fund (TWPF) established in 1974 to assist Wildlife Department manage wildlife throughout Tanzania; initially, TWPF retained 25 per cent of game fees; since 1989 TWPF retained 100 per cent of observer, conservation, permit and trophy hunting fees paid by tourist hunters; custom-made, computerised hunting management system not implemented as computer-generated receipts said to be not in line with Government regulations; Wildlife Department unable to properly account for fees worth millions of dollars each year.	PAWM (1996a, 1996b); Baldus & Cauldwell (2004)

(Continued)

Table 18.2   Continued.

Issue (1), level (2) and management concern (3)	Constraints to effective management	Source
(1) Lack of competition in allocating hunting blocks and fixed right-to-use fees (2) Minister, Director of Wildlife (3) Financial	Tanzania's non-competitive system of allocating hunting blocks produces low returns; despite introduction of fixed block fees in 1993, returns in Tanzania are c.10 times lower than the competitive US$100,000 to 250,000 right-to-use fees paid by outfitters in other southern African hunting destinations; WD and TAHOA, the professional hunters' association, block further competition.	Price Waterhouse (1996); Saiwana (1996); Nshala (2001); Barnett & Patterson, (2006)
(1) Abuse of Presidential Licence (2) Minister, Director of Wildlife (3) Financial	Presidential licences regularly issued to friends of the Minister or Director, or to former holders of these posts; does not require any licence fees to be paid for animals killed.	Nshala (2001)
(1) Patronage and nepotism in allocating hunting blocks (2) Senior politicians, Minister, Director of Wildlife	Many concessions allegedly leased to smaller national companies silently owned by senior public officials with political links in Ministry or Wildlife Department; most national safari outfitters acquiring	Nshala (2001)

(3) Financial and biological	concessions through patronage lack capacity to effectively market hunting opportunities, nor employ qualified professional hunters; concessions allegedly sub-leased, mostly to non-resident foreigners, and income generated does not enter Tanzania; revenue authorities cannot assess funds due for taxation; foreign professional hunters who sub-lease blocks only take a short-term view, and maximise off takes during their lease.	
(1) High-level political decisions to offer foreigners exclusive hunting opportunities (2) Former President (3) Ethical, biological and financial	Administration of former President granted hunting block in Loliondo to group from United Arab Emirates for 10 years, twice as long as previous maximum length for block concessions in Tanzania; lease agreed allegedly. without consulting Director of Wildlife or local communities living in Loliondo; concession re-allocated to same group without competition in 2002.	Nshala (2001)

pay right-to-use fees, some outfitters have leased hunting blocks at well below market prices for long periods, in turn leading to massive losses of revenue for the Government over the years. Furthermore, these losses are not compensated by an increasingly complex system of pay-as-use fees levied on the hunter (Baker, 1997). Nevertheless, senior staff in the Wildlife Department who allocate the blocks have continued to defend the system, together with those outfitters who benefit from it (Nshala, 2001). Meanwhile, the Presidential decision to grant a group from the United Arab Emirates exclusive hunting rights in Loliondo may have some parallels with the decision to offer bustard hunting opportunities in Pakistan to neighbouring Arab countries (Table 18.1).

Impacts of alleged corruption on tourist hunting

The impacts of alleged corruption on the outcome of tourist hunting will be complex (Table 18.2), as with other aspects of conservation (Smith & Walpole, 2005). For example, the biological impact of not enforcing hunting regulations might reduce the sustainability of tourist hunting, such as if male lions *Panthera leo* are drawn out of national parks by baiting too close to their boundary (Loveridge *et al.*, 2006) or if female leopards *Panthera pardus* are shot (Spong *et al.*, 2000). However, outfitters who retain the same hunting blocks, by whatever means, probably take a long-term view over husbanding hunting opportunities in their blocks. Thus a study of lions in an area of Selous Game Reserve retained by the same outfitter since 1967 suggested that existing quotas were too high, but that actual off takes were much lower than those allowed by the quota and appeared sustainable (Creel & Creel, 1997). Therefore, we here seek to generalise some of the biological and financial impacts of the alleged role of corruption for tourist hunting, first for an unoccupied game reserve and, second, for people living inside hunting areas such as game controlled areas, or whose lands border game reserves.

Conservation impacts in the Selous Game Reserve

The 50,000 km^2 Selous Game Reserve (SGR) is the best known hunting area in Tanzania. Currently a World Heritage Site and an IUCN Category IV protected area, SGR has long been supported through revenue from tourist hunting (Nicholson, 1970, 2001). The management of SGR collapsed following the hunting ban in the mid-1970s, and by the late 1980s its annual operating budget from the Treasury was US$150,000, equivalent to US$ 3 per km^2. Meanwhile, commercial poaching for ivory, rhino horn and meat reduced SGR's elephant population from 110,000 in the mid-1970s to less than 30,000 in 1989, while black rhinos were reduced from over 3000 to probably less than 100 (Siege, 2000). Wildlife staff were allegedly instrumental in the poaching, acting on orders from their superiors or from politicians.

Tanzania requested German development assistance in the mid-1980s to restore the SGR. The Selous Conservation Programme (SCP) began at the end of 1987, and sought to achieve two long-term goals: first, to reintroduce management systems and rehabilitate the infrastructure of SGR; and second, to encourage community-based natural resources management in the 15,000 km^2 of buffer zones around SGR. To secure these goals long-term once German funding ended, SCP sought a sustainable income for SGR, and succeeded in difficult negotiations to retain 50 per cent of the Treasury's share of game fees to use directly for SGR management (PAWM, 1996a; Baldus, 2003). The funding allocated by the Treasury to SGR increased sixfold to US$900,000 when the retention scheme started in 1994, and had further increased to US$2.8 million, equivalent to US$ 60 per km^2, by 2003. Tourist hunting provided around 90 per cent of all SGR's retained revenue, while more than 100 photographic tourists were needed to achieve the returns derived from a single tourist hunter (Planning and Assessment for Wildlife Management (PAWM), 1996d; Baldus *et al.*, 2003). These increases in retained revenue allowed SGR to top up staff salaries and introduce proper allowances for game scouts, and to improve management and infrastructure. In turn, elephant numbers have since recovered to around 70,000 and continue to increase (URT, 2007), while most other wildlife populations including black rhinos have also begun to recover. Thus, the changes in fortune of SGR appear closely tied to the potential of retained revenue from tourist hunting to develop a sustainable funding base. Unfortunately, once the SCP ended in 2003, the Government of Tanzania has not since fully honoured the retention scheme agreed in 1994, and disbursements have been much less than the previously agreed 50 per cent share of fees due to the Treasury (Baldus, 2006).

Impacts on benefit-sharing with local communities

Various measures have been officially agreed by the Government of Tanzania to allow local communities to share benefits from hunting carried out either on their own land, or in protected areas bordering their land. The then Prime

Minister issued a directive in 1992 stating that district councils on whose land hunting took place should receive 25 per cent of the Treasury share of game fees, in order to compensate local people for the costs of living with wildlife (PAWM, 1996a; Nshala, 2001). If appropriately and equitably shared, this would have equated to significant sums even in the early 1990s (Planning and Assessment for Wildlife Management (PAWM), 1996c). Nevertheless, it remains unclear if this revenue was always distributed to district councils, let alone shared by district councils with local communities (Nelson *et al.*, 2007), as with revenue accruing to district councils from tourism in the Masai Mara, Kenya (Thompson & Homewood, 2002).

In addition to direct sharing of revenue, workshops on tourist hunting and community-based conservation held in the 1990s agreed that local communities should control resource use on their land in buffer areas outside more strictly protected areas (Leader-Williams *et al.*, 1996a, 1996b). Indeed, the resulting 'Policy for Wildlife Conservation and Utilisation' subsequently endorsed the establishment of Wildlife Management Areas (WMA) under community control (Ministry of Tourism, Natural Resources and Environment (MTNRE), 1998). Furthermore, WMAs were also mandated by official policies in other sectors, including the Poverty Reduction Strategy of Tanzania (Baldus *et al.*, 2004). These policies recognised that the best-suited land use option to generate funds for WMAs in many parts of Tanzania would be tourist hunting (PAWM, 1996c, 1996d), but the development of WMAs has been seriously delayed (Nelson *et al.*, 2007). There is still no clear schedule for sharing financial benefits from tourist hunting with local communities, nor an agreed way for local communities to decide which outfitter hunts on their land or to agree quotas for such hunting (Nelson *et al.*, 2007). Officially, it has been noted that reforms such as establishing WMAs should not be rushed, but the reform process has been ongoing for over 15 years, leading to speculation that opportunities for private gain by senior public officials and officers of TAHOA underlie such delays (Nshala, 2001; Baldus *et al.*, 2004). Indeed, the recent Wildlife Conservation (Non-Consumptive Wildlife Utilization) Regulations 2007 have concentrated all management powers and revenues centrally, instead of devolving such powers to, and sharing benefits with, local communities.

In the meantime, pending implementation of WMAs, outfitters are required both to contribute towards wildlife protection and to support local communities. However, these requirements are only vaguely specified and cannot be effectively evaluated (Nshala, 2001). Hence only a few outfitters voluntarily support communities through schemes such as the Cullman Wildlife

Project (Robin Hurt Safaris, 1996), while most hesitate over community empowerment, feeling greater security in perpetuating the state-controlled monopoly over wildlife, compared with facing the unknowns of democratically elected village committees (Nelson *et al.*, 2007). Outfitters also fear the advent of competition from other operators if communities are empowered to develop a market-based approach to concession lease fees in WMAs (Baldus & Cauldwell, 2004). Therefore, it is also widely speculated that the delay in implementing WMAs has in part arisen from high-level influence by leading outfitters through TAHOA, the professional association for outfitters in Tanzania (Nshala, 2001).

A special case where decisions have allegedly been taken out of the hands of local communities by even higher authority is Loliondo Game Controlled Area (Table 18.2), with its prime position within the migration of ungulates through the Serengeti-Mara ecosystem. Reports have alleged breaches of hunting regulations and human rights abuses against local Masai,[1] while a long airstrip has been built and allows military aircraft to fly in hunting parties with their vehicles and equipment, and to fly out trophies, meat and live animals, apparently with no reference to local communities.[2] The Government of Tanzania defends its decision to allocate the block in this way, citing the block fees of US$300,000 and payment in full for the quota.[3]

For all these various reasons, local communities are increasingly frustrated that promised benefits from tourist hunting, and the promised reform of wildlife policy to establish WMAs, have not so far been forthcoming. Such frustration may in turn encourage local communities to revert to poaching or habitat conversion, or to convert hunting blocks into photographic tourism areas (Nelson *et al.*, 2007), unless reforms are forthcoming.

The need for reform

Tanzania could reform its tourist hunting industry by implementing the *Policy and Management Plan for Tourist Hunting* (Table 18.3) developed following workshops held in 1993 and 1994 (Leader-Williams *et al.*, 1996a, 1996b). This policy was accepted by the then Director of Wildlife, but was never subsequently signed by the Minister, nor implemented. Political will to reform the tourist hunting industry has since been lacking (Nelson *et al.*, 2007), despite empirical studies showing the benefits that could accrue to the national exchequer, to the

Table 18.3   **Key aims and objectives of Tanzania's** *Policy and Management Plan for Tourist Hunting* **(Ministry of Tourism, Natural Resources and Environment (MTNRE), 1995).**

Overall objective: *to develop a tourist hunting industry that makes significant contributions to conservation objectives and to Tanzania's economy*

Issue	Aim
Hunting blocks	to allocate hunting blocks through market-based competition that does not compromise existing high standards of many outfitters nor prejudice long-term returns from hunting.
Fee structure	to adopt simple fee structure that combines right-to-use concession fee paid by the outfitter in return for a long-term lease of that block, and a pay-as-you-use trophy fee per animal shot.
Quotas	to set and monitor sustainable quotas that promote trophy quality on a scientific basis.
Codes of conduct and examinations	to adopt codes of conduct for outfitters and examinations for professional hunters.
Management of GRs	to reinvest funds from tourist hunting to better manage GRs that serve as core areas for the industry.
Establishment of WMAs	to encourage local communities to become principal decision-makers for allocating concessions and setting quotas for hunting on their land, from which they receive and manage funds so generated.
Wildlife legislation	to update, and where necessary amend, wildlife legislation related to the conduct of hunting, and to hunting in the context of community-based conservation.

management of individual game reserves and to local communities (PAWM, 1996c, 1996d; Baldus & Cauldwell, 2004). Since 2006, reform of the tourist hunting industry has again become an issue for the Government of Tanzania. A technical committee prepared a draft proposal to reform the administration of hunting (Ministry of Natural Resources and Tourism (MNRT), 2006), the discussion of which has involved public accusations of institutionalized corruption that has lead to massive losses of revenue for Tanzania's wildlife sector.

Future challenges for recreational hunting globally?

The struggle for good governance is a universal truth. In terms of conservation, corruption is neither restricted to recreational hunting, nor to Tanzania. However, debates over the role of corruption in conservation (Smith, Muir *et al.*, 2003) remain heavily contested (Barrett *et al.*, 2006). While further research is needed, the negative consequences of endemic and systemic corruption are well enough understood to initiate some actions. However, reform of alleged corruption in recreational hunting (Tables 18.1 and 18.2) will prove easier to articulate than to implement, as in Tanzania. Indeed, senior officials and elected politicians will resist changes to the status quo because of the wealth they accrue from current practices in recreational hunting. There appear parallels with 'the curse of oil', where oil wealth in poorly governed countries helps entrench powerful elites (Shaxson, 2007). Given their power, reform of corrupt practices is unlikely to come from public officials and elected politicians within countries with poor governance. Therefore, what avenues are open to reform the governance of recreational hunting?

First, many expect the international donor and NGO community to follow a policy of no-tolerance and encourage appropriate and stepwise reforms in conservation, as has occurred in other sectors (Smith, Muir *et al.*, 2003). Currently, many donors and NGOs bemoan institutionalised and systemic corruption, but continue to provide unconditional budgetary support. However, persuading donors and NGOs not to fund projects in stable countries with diversifying and growing economies, may be naïve. If donors stand off from funding projects in favoured countries such as Tanzania and Mozambique (Hanlon, 2004), then whom should they support?

Second, hunters could deflect opposition to recreational hunting and adopt a consumer-based policy of no-tolerance. While hunting bans are widely advocated, they may remove incentives to retain land under wildlife management, whether in formally protected areas or in areas occupied by people outside more strictly protected areas (Child, 1995). Recreational hunting can be a powerful tool to finance conservation and to generate income for rural communities through nature-friendly ecotourism, even when sub-optimally managed in countries like Tanzania. Therefore, the hunting industry itself should develop principles and guidelines for improving the sustainability of recreational hunting, and increase pressure on countries, wildlife administrations and hunting industries that perform against such principles (Baldus *et al.*, 2008).

Finally, local communities in many countries like Tanzania feel betrayed that benefits promised from recreational hunting have failed to materialise. Ironically, Tanzania's recent policies have articulated devolving responsibility for wildlife management to local communities (MTNRE, 1998). In practice, the opposite has occurred, and its benefits have instead been centralised into the hands of elites (Nelson *et al.*, 2007). Consequently, local people and civil society should be encouraged to press their democratically elected representatives for appropriate reforms. For recreational hunting, the critical link lies between land rights and wildlife management (Nelson *et al.*, 2007), which local community-based organisations are particularly well suited to articulate.

All these suggested avenues to reform the governance of recreational hunting will prove difficult, and none should be considered in isolation. Successful reform of endemic and systemic corruption, though likely to be slow, will lie in judiciously combining approaches in a stepwise and probably situation-specific fashion. Hence, lessons learned in one situation should be made available elsewhere (Smith & Walpole, 2005). Without such reforms, proponents of recreational hunting will continue to be challenged when extolling its benefits by increasingly well organised opposition.

Acknowledgments

We thank Ludwig Siege, Andrew Cauldwell and in particular Gerhard Damm for valuable advice. However, they do not bear any responsibility for the article.

Notes

1 http://www.awionline.org/pubs/Quarterly/su02/loliondo.htm
2 http://www.ntz.info/gen/n01526.html
3 http://www.tanzania.go.tz/wildlife2503eng.htm

References

Azfar, O., Lee, Y. & Swamy, A. (2001) The causes and consequences of corruption. *Annals of the American Academy of Political and Social Science*, 573, 42–56.

Baker, J.E. (1997) Development of a model system for touristic hunting revenue collection and allocation. *Tourism Management*, 18, 273–286.

Baldus, R.D. (2003) African wildlife: must it be subsidized? *Internationales Afrika Forum*, 39, 387–392.

Baldus, R.D. (2006) The crucial role of governance in ecosystem management: results and conclusions of the Selous Conservation Programme. In *Managing Africa's Natural Ecosystems: Report of the Information Sharing and Best Practices Workshop*, ed. TANAPA, FZS & BMU, pp. 85–92. Serengeti National Park, Tanzania.

Baldus, R.D. & Cauldwell, A.E. (2004) *Tourist Hunting and Its Role in Development of Wildlife Management Areas of Tanzania*. Paper to Sixth International Game Ranching Symposium, Paris, 6–9 July 2004.

Baldus, R.D., Damm, G.R. & Wollscheid, K. (2008) *Best Practices in Sustainable Hunting*. CIC Technical Series Publication No. 1, Budakeszi, 3–4.

Baldus, R.D., Kaggi, G.Th & Ngoti, P.M. (2004) Community based conservation (CBC): where are we now? Where are we going? *Kakakuona*, 35, 20–22.

Baldus, R.D., Kibonde, B. & Siege, L. (2003) Seeking conservation partnerships in the Selous Game Reserve, Tanzania. *Parks*, 13, 50–61.

Bardhan, P. (1997) Corruption and development: a review of issues. *Journal of Economic Literature*, 35, 1320–1346.

Barnett, R. & Patterson, C. (2006) *Sport Hunting in the Southern African Development Community (SADC) Region: an Overview*. TRAFFIC East/Southern Africa, Johannesburg.

Barrett, C.B., Brandon, K., Gibson, C. & Gjertsen, H. (2001) Conserving tropical biodiversity amid weak institutions. *Bioscience*, 51, 497–502.

Barrett, C.B., Gibson, C.C., Hoffman, B. & McCubbins, M.D. (2006) The complex links between governance and biodiversity. *Conservation Biology*, 20, 1358–1366.

Child, G. (1995) *Wildlife and People: the Zimbabwean Success*. Wisdom, Harare and New York.

Creel, S. & Creel, N.M. (1997) Lion density and population structure in the Selous Game Reserve: evaluation of hunting quotas and offtake. *African Journal of Ecology*, 35, 83–93.

Ferraro, P. (2005) Conservation and corruption: the need for empirical analyses. *Oryx*, 39, 257–259.

Hanlon, J. (2004) Do donors promote corruption? The case of Mozambique. *Third World Quarterly*, 25, 747–763.

Hurt, R. & Ravn, P. (2000) Hunting and its benefits: an overview of hunting in Africa with special reference to Tanzania. In *Conservation of Wildlife by Sustainable Use*, eds. H.H.T. Prins, J.G. Grootenhuis & T.T. Dolan, pp. 295–313. Kluwer Academic Publishers, Boston.

Kaufmann, D. (1997) Corruption: the facts. *Foreign Policy*, 107, 114–131.

Laurence, W.F. (2004) The perils of payoff: corruption as a threat to global biodiversity. *Trends in Ecology and Evolution*, 19, 399–401.

Leader-Williams, N. (2000) The effects of a century of policy and legal change upon wildlife conservation and utilisation in Tanzania. In *Conservation of Wildlife by Sustainable Use*, eds. H.H.T. Prins, J.G. Grootenhuis & T.T. Dolan, pp. 219–245. Kluwer Academic Publishers, Boston.

Leader-Williams, N., Albon, S.D. & Berry, P.S.M. (1990) Illegal exploitation of black rhinoceros and elephant populations: patterns of decline, law enforcement and patrol effort in the Luangwa Valley, Zambia. *Journal of Applied Ecology*, 27, 1055–1087.

Leader-Williams, N., Kayera, J.A. & Overton, G.L. (eds.) (1996a) *Tourist Hunting in Tanzania*. IUCN – The World Conservation Union, Gland and Cambridge.

Leader-Williams, N., Kayera, J.A. & Overton, G.L. (eds.) (1996b) *Community-based Conservation in Tanzania*. IUCN – The World Conservation Union, Gland and Cambridge.

Loveridge, A.J., Searle, A.W., Murindagomo, F. & Macdonald, D.W. (2006) The impact of sport-hunting on the population dynamics of an African lion population in a protected area. *Biological Conservation*, 134, 548–558.

Maliyamkono, T.L. & Bagachawa, M.S.D. (1990) *The Second Economy in Tanzania*. James Currey, London.

Ministry of Tourism, Natural Resources and Environment (MTNRE) (1995) *Policy and Management Plan for Tourist Hunting*. Dar es Salaam.

Ministry of Tourism, Natural Resources and Environment (MTNRE) (1998) *The Wildlife Policy of Tanzania*. Dar es Salaam.

Ministry of Natural Resources and Tourism (MNRT) (2006) *Taarifa ya Tasnia ya Uwindajiwa Kitalii* [*Report on Tourism Hunting Reform*]. Mimeo, Dar es Salaam.

Nelson, F., Nshala, R. & Rodgers, W.A. (2007) The evolution and reform of Tanzanian wildlife management. *Conservation and Society*, 5, 232–261.

Nicholson, B. (1970) *The Selous Game Reserve: Recommendations for a Ten-year Development Programme*. Report to the Director of Natural Resources, Dar es Salaam, Tanzania. Typescript, 24 pp.

Nicholson, B. (2001) *The Last of Old Africa*. Long Beach, California.

Nshala, R. (2001) *Granting Hunting Blocks in Tanzania: the Need for Reform*. Policy Brief No 5, Lawyers' Environmental Action Team, Dar es Salaam.

Planning and Assessment for Wildlife Management (PAWM) (1996a) The structure of Tanzania's tourist hunting industry. In *Tourist Hunting in Tanzania*, eds. N. Leader-Williams, J.A. Kayera & G.L. Overton, pp. 23–37. IUCN – The World Conservation Union, Gland & Cambridge.

Planning and Assessment for Wildlife Management (PAWM) (1996b) Returns from tourist hunting in Tanzania. In *Tourist Hunting in Tanzania,* eds. N. Leader-Williams, J.A. Kayera & G.L. Overton, pp. 71–80. IUCN – The World Conservation Union, Gland & Cambridge.

Planning and Assessment for Wildlife Management (PAWM) (1996c) Potential benefits from tourist hunting available for local communities in Tanzania. In *Tourist Hunting in Tanzania,* eds. N. Leader-Williams, J.A. Kayera & G.L. Overton, pp. 97–101. IUCN – The World Conservation Union, Gland & Cambridge.

Planning and Assessment for Wildlife Management (PAWM) (1996d) Options for community-based conservation in Tanzania, with special reference to possible benefits and village title. In *Community-based Conservation in Tanzania,* eds. N. Leader-Williams, J.A. Kayera & G.L. Overton, pp. 169–194. IUCN – The World Conservation Union, Gland & Cambridge.

Price Waterhouse (1996) The trophy hunting industry: an African perspective. In *Tourist Hunting in Tanzania,* eds. N. Leader-Williams, J.A. Kayera & G.L. Overton, pp. 12–13. IUCN – The World Conservation Union, Gland & Cambridge.

Robin Hurt Safaris (1996) The Cullman Wildlife Project. In *Community-based Conservation in Tanzania,* eds. N. Leader-Williams, J.A. Kayera & G.L. Overton, pp. 97–107. IUCN – The World Conservation Union, Gland & Cambridge.

Saiwana, L. (1996) Granting of safari hunting rights in Game Management Areas in Zambia. In *Tourist Hunting in Tanzania,* eds. N. Leader-Williams, J.A. Kayera & G.L. Overton, pp. 50–52. IUCN – The World Conservation Union, Gland & Cambridge.

Severre, E.L.M. (1996) Setting quotas for tourist hunting in Tanzania. In *Tourist Hunting in Tanzania,* eds. N. Leader-Williams, J.A. Kayera & G.L. Overton, pp. 57–58. IUCN – The World Conservation Union, Gland & Cambridge.

Shaxson, N. (2007) Oil, corruption and the resource curse. *International Affairs,* 83, 1123–1140.

Siege, L. (2000) *From Decline to Recovery: the Elephants of the Selous.* Tanzania Wildlife Discussion Paper No. 27, Dar es Salaam.

Smith, J., Obidzinski, K., Subarudi & Suramenggala, I. (2003) Illegal logging, collusive corruption and fragmented governments in Kalimantan, Indonesia. *International Forestry Review,* 5, 293–302.

Smith, R.J. & Walpole, M.J. (2005) Should conservationists pay more attention to corruption? *Oryx* 39, 251–256.

Smith, R.J., Muir, R.D.J., Walpole, M.J., Balmford, A.P. & Leader-Williams, N. (2003) Governance and the loss of biodiversity. *Nature,* 426, 67–70.

Spong, G., Hellborg, L. & Creel, S. (2000) Sex ratio of leopards taken in trophy hunting: genetic data from Tanzania. *Conservation Genetics,* 1, 169–171.

Thompson, M. & Homewood, K.M. (2002) Entrepreneurs, elites and exclusion in Maasailand: trends in wildlife conservation and pastoralist development. *Human Ecology*, 30, 107–138.

Transparency International (2007) *Transparency International's Corruption Perceptions Index 200**. http://www.transparency.org.

United Republic of Tanzania (URT) (1996) *Commission Report on the State of Corruption in the Country*. PUBD.121996.001 COR. Dar es Salaam.

United Republic of Tanzania (URT) (2007) *Proposal to downlist the elephant population from Appendix I to Appendix II of CITES*. Dar es Salaam.

Wright, S.J., Sanchez-Azofeifa, G.A., Portillo-Quintero, C. & Davies, D. (2007) Poverty and corruption compromise tropical forest reserves. *Ecological Applications*, 17, 1259–1266.

Regulation and Certification

19

Regulation and Recreational Hunting

Alison M. Rosser

Durrell Institute of Conservation and Ecology,
University of Kent, Canterbury, UK

Introduction

Recreational hunting occurs in both the biodiversity-rich and biodiversity-poor countries and involves native and non-native species, the latter often specially reared for hunting. Hunters from the biodiversity-poor countries often seek the challenge of hunting native fauna elsewhere and thus support a hunting tourism industry with its associated cross-border movements of hunting trophies. Such recreational hunting can generate considerable incentives for conservation. In the southern African countries, for example, recreational hunting generated over US$96 million annually in the late 1990s (Barnett & Patterson, 2005). However, recreational hunting may result in ecological problems such as loss of heterozygosity through targeting prime males or over-harvesting; and it may also result in welfare problems, through the inhumane treatment of quarry species, and corrupt practices among hunting operators (Gunn, 2001; Holmes, 2007). There may also be problems with the international movements of hunting trophies. Thus, to improve its conservation outcomes, recreational

Recreational Hunting, Conservation and Rural Livelihoods: Science and Practice, 1st edition.
Edited by B. Dickson, J. Hutton and B. Adams. © 2009 Blackwell Publishing,
ISBN 978-1-4051-6785-7 (pb) and 978-1-4051-9142-5 (hb).

hunting needs to stress and strengthen its positive aspects, while eliminating its controversial aspects – but how?

Historic use of regulation

Regulation, which generally refers to 'a rule prescribed for the management of some matter, or for the regulating of conduct' (Oxford English Dictionary, 1989) is the most commonly followed approach to ensure that recreational hunting meets certain standards. In many countries, the regulation of recreational hunting has a long history. The New Forest in southern England, recently designated as a national park, was originally set aside for conservation in 1066, when it was established as a royal hunting preserve from which subsistence hunters were excluded on sentence of death for poaching (Tubbs, 1986). Likewise, the first written hunting laws in Finland date from 1347 (Mykra et al., 2005). Similar ancient approaches to the regulation of recreational hunting have helped conserve Poland's Bialowecza Forest, India's Sariska Tiger Reserve and Nepal's Chitwan National Park, as well as many other areas around the world (see Allsen, 2006). Based on such approaches, some Latin American and African countries responded to evident over-harvests of wildlife in the 1960s and poor practices among the operators by heavily restricting hunting, in some cases including bans on sport hunting (Ojasti, 1996; Price Waterhouse, 1996a). More recently, such protectionist policies have been revisited and reintroduction of sport hunting has been considered in Brazil and in Kenya (UPI, 2007; Kameri-Mbote, 2005). Meanwhile, questionable practices such as canned hunting and disregard for hunting regulations compromise the continued existence of recreational hunting in the developed and developing world alike (Holmes, 2007).

This chapter provides an overview of some aspects of the regulation of recreational hunting, and highlights some areas that require further consideration by regulators and hunters alike. Recreational hunting is generally regulated through overarching national legislation, which designates one or several appropriate authorities for more detailed management. National legislation may be further interpreted through local by-laws or in federal systems through individual state provisions. Beyond domestic jurisdictions, foreign national legislation and regional policy may also affect hunting management through regulations pertaining to the control of international movements of hunting trophies from certain species. The first part of this chapter will consider sub-national and national legislation, while the second part will discuss international regulations such as the Convention on International Trade in Endangered Species of Wild Fauna and Flora (CITES) and examples of national legislation that restrict imports, so called 'stricter domestic measures'.

Current national regulations

For recreational hunting to contribute positively to conservation, it is important that countries have clear regulations and well-managed agencies to implement and enforce that legislation. Yet hunting regulations are often complex and promulgated in myriad acts, statutes and decrees in different countries throughout the world. For example, legislation on wildlife conservation or environmental protection generally deals with quarry species and the sustainability of hunts (e.g. New Zealand's Wildlife Act of 1953; Australia's Environmental Protection and Wildlife Conservation Act 1999; and the 1981 Wildlife and Countryside Act of Great Britain). Meanwhile, access to hunting areas is regulated both through legislation relating to property as well as through wildlife conservation and game legislation. Although regulations pertaining to animal welfare are mainly found in conservation and game legislation, there is now an increasing trend towards standalone legislation relating to animal welfare, particularly in developed countries (e.g. Germany's Animal Welfare Act 1998). In contrast, the ownership of hunting weapons is generally regulated through firearms legislation with little reference to conservation law (e.g. Argentina's Act on Weaponry and Explosives 1973 and subsequent decrees). Furthermore, in countries such as the United States, Australia and South Africa that have a decentralised or federal governance structure, there may be both federal and state regulations on many of these issues (see Cirelli, 2002).

Consequently, this section provides an overview of national regulations pertaining to hunting in five broad categories as follows:

(a) Those that deal with sustainability. (What to hunt or more likely what not to hunt? How many animals can be hunted and when?)
(b) Those that deal with access and ownership. (Where to hunt? Who benefits?)
(c) Those that deal with the skills of hunting. (What knowledge is required to hunt? Which hunting methods to use?)

(d) Those that deal with welfare of the quarry. (Which hunting methods to use?)

(e) Those that deal with exports and imports of hunting trophies.

Quarry species are often divided into native and non-native species within hunting regulations. Native species usually enjoy different levels of protection in national legislation, based both historically on their perceived usefulness, and currently on the perceived or assigned level of threat that each currently faces (Mykra et al., 2005). A basic management goal for native species is to ensure that hunting does not threaten their populations, while an enlightened management goal would seek to ensure that hunting enhances their population status and the conservation of their habitats. However, much recreational hunting as it is currently understood in popular culture often targets pest and/or introduced species, particularly in developed countries, for which conservation authorities have very different management goals (Dryden & Craig-Smith, 2004).

This distinction is important, as some concerns regarding hunting and its regulation pertain to the less restrictive practices associated with hunting non-native species and to the ethical and welfare concerns of recreational hunting of non-native species. For example, New Zealand has separate regulations for: native species that are generally protected; game species which include a few native species that are well-managed through bag limits; and so-called 'noxious species' for which there are no limits. The noxious species comprise introduced species, many of interest to recreational hunters, including the exotic deer species sometimes involved in 'put and take' hunts (where the quarry species is released into the hunting area just prior to the hunt) in New Zealand (Eggleston et al., 2003). Similarly, in the UK, many popular game species are non-native, and hunted populations are often supplemented by careful management each season, involving stocking and predator control. This management in turn raises welfare questions for the stocked species as well as conservation concerns for native species. In Latin America, the majority of recreational hunting targets field birds, but big game is also hunted and a regulatory distinction is made between hunting of native and non-native mammal species (Ojasti, 1996). Therefore, to facilitate hunter understanding, the Wildlife Department in Chile has produced an illustrated guide that combines information on species that can be hunted and on fully protected species (SAG, 2006). However, given that hunting of non-native species may contribute to habitat conservation, as shown recently in lowland England

(Oldfield *et al.*, 2003), questions can be asked about simplistic native/non-native distinctions.

Authority for managing the sustainability of hunting is generally delegated to the appropriate national or federal wildlife authority, which in turn determines the methods for regulating numbers hunted and when to hunt (Ojasti, 1996; Cirelli, 2002; Barnett & Patterson, 2005; FAO, 2006). Sustainability of recreational hunting, particularly of native species, can be managed either by regulating the number of animals to be hunted each year through the establishment of a fixed quota, or by regulating hunting effort, or by regulating the proportion of the population that is hunted (see Milner-Gulland & Mace, 1998).

In general in Africa, where trophy hunting is promoted as a form of land use for conserving natural habitats, many native species can be hunted and harvests are often regulated through hunting quotas, whether they be national hunting quotas or trophy export quotas under CITES (Price Waterhouse, 1996a; Barnett & Patterson, 2005). In contrast, sustainability in Scandinavia and the US is usually delivered by managing hunting on the basis of controlling effort, or of taking a constant proportion of the population (see Robertson & Rosenberg, 1988; Bregenballe *et al.*, 2006). In the US the hunting of native species is generally managed through licences that specify how many carcasses a hunter may have in their possession on any one day (daily bag limits). In turn, the licences issued in any season are generally allocated by lottery and hunters are required to complete data forms and attach a tag immediately when they complete a hunt. These hunter returns for water-birds are processed as part of the Hunter Inventory Project, the results of which determine the numbers of licences allowed in subsequent years (USFWS, undated). In contrast, responsibility for managing hunting in the UK lies more closely with the landholder than in the USA, and the bag size for a day (or season) is determined largely by the landholder (Parrott *et al.*, 2003).

Many countries also regulate hunting effort through defining closed seasons, in which quarry species cannot be hunted. Closed seasons are often laid out in national legislation and are usually based on a combination of ecological and welfare concerns, such as avoiding hunting during breeding seasons. Long closed seasons may serve to limit the numbers of individuals that can physically be hunted in any one year, an example being the US state of Kansas, where the open season for particular taxa is very short (see KDWP, undated).

Although many countries regulate hunting off take through licences, quotas, bag limits and restrictions on hunting areas, monitoring populations and off take levels for many species may not be sufficiently robust to ensure that harvests are sustainable (Parrott *et al.*, 2003; Barnett & Patterson, 2005). In Eurasia, several Central Asian countries noted problems with the regulation of hunting because there is insufficient information to set quotas (FAO, 2006). Under these conditions harvest regulations may fail to take full account of the behavioural ecology of the species, which in turn can affect productivity (Caro, 2000; Whitman *et al.*, 2004). In some cases, the national legislation designed to protect wildlife may give those administering the hunting regulations too much discretion (Nshala, 1999). To deal with these problems it has been suggested that harvest levels should be set very low initially, and be fine-tuned upwards through transparent adaptive management (Caughley & Sinclair, 1994), with appropriate independent monitoring of wildlife authorities. There is also a role for hunters and operators to play in ensuring that their sport only takes harvests that are sustainable and transparently monitored to ensure that the industry promotes conservation.

Hunting regulations generally specify the types of land upon which hunting can take place. Land ownership can be broadly divided into state, private and communal lands of various types. For example, in 2000 the countries of southern Africa had allocated land in different categories of ownership to sport hunting as follows: 420,000 km^2 of communal land; 188,000 km^2 of commercial land; and 420,089 km^2 of state land (Barnett & Patterson, 2005). In many areas of Africa, state-run game reserves where hunting was allowed were established by the colonial and post-colonial regimes as an adjunct to protection reserves. More recently in Tanzania, Zambia, Zimbabwe and Namibia amongst others, benefits are shared with different combinations of state and local communities running hunting concessions (Leader-Williams *et al.*, 1996; Kameri-Mbote, 2005). Similarly in the Carpathian area of Europe, the majority of forested land belongs to the state, and these lands are divided into hunting grounds that are managed by various governmental agencies and NGOs (Salvatori *et al.*, 2002). In contrast, hunting in the UK is only allowed on private lands and is generally organised by the landholder. In Latin America the situation varies depending on regulations regarding ownership of animals (Ojasti, 1996; Cirelli, 2002).

In most jurisdictions, hunting on private land requires the permission of the landowner (see Australian Act; S.A.G., 2006). However, the access for hunting can be extremely complex. For example in Alberta, Canada, hunting guides refer to three pieces of legislation when outlining access arrangements for hunting, comprising: the Wildlife Act 1994; the Petty Trespass Act 2004;

and the Federal Criminal Code section 41. Likewise, the situation varies across the USA where 21 states have statutes that require hunters to gain landowner permission before hunting, while a further 29 states have 'posting statutes' that require the landowner to post 'no hunting' notices if they do not wish to grant hunters access to their land (Sigmon, 2004).

Ownership of the quarry species is not simple either, as wildlife may be variously owned by the state, the landowner or the hunter in different countries. Countries with common law systems, including many British Commonwealth countries, generally regard wild animals as *res nullius*, and as not being the subject of private property until reduced into possession by being killed or captured (Cirelli, 2002). Meanwhile, *ratione soli* systems give rights over the animals to the landowner; whilst *ratione privilegii* systems retain rights under government control which can be given out by licence. Access and ownership issues may be important in contributing to the assessments of sustainability of the hunt, and to the right to hunt, such that considerable clarification is required of whoever has responsibility to manage the hunt.

Differences in the ownership of land and wildlife are both critical to how benefits from hunting may be distributed. There is broad agreement that recreational hunting is a sufficiently lucrative industry to generate considerable revenue for conservation. However, it is considered important that appropriate shares of that revenue help provide benefits to local communities as well as support conservation objectives. Such benefits are particularly important for those communities that suffer the costs of living among wildlife, for example through loss of life and livelihood, through crop-raiding and/or through livestock depredation. Increasingly, local communities in southern Africa are managing hunting concessions where they have been granted usufruct rights, or are included in benefit sharing arrangements from land and hunting concessions that are not under their direct management (Child, 1996; O'Gorman, 2006). In some countries, the proportions of revenue shared with their different community-based conservation programmes may be mandated by regulation: for example, in Zambia's ADMADE and Namibia's Conservancy programmes. In Tanzania, regulations stipulate the sharing of benefits from hunting in unoccupied state-owned game reserves. In other countries, the arrangements for revenue-sharing may be informal (Barrow *et al.*, 2001).

Recreational hunting is controlled by the issuance of licences in most countries, but what knowledge is required to obtain such permits? In some countries, these licences are divided into those for residents and non-residents, and are associated with different requirements and rights. In Latin America, the usual requirements for a sport hunting licence include: (a) a minimum age of 18 years; (b) a request with personal history and photograph; (c) a valid licence to carry a gun; and (d) payment of the fees entailed (Cirelli, 2002). In Brazil and Mexico, the applicant must also belong to a hunting club, and Venezuela extends preferential treatment to organised hunters. Similarly, Colombian legislation also requires familiarity with the use of firearms and knowledge of the regulations in force (Ojasti, 1996). In Europe, the European Parliament has recommended that voluntary education programmes be established to guarantee that hunters have the necessary expertise to ensure that quarry is despatched in a humane and sustainable manner (EU, 2005). In North America, hunters must pass exams as a prerequisite to issuing hunting permits in a number of states (ADF&G, 2006) and a number of countries in Africa have similar requirements, especially for professional hunters (Price Waterhouse, 1996b). Increasingly, hunters are required to demonstrate the following: safe handling of weapons; an intimate knowledge of local, national, regional and international regulation; understanding of the rules of fair chase operating in the area where they wish to hunt; and an understanding of regulations developed to ensure sustainability of the hunt. However, there are concerns that education and training programmes are not mandated in all countries that allow hunting. A recent report on the industry in southern Africa, recommends that such training be administered through professional hunters' organisations and become mandatory (Barnett & Patterson, 2005).

Basic animal welfare aspects of hunting have been addressed through regulations for some time, but are now receiving greater scrutiny. Many countries regulate the type and calibre of firearm or bow that can be used for specific species, thereby seeking to ensure a humane end for the quarry. For example, the Alaskan bow hunting regulations allow the use of longbow, recurve and compound bow for use with game animals, but restrict the use of crossbows to designated big game species. Likewise, many states in the US lay down power requirements for bows and designate what types of arrowhead can be used for particular species. Similarly, closed seasons also encompass humane considerations, by ensuring that females of the quarry species are not hunted or disturbed during the period in which they are carrying or nursing young. However, although the concept of fair chase is associated with many hunting cultures and regulations, it is unfortunately not practised by everyone. Thus, practices such as canned hunting or put-and-take hunting, in which captive reared animals are released briefly for hunting purposes, have given rise to genuine animal welfare concerns. In response, South Africa has recently committed that its hunting regulations will require that animals to be hunted must have lived *in situ* for two years in order to outlaw such practices (IOL, 2007).

International movements of hunting trophies – the role of CITES

CITES regulates trade[1] in specimens of species that are threatened with extinction, or may become so unless such movements are strictly regulated in order to avoid utilisation incompatible with their survival (Wijnstekers, 2006). Thus CITES regulates the cross-border movement of hunting trophies of certain species listed in appendices to the Convention. CITES has proven to be amongst the most innovative and effective of the multilateral environmental agreements (Reeve, 2002). Although it entered into force on 1 July 1975, the Convention has adapted to the changing world through flexible decision-making and policy development. It continues to attract new state members and currently has 173 (October 2008) Parties, which means that its membership is now nearly universal. The Convention establishes an international legal framework with common procedural mechanisms for the prevention of international commercial trade in endangered species, and for an effective regulation of international trade in others (Table 19.1).

CITES regulates international trade through a system of export and import permits and certificates which are issued only when certain conditions are met, and which must be presented when leaving and entering a country. Each Party to the Convention must designate one or more Management Authorities to administer the licensing system and one or more Scientific Authorities to advise them on the effects of trade on the status of the species.

The species covered by CITES are listed in three Appendices according to the degree of protection they need. CITES Appendix I contains species threatened with extinction and commercial international trade in specimens of these species is generally prohibited,[2] except for some very specific purposes. Appendix I includes a number of animal species targeted for hunting.

Appendix II contains species that are not currently threatened with extinction but for which the international trade must be controlled to avoid their becoming threatened. It also contains the species that visually resemble other species already included in Appendix I or II, so as to simplify the job of enforcement officers. International trade in specimens of species included

Table 19.1   **Requirements for permits for import/export of specimens of species included in CITES Appendices I and II together with conditions to be fulfilled before issuance.**

Import/export	Appendix I	Appendix II
Import	An import permit is required and may be issued only if: • The purpose of the import will not be detrimental to the survival of the species; • The proposed recipient is suitably equipped to house and care for any live specimen; and • The specimen is not to be used for primarily commercial purposes.	*The Convention does not require an import permit – any requirement by a Party for an import permit is a stricter domestic measure (which is permitted under the terms of the Convention).*
Export	An export permit is required and may be issued only upon presentation of the import permit, and only if: • The export will not be detrimental to the survival of the species; • The specimen was acquired in accordance with national wildlife protection laws; and • Any live specimen will be shipped in a manner which will minimise the risk of injury, damage to health or cruel treatment.	An export permit is required and may be issued only if: • The export will not be detrimental to the survival of the species; • The specimen was acquired in accordance with the national wildlife legislation; and • Any live specimen will be shipped in a manner which will minimise the risk of injury, damage to health or cruel treatment.

in Appendix II is permitted but regulated. More than 4400 animal species are included in this appendix, of these several hundred are attractive to hunters. Appendix III includes species that are listed unilaterally by their country of origin because it seeks assistance in regulating their trade; this Appendix will not be considered further.

The responsibility for ensuring that the trade in a CITES-listed species is not detrimental to the survival of that species largely rests with the exporting country (Table 19.1). This reflects one of the fundamental principles of CITES; namely that peoples and states are and should be the best protectors of their own wild fauna and flora.

The regulation of international trade in hunting trophies has been discussed by the CITES Parties regularly at meetings of the Conference of the Parties (Figure 19.1). For Appendix I hunting trophies, resolutions adopted by the Conferences of the Parties have introduced a number of measures that facilitate the cross-border movement of hunting trophies from these species.[3] Although these resolutions are not binding, they are widely respected and applied and reflect the will of the Parties. In particular, it was agreed as long ago as 1979 that provided that any quota agreed by the Conference of the Parties is not exceeded then the Scientific Authority of the importing country should accept the finding of the exporting country that the export of the

Figure 19.2 **The fourteenth meeting of the Conference of the Parties to the Convention on International Trade in Endangered Species, The Hague, June 2007.**

hunting trophy is not detrimental to the survival of the species unless there are data to indicate otherwise (Resolution Conf. 2.11 (Rev.)).

In addition to this generic principle, the Parties have decided on specific measures to apply to the international movement of hunting trophies for a number of species included in Appendix I. In the case of the leopard *Panthera pardus* the CITES parties in 1983 recognised that the killing of specimens of leopard may be sanctioned by countries of export in defence of life and property and that the species is in no way endangered in certain countries (Resolution Conf. 10.14 (Rev. CoP14)). At the same time, they reaffirmed that the commercial market for leopard skins should not be reopened and recommend that:

- the Scientific Authority of the state of import approve applications for import permits if it is satisfied that the skins being considered for import are from one of the states having a quota approved by the Conference of the Parties;
- the Management Authority of the state of import be satisfied that the transaction will not be for primarily commercial purposes (if acquired in and legally exported from the country of origin and the owner imports no more than two skins per year and does not resell them); and
- skins should be tagged.

Export quotas have been agreed on the basis of this resolution over the years for a number of Parties (see Table 19.2).

As the approach taken for leopard seemed to be working satisfactorily, the Parties developed a similar set of procedures for markhor goats *Capra falconeri* in 1997 (Resolution Conf. 10.15 (Rev. CoP14)). In this instance an export quota of 12 hunting trophies per year has been approved for Pakistan and exports are limited to one specimen per person per year. The most recently agreed procedure on this model agreed on export quotas for black rhinoceros *Diceros bicornis* hunting trophies in 2004 (Resolution Conf. 13.5 (Rev.Co.P14)). Two countries of export are involved, Namibia and South Africa, each with five adult male specimens per year. The specification of adult male specimens only and the requirement that the export quotas be part of a national conservation and management plan or programme are a reflection of the fact that the conservation status of this species is more precarious than that of the markhor or leopard.

Resolutions are one method for the CITES Parties to promulgate their policies but another can be an amendment to the list of CITES species itself. In the

Table 19.2   **Current export quotas for leopards following Resolution Conf. 10.14 (Rev. CoP14).**

State	Annual export quota[*]
Botswana	130
Central African Republic	40
Ethiopia	500
Kenya	80
Malawi	50
Mozambique	120
Namibia	250
South Africa	150
Uganda	28
Tanzania	500
Zambia	300
Zimbabwe	500

*For whole skins or nearly whole skins.

Table 19.3   **Examples of voluntary export quotas for Appendix I hunting trophies of African elephant *Loxodonta africana* in 2008.**

Country	Quota
Cameroon	80 animals (160 tusks)
Gabon	50 animals (100 tusks)
Mozambique	40 animals (80 tusks)
Tanzania	200 animals (400 tusks)
Zambia	20 animals (40 tusks)

case of the cheetah *Acinonyx jubatus*, the Parties decided to establish hunting trophy export quotas by the latter route – probably a historical accident. Thus, in 1992, the CITES Appendices were amended to note annual export quotas for live specimens; hunting trophies for three countries were granted and are still in force today.

For all other hunting trophies from Appendix I species, the generic rule laid out in Resolution Conf 2.11 (Rev.) applies as described above. Some Parties have established voluntary export quotas for such hunting trophies (Table 19.3). These voluntary export quotas serve largely to assist management

Table 19.4 **Examples of voluntary export quotas for Appendix II hunting trophies.**

Species	Latin name(s)	Country with voluntary export quota
Primates	*Cercopithecus mitis/Colobus guereza/Papio hamadryas/ Theropithecus gelada*	Ethiopia
Caracal	*Caracal caracal*	Ethiopia
Wild cat	*Felis silvestris*	Ethiopia, Romania
Serval	*Leptailurus serval*	Ethiopia
Eurasian lynx	*Lynx lynx*	Romania
Lion	*Panthera leo*	Ethiopia
Wolf	*Canis lupus*	Romania
Brown bear	*Ursus arctos*	Romania
African elephant	*Loxodonta africana*	Botswana, Namibia, South Africa and Zimbabwe
Nile crocodile	*Crocodylus niloticus*	Botswana, Ethiopia and Namibia
Hippopotamus	*Hippopotamus amphibius*	Ethiopia and Tanzania

controls in exporting countries. As any export permit associated with these trophies is annotated to show that the specimen(s) refer to the quota in question, this also helps importing states ascertain that a particular trophy being imported is from the quota established. For Appendix II species several Parties have chosen to establish voluntary export quotas for hunting trophies for a variety of (mostly mammalian) species (Table 19.4).

In addition to the procedures for importing and exporting hunting trophies, the CITES Parties have, in one instance, formally amended the level of CITES protection afforded to a species on the basis of recreational hunting considerations. In 1994, the South African and Swazi populations of southern white rhinoceroses *Ceratotherium simum simum* were transferred from Appendix I to Appendix II 'for the exclusive purpose of allowing international trade in live animals to appropriate and acceptable destinations and hunting trophies'. Thus, under CITES rules, hunting trophies of southern white rhinoceroses from these countries can be imported into the country of residence of the hunting tourist without the requirement for an import permit.

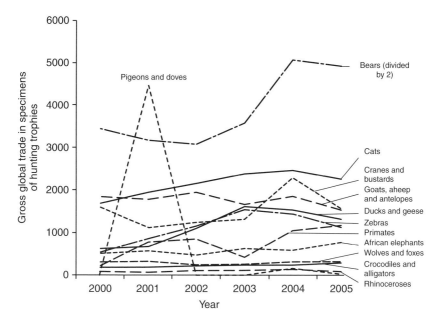

Figure 19.2 **Gross global trade in specimens of hunting trophies reported by CITES Parties for principal animal groups during 2000–2005 (CITES trade statistics derived from the _CITES Trade Database,_ UNEP World Conservation Monitoring Centre, Cambridge, UK).**

A considerable volume of international trade in hunting trophies has occurred in recent years, using the special provisions described earlier, or simply under the normal permitting procedures (Figure 19.2). The figure shows the principal groups of CITES-listed animals of interest in hunting tourism and the fact that broadly, the number of such transactions shows a steady or increasing trend over recent years. However, the CITES trade data in Figure 19.2 need to be treated with some caution. Gross trade is the sum of all reported imports, exports and re-exports in a particular commodity or species and may overestimate the total number of actual specimens in trade as re-exports are not deducted from the total. Additionally, the number of hunting trophy specimens cannot be equated to the number of animals involved as sometimes a number of hunting trophy parts will be derived from a single specimen (skin, tusks, horns, skull, etc.). Finally, the data for 2005 might not be complete as some Parties have yet to submit their records.

Imports of trophies – stricter domestic measures and regional regulations

In addition to their CITES implementing legislation, a number of countries have national legislation which regulates the import of hunting trophies and is stronger than multilateral measures imposed under CITES. Prime examples of these so called 'stricter domestic measures' are the US Endangered Species Act 1973 and Lacey Act Amendments 1981 (Ortiz, 2005) and, in the European Union (EU) (2005), the CITES enabling legislation (Council Regulation (EC) No. 338/97).[4] Both the US and EU can refuse the import of certain hunting trophies under their domestic measures. This has caused considerable concern to programmes that aim to conserve species and habitat through income derived from hunting a small number of high value specimens (Hutton & Dickson, 2000; Rosser *et al.*, 2005). Hunting programmes that have been impacted by the Endangered Species Act include those for straight-horned markhor, wood bison, Namibian cheetah and black-faced impala.

Encouragingly, a number of regional overviews of the industry have either been completed recently or are currently underway and have the potential to consider the issue of stricter measures. For example, the European Parliament has recently approved a study to examine regional approaches to regulation of hunting. The FAO is also facilitating reviews of hunting provisions in a number of Central Asian countries and USAID has just supported a similar review in Southern Africa (Barnett & Patterson, 2005; FAO, 2006). The latter makes a number of recommendations that may provide a useful framework for use in strengthening the conduct of the industry in other regions (see Box 19.1).

Conclusions

Recreational hunting can provide benefits for conservation (Wall & Child, this volume, Chapter 15; Mahoney, this volume, Chapter 16) and from this brief review it is clear that national regulation is important in providing a framework for the industry. However, in many cases regulations are complex and

Box 19.1 Recommendations for improving the regulation of trophy hunting in southern Africa

Maintaining quality and standards of the sport hunting industry
- Set minimum trophy quality sizes and standards
- Enact and enforce wildlife hunting regulations
- Form professional hunting associations
- Hold professional hunting training courses
- Establish professional hunter standards

Monitoring and administration of the sport hunting industry
- Develop and implement sound monitoring systems
- Standardise data collection forms
- Submit and analyse hunt return registers

Quota setting
- Establish quota setting processes and procedures
- Demonstrate compliance with CITES
- Demonstrate effective management capacity
- Collect and analyse information and data
- Agree and standardise information sources
- Establish effective monitoring systems
- Record and analyse trophy quality data
- Establish oversight and approval process for quotas

Maximising economic and social benefits from the sport hunting industry
- Adopt transparent mechanism for allocation of hunting concessions
- Develop and use screening criteria for hunting operators
- Mandate annual reporting and accounting of revenues
- Market hunting packages effectively
- Revise government hunting fees periodically
- Set longer-term arrangements for hunting revenues
- Establish equitable revenue retention guidelines

Source: Extracted from Barnett & Patterson (2005).

might benefit from consolidation and from interpretive guides. Nevertheless, it is also important to recognise that regulation alone cannot ensure the sustainability of hunting, and that resources are required for data collection to support the setting of precautionary hunting limits. With regard to ownership and benefit sharing, some regulations may require updating to reflect current social, conservation and welfare policy. In this respect, much can be learned from the CITES experience, where Parties have recognised that in the right circumstances, international trade in hunting trophies is not inimical to the conservation of the species. As a reflection of this fact, whilst maintaining strict controls on this trade, the CITES rules have been amended in a variety of ways to facilitate it. Given the recognition of national sovereignty over biodiversity in forums such as the Convention on Biological Diversity and CITES, the use of stricter domestic measures remains a bone of contention that will not easily be addressed.

Once hunting regulations have been revised, they need to be implemented. Such implementation can be encouraged through a mixture of incentives and enforcement. Hunting organisations are well placed to develop hunter education packages and ensure that members have appropriate knowledge of regulations, the rules of fair chase and how to operate their weapons. Individual tourist hunters too have a responsibility to ensure that the outfitters with whom they do business adhere to regulations and there may even be value in a common code of conduct. In terms of enforcement, resources are required to aid the implementation of hunting regulations in many jurisdictions, both to allow sufficient data collection to assist management authorities to set appropriate harvest limits and also to support effective investigative and enforcement services.

This review has indicated some aspects of hunting regulations that might benefit from review and revision, and projects under way in southern Africa, Central Asia and Europe look set to take these issues forward. Finally, proponents of hunting are well placed to encourage positive change and improve the conservation benefits of hunting by demanding best practice in hunting regulation, whether it be local, national or international.

Acknowledgements

I would like to acknowledge a major contribution to this chapter by David Morgan, Chief of the Scientific Support Unit at the CITES Secretariat in

Geneva, Switzerland, who provided information on the regulations pertaining to international trade.

Notes

1 The term 'trade' in CITES means any export, re-export, import and introduction from the sea.
2 In the context of this chapter, the term 'specimen' under CITES means any animal alive or dead and any specimen which appears from an accompanying document, the packaging or a mark or label, or from any other circumstances, to be a part or derivative of such an animal.
3 Resolution Conf. 2.11 (Rev.) on trade in hunting trophies of species listed in Appendix I; Resolution Conf. 10.14 (Rev. CoP14) on quotas for leopard hunting trophies and skins for personal use; Resolution Conf. 10.15 (Rev. CoP14) on establishment of quotas for markhor hunting trophies; Resolution Conf. 13.5 (Rev. CoP14) on establishment of export quotas for black rhinoceros hunting trophies.
4 Council Regulation (EC) No. 338/97 deals with the protection of species of wild fauna and flora by regulating the trade in these species. The full text is published in the *Official Journal of the European Communities*, L 61, Vol. 40, 3 March 1997 (ISSN 0378-6978). Commission Regulation (EC) No. 1808/2001 lays down detailed rules for Member States on the implementation of Council Regulation (EC) No 338/97, as described above. The Regulation is published in the Official Journal n° L 250 of 19.09.2001. Commission Regulation (EC) No. 865/2006 adopted on 4 May 2006 lays down detailed rules for Member States on the implementation of Council Regulation (EC) No 338/97. The Regulation is published in the Official Journal n° L 166 of 19.06.2006.

References

Alaska Department of Fish and Game (ADF&G) (2006) *2006–7 Alaska Hunting Regulations*. Alaska Department of Fish and Game, Publication No. 47, http://www.wildlife.alaska.gov/regulations/pdfs/regulations_complete.pdf viewed 2/3/2007.

Allsen, T.T. (2006) *The Royal Hunt in Eurasian History*. University of Pennsylvania Press, Philadelphia.

Barnett, R. & Patterson, C. (2005) *Sport Hunting in the Southern African Development Community (SADC) Region: An Overview*. TRAFFIC East Southern Africa, Johannesburg.

Barrow, E., Gichohi, H. & Infield, M. (2001) The evolution of community conservation policy & practice in East Africa. In *African Wildlife & Livelihoods: The Promise and Performance of Community Conservation*, eds. D. Hulme & M. Murphree. James Currey, Oxford.

Bregenballe, T., Noer, H., Christensen, T.K., Clausen, P., Asferg, T. & Delany, S. (2006) Sustainable hunting of migratory waterbirds: the Danish approach. In *Waterbirds Around the World: A Global Overview of Conservation, Management and Research of the World's Waterbird Flyways*, eds. G.C. Boere, C.A. Galbraith & D.A. Stroud. The Stationery Office, Edinburgh.

Caro, T. (ed.) (2000) *Behavioural Ecology and Conservation Biology.* Oxford University Press, Oxford.

Caughley, G. & Sinclair, A.R.E. (1994) *Wildlife Ecology and Management.* Blackwell Science, Oxford.

Child, G. (1996) The role of community-based wild resource management in Zimbabwe. *Biodiversity and Conservation*, 5(3), 355–367.

Cirelli, M.T. (2002) *Legal Trends in Wildlife Management.* FAO Legislative Study 74. FAO, Rome.

Dryden, G. & Craig-Smith, S.J. (2004) *Safari-hunting of Australian Exotic Wild Game.* A report for the Rural Industries Research and Development Corporation. RIRDC, Barton, ACT.

Eggleston, J.E., Rixecker, S.S. & Hickling, G.E. (2003) The role of ethics in the management of New Zealand's wild animals. *New Zealand Journal of Zoology*, 30, 361–176.

European Union (EU) (2005) *Hunting and Europe's Environmental Balance, Recommendation 1689, 2004.* Reply from the Committee of Ministers, 10 June 2005. Doc. 10576.

Food and Agriculture Organisation (FAO) (2006) Joint FAO/Czech Republic Workshop on Wildlife Policy and Institutions for Sustainable Use and Conservation of Wildlife Resources, 11–15 September 2006, Prague, Czech Republic. www.fao.org/forestry/site/35813/en viewed 2/3/2007.

Gunn, A.S. (2001) Environmental ethics and trophy hunting. *Ethics & the Environment*, 6, 68–95.

Holmes, B. (2007) Bag a trophy, save a species. *New Scientist*, 193, 2585, 6–7.

Hutton, J. & Dickson, B. (2000) *Endangered Species, Threatened Convention: The Past, Present and Future of CITES.* Earthscan, London.

Independent on Line (I.O.L.) (2007) Lion hunting dealt death blow. Independent on Line, South Africa, Feb. 20, 2007. http://www.iol.co.za/index.php?click_id=31&set_id=1&art_id=nw20070220140200685C795016 viewed 2/3/2007.

Kameri-Mbote, P. (2005) *Sustainable Management of Wildlife Resources in East Africa: A Critical Analysis of the Policy, Legal and Institutional Frameworks.* International Environmenetal Law Research Centre working paper no. 5, Geneva. http://www.ielrc.org/content/w0505.pdf viewed 2/3/2007.

Kansas Department of Wildlife and Parks (KDWP) (undated) *Kansas Sportsman's Calendar 2007*. Kansas Department of Wildlife and Parks. http://www.kdwp.state.ks.us/news/hunting/when_to_hunt/sportsmen_s_calendar viewed 7/3/2007.

Leader-Williams, N., Kayera, J.A. & Overton, G.L. (eds.) (1996) *Tourist Hunting in Tanzania*. Occasional Paper of the IUCN Species Survival Commission, No. 14. IUCN, Cambridge.

Milner-Gulland, E.J. & Mace, R. (1998) *Conservation of Biological Resources*. Blackwell Science, Oxford.

Mykra, S., Vuorisalo, T. & Pohja-Mykra, M. (2005) A history of organized persecution and conservation of wildlife: species categorizations in Finnish legislation from medieval times to 1923. *Oryx*, 39, 275–283.

Nshala, R. (1999) *The Wildlife Conservation Act of Tanzania 2001*. Lawyers' Environmental Action Team. http://www.leat.or.tz/publications/hunting.blocks viewed 2/7/2007.

Oxford English Dictionary (1989) *Oxford English Dictionary*, 2nd edn. Oxford University Press, Oxford.

O'Gorman, T.L. (2006) *Species and People: Linked Future*. WWF, Gland.

Ojasti, J. (1996) *Wildlife Utilization in Latin America: Current Situation and Prospects for Sustainable Management*. FAO Conservation Guide 25. FAO, Rome.

Oldfield, T.E.E., Smith, R.J., Harrop, S.R. & Leader-Williams, N. (2003) Field sports and conservation in the United Kingdom. *Nature,* 423, 531–533.

Ortiz, P.A. (2005) An overview of the Lacey Act Amendments of 1981 and a proposal for Model Port State Fisheries Enforcement Act. US NOAA report. http://www.illegal-fishing.info/uploads/Lacey_Act_Paper.pdf viewed 2/3/2007.

Parrott, D., Moore, N., Browne, S. & Aebischer, N. (2003) *Provision of Bag Statistics for Huntable Birds*. Defra Project CRO281. http://www.defra.gov.uk/science/Project_Data/DocumentLibrary/WC04001/WC04001_1244_FRP.doc viewed 2/3/2007.

Price Waterhouse (1996a) The trophy hunting industry: an African perspective. In *Tourist Hunting in Tanzania,* eds. N. Leader-Williams, J.A. Kayera & G.L. Overton. Occasional Paper of the IUCN Species Survival Commission, No. 14. IUCN, Cambridge.

Price Waterhouse (1996b) Training of professional hunters in Southern Africa. In *Tourist Hunting in Tanzania,* eds. N. Leader-Williams, J.A. Kayera & G.L. Overton. Occasional Paper of the IUCN Species Survival Commission, No. 14. IUCN, Cambridge.

Robertson, P.A. & Rosenberg, A.A. (1988) Harvesting gamebirds. In *Ecology and Management of Game Birds*, eds. P.J. Hudson & M.R.W. Rands, pp. 177–201. BSP Professional Books, Oxford.

Reeve, R. (2002) *Policing International Trade in Endangered Species: the CITES Treaty and Compliance*. Royal Institute of International Affairs & Earthscan Publications Ltd., London.

Rosser, A.M., Tareen, N. & Leader-Williams, N. (2005) The Precautionary Principle, uncertainty and trophy hunting: a review of the Torghar population of central Asian markhor *Capra falconieri*. In *Risk and Uncertainty in Conservation and Sustainable Use*, eds. R. Cooney & B. Dickson. Earthscan, London.

S.A.G. (2006) *Cartilla Para Cazadores*. Wildlife Division Ministry of Agriculture, Santiago.

Salvatori, V., Okarma, H., Ionescu, O., Dovhanych, Y., Find'o, S. & Boitani, L. (2002) Hunting legislation in the Carpathian Mountains: implications for the conservation and management of large carnivores. *Wildlife Biology*, 8, 3–10.

Sigmon, M.R. (2004) Hunting and posting on private land in America. *Duke Law Journal*, 54, 549–585.

Tubbs, C.R. (1986) *The New Forest*. Collins, London.

United Press International (UPI) (2007) Kenya considers sport hunting legislation. United Press International, 25 Feb. 2007, http://www.upi.com/NewsTrack/Top_News/Kenya_considers_sport_hunting_legislation/20070225–125602-1858r/ viewed 3/3/2007.

United States Fish and Wildlife Service (USFWS) (undated) *Hunting*. http://www.fws.gov/hunting/. viewed 1/2/2006.

Whitman, K., Starfield, A.M., Quadling, H.S. & Packer, C. (2004) Sustainable trophy hunting of African lions. *Nature*, 428, 175–178.

Wijnstekers, W. (2006) *The Evolution of CITES*. Bernan Press, Lanham, MD.

(20)

The Application of Certification to Hunting: A Case for Simplicity

Brian Child[1] and Bill Wall[2]

[1]Center for African Studies & Department of Geography,
University of Florida, Gainesville, FL, USA
[2]Independent Consultant, Alaska, USA

Introduction

This chapter focuses on systems of recreational hunting where the hunter travels, often between continents, and pays for hunting opportunities to secure a hunting 'trophy' (durable parts of the animal such as horns). Consequently, this form of hunting is commonly called trophy hunting or, especially in Africa, 'safari hunting'. The quarry species are often referred to, even in legislation, as 'game'.

Most current initiatives to establish criteria for sustainable hunting focus on biological criteria. Our contribution spotlights governance because ecological management is invariably an outcome of incentive and governance structures and, moreover, certification is essentially a tool for improving governance. However, governance is one of the least understood aspects of sustainability. We therefore try to define governance conditions that encourage sustainable hunting, ascertain if they are measurable, and judge if certification can play a role in bringing them about. We argue that certification can encourage markets for trophy hunting to promote sound practices and discourage those that

Recreational Hunting, Conservation and Rural Livelihoods: Science and Practice, 1st edition.
Edited by B. Dickson, J. Hutton and W.M. Adams. © 2009 Blackwell Publishing,
ISBN 978-1-4051-6785-7 (pb) and 978-1-4051-9142-5 (hb).

are unethical or unsustainable. We suggest that certification should be kept simple.

The primary threat to wildlife, including quarry species, is habitat loss and illegal hunting. This happens where landholders have inadequate incentives and authority to husband and protect wildlife. Well-managed recreational hunting is very effective at providing these conservation benefits (Lindsey *et al.*, 2007). In developing countries hunting often fails to benefit conservation because that national or global elites and agencies capture the lion's share of the benefits, yet landholders and communities bear the costs of living with wildlife. This imposes an unfavourable cost/benefit on landholders so they replace wildlife with land uses that are more economically viable to them, like the growing of crops or the ranching of cattle. This situation is also common in wealthy industrialised countries. Hunting is treated as a public good, and is allocated fairly (but cheaply) to the general public, so landholders can end up bearing the costs of wildlife without sharing in the benefits.

Certification, power and governance

Certification in the natural resources sector was initiated largely to address deforestation caused by unchecked Government collusion with unscrupulous business interests in biodiversity-rich developing countries. The core problem is that natural resources like forests and wildlife are subjected to excessive government authority in the absence of mechanisms of accountability. Certification uses information to rework power relationships, increasing the range of oversight stakeholders to include consumers, corporations, and transnational NGOs (Klooster, 2006). Its underlying assumption is that where consumer interest is strong, information can be deployed to influence consumer spending to bring about more accountability in the governance of wild resources (Upton & Bass, 1995). Essentially, certification strives to enlarge the criteria that influence resource purchasing decisions to include processes like governance, sustainability and justice. Certification is usually voluntary and lacks enforcement, relying instead on information and influence.

Government-led conservation approaches are particularly prone to mismanagement in weak states where policy agencies become more political than technical (Grindle & Thomas, 1991), where the checks and balances provided by civil society are weak, and where political actors and elites maximise their returns by increasing institutional uncertainty and confusion (Chabal &

Daloz, 1999). Because historically the benefits flowing from the management and regulation of wildlife and wild resources have been highly centralised, they are especially prone to problems associated with imperfect state management, or corruption. This problem is magnified because state forestry and wildlife agencies often combine both managerial/commercial and regulatory functions, introducing a significant conflict of interest. This can create the dangerous combination of a trading monopoly with regulatory power. Certification attempts to address the highly dysfunctional manner in which the governance of wild resources is often structured by establishing new rules, norms and information that reflect the process of production, and by increasing the oversight role of civil society.

A number of initiatives are currently developing criteria and indicators for sustainable hunting. Most of these are 'developing operational concepts of sustainable use [that require] adequate tools for assessment, monitoring and adaptive management' (Lexer *et al.*, 2005). In effect these initiatives are often engaged in two tasks that should be kept separate. First, seeking to identify the principles and criteria of sustainable use as applied to recreational hunting. Second, exploring the use of such criteria to address the (mis-) governance of conservation hunting through certification.

Certification and hunting

Hunters buy trophy animals, outfitting services and access to the scenic environment in which the hunt occurs. However, the organisation of hunting is a highly complex process much of which is hidden from the customer, so it is all too easy for a hunter to buy a product that ultimately will not contribute to conservation.

Wall & Child (this volume, Chapter 15) have established a typology for 'conservation hunting'. They suggest that public hunting is effective in 'impersonal economies' with competitive political and economic transactions governed by property rights, rule of law, reliable justice and effective civil society. However, a more devolved model based on landholder incentives produces better conservation outcomes in 'personalised economies' with weak bureaucracies, weak property rights and legal systems, and ineffective civil society. It is in highly personalised economies where state actors often extract the rights and benefits from communities (i.e. undermining the necessary conditions for sustainable hunting) that certification has the greatest potential to contribute to conservation and other desirable social values. By rechannelling to landholders the large amounts of money that tourist hunters spend in developing countries, certified hunting could become a powerful force for conservation. If this revenue is governed properly it also promotes social justice, improved livelihoods and even rural democratisation through soundly structured community-based hunting programmes (see Frisina & Tareen, this volume, Chapter 9).

Public hunting in free market democracies is generally well managed. We need to assess carefully the purpose of certification here. Initiatives to develop certification are currently being encouraged by some hunters in Europe to gain public acceptance and avoid the perceived need for regulation which many fear would be stifling. Similarly, there is growing feeling in North America that some form of certification might strengthen the case for hunting to a public that is often unsupportive, and improve the objectivity and effectiveness of civic action. A reliable international certification process would also streamline the importation of hunting trophies from legitimate suppliers in Africa, Latin America and Asia into Europe and the United States, reducing the influence of special interest groups (Hutton, 1995) which often have little accountability for their actions. Reliable certification protects legitimate conservation hunting against anti-hunting interests and campaigns in the first world, as well as weak governance in the third world.

Certification of recreational hunting has several objectives:

(a) To improve conservation practice by creating greater transparency around the principles that lead to conservation hunting – especially appropriate benefit, power, and responsibility sharing arrangements that prioritise landholder benefit.
(b) To improve business practice within industry, including reducing practices that are unethical or damaging to conservation.
(c) To encourage acceptance and application of principles for sustainable use by the conservation community by providing greater clarity about these principles and checks on cases of misuse.
(d) To improve the linkages between the regulation of trophy imports into industrialised countries with good practice in developing countries.
(e) To reduce the pressure for increased or inappropriate regulation that harms good practice.
(f) To improve public acceptance of hunting in rich, urbanised societies.

Can certification work for hunting?

Converting the promise of certification into practical results has proven difficult in the forestry sector where it was pioneered, particularly in the context of developing countries and rural communities. Are there reasons to think it can be more effective with wildlife and specifically trophy hunting?

The conditions under which conservation hunting and community conservation are most effective have been detailed by Wall & Child (this volume). In short, the landholder must have a high degree of authority, responsibility and accountability over the hunting *and* receive the full benefits that flow from it. However, these principles are often undermined by poor or corrupt management (Leader-Williams *et al.*, this volume, Chapter 18). Symptoms include the corrupt allocation of hunting blocks, biologically unrealistic quotas, a failure to reinvest in the resource base and, most insidiously, a refusal to return benefits to the communities on whose land the animals are living.

To establish effective certification we need to meet two challenges. First, we have to define the mechanisms that lead to sustainable hunting in at-risk areas in clear, measurable terms. Second, we need to establish conditions where knowledge and information about these mechanisms influences how hunters spend money.

The marketing chain for the trophy hunting industry favours certification. First, the customer is highly informed about the animals he or she wants to hunt and, in visiting the site of production to savour a wilderness experience is exposed to the on-the-ground realities in a way that a person who visits a warehouse to buy certified timber from a tropical forest can never be. Thus, hunting purchases are high cost, highly personal and highly differentiated. This contrasts with forest certification where marketing chains between forest and consumer are much longer, and where the final product bears little similarity to the forest from which it comes.

Second, hunting involves a great deal of personal expense and time, so hunters research their purchasing decisions carefully. They validate their purchases through numerous clubs, magazines, detailed comparisons on websites or newsletters (e.g. the 'Hunting Report'[1]) hunting shows, and especially through personal contacts with outfitters and previous customers. Being involved in a recreational choice, hunters are more likely to understand the underlying production process than even a responsible individual buying timber or furniture.

Third, the hunting product is far more homogeneous and globally comparable than one would, at first glance, believe. The species hunted differ between North America, Africa and Asia, but the basic components of a hunt are remarkably constant. The customer buys two things: trophy animals and outfitting services. The quality of the trophy is highly comparable through detailed registers (e.g. those of Safari Club International, Boone and Crockett and Rowland Ward), with North American hunters tending to seek large horns and Europeans preferring animals of age and 'character'. The customer also considers the wilderness quality of the experience, and the quality of outfitting services, including guiding, tracking, accommodation, vehicles and so forth. In general, the landholders' production (i.e. the animals plus a wilderness experience) is captured in the 'concession fee' and the 'trophy fee' while a 'daily rate' covers outfitter services. These characteristics suggest that certification could work well for trophy hunting.

How could we set up workable criteria?

Principles and criteria for a workable certification system should be based on an explanatory model to avoid the all-too-common proliferation of extensive and non-prioritised 'lists'. The Addis Ababa 'Principles and guidelines for the sustainable use of biodiversity', formulated under the Convention on Biological Diversity[2] contain fourteen 'practical principles' and no less than eighty 'operational guidelines'. This may reflect the complexity of socio-ecological systems, but it is quite useless for certification purposes. Instead, we should be guided by three maxims:

- if you cannot describe it, you cannot measure it;
- if you cannot measure it, you cannot manage it;
- if you have too many controls (e.g. indicators or principles) you risk losing control altogether.

Thus, the number of metrics should be strictly limited. It is generally recognised that managers cannot cognitively process more than twenty metrics, or seven 'key performance areas', so we need to keep the certification process within these bounds. Moreover, the people buying hunting do so for recreation and not to fulfil any basic need, so we suggest that certification should be simple and aim for as few metrics as possible to avoid creating any unnecessary disincentives. We propose four metrics as the starting point for a certification initiative:

1 The most important indicator is 'who gets the lion's share of the income', given that incentive is the fulcrum for responsibility (Murphree, 2000). Tracking the value chain is intuitively easy to understand, as is favouring purchases where payments are reinvested in the resource base, and discouraging those where payments are captured by others. Changing the incentive structure in favour of landholders and communities has profound governance and investment implications that it, alone, makes such certification worthwhile.

2 We need a measure of harvest sustainability. Trophy quality is an excellent indicator. It measures the effects on the harvested population directly, is easy to collect and can be presented as a single number. It would be useful to supplement trophy quality with a measure of habitat quality and fragmentation, but measuring this in a standard way has so far proved difficult and/or prohibitively expensive.

3 We suggest the need for an indicator concerned with ethics. This will be very difficult to design, but it is necessary in order to exclude poor hunting practice and outfitters who so greatly damage the image of hunting despite being uncommon. It could take the form of a blacklist rather than a certification indicator.

4 The last indicator would, we suggest, deal with the quality of the hunting experience (i.e. the quality of the outfitter, hunt and hunting area). While not theoretically necessary as part of a certification scheme, this could provide the vehicle for integrating the other three process-orientated parameters into purchasing decisions.

Metric 1: Governance and incentives

A wealth of literature describes the key factors that underpin the sustainable use of natural resources in a common-property regime (e.g. Ostrom, 1990; Hulme & Murphree, 2001) and a number of principles have emerged (Murphree, 2000). However, it is still easy to get lost in the complex discussion about the rights and benefits and the way these influence sustainable outcomes. Fortunately, Bond (1999) has developed a very useful model which suggests that the probability of successful resource management increases as

(a) its tangible value to landholders increases and (b) as they gain more proprietary control over the resource. People will conserve a resource that is not
very valuable simply because it is theirs. They will also conserve a valuable
resource even if they only have diluted rights to manage it. But the highest
probability of successful management is where there are both tangible benefits and strong proprietorship. Bond's model reflects a philosophy of sustainable use based on the so-called price-proprietorship-subsidiarity hypothesis
(Figure 20.1): *If* the resource is valuable, *and* the landholder has the rights to
use, benefit and manage it, *and* rights are accumulated by upward delegation,
then there is a high probability of improved resource management (Child G.,
1995; Child, B., 2002).

It is equally easy to get lost in complex arguments, and even more complex
indicators, about economic power, scale, biological and organisational factors.
But for the maximum conservation gain from hunting we can say simply that
the person on whose land wildlife occurs should be the primary beneficiary,
with the right to receive and control the income generated by wildlife. Assuming
that money follows power we can incorporate much of this complexity into

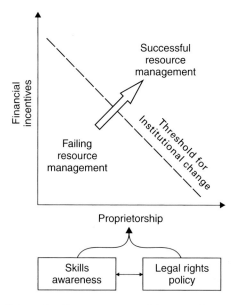

Figure 20.1 **Conditions leading to successful natural resource management
institutions.**

a single variable – the amount and proportion of money getting to the land-holder. Figure 20.2 shows this value chain in the context of community-based hunting in Africa, and provides a strong visual analysis of where power/money accumulates in the managerial hierarchy. This suggests that we can use a single metric to capture much of the complexity of governance: Who gets the lion's share of benefits, and how much is it? Measuring the magnitude of private benefit (i.e. the discretionary income reaching the landholder/community) provides important information about how effectively wildlife conservation is incentivised. Measuring the proportion of benefits getting to landholders provides a handy measure of proprietary rights and responsibilities.

Reducing a range of complex socioeconomic factors to one index in this way is likely to be viewed by many as a radical step, but arguably it is simple, powerful, effective and workable. We fully acknowledge that wildlife management

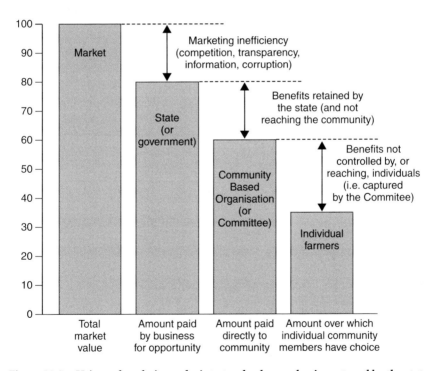

Figure 20.2 **Using value chain analysis to track where value is captured by the state or community so it reduces individual incentives for wildlife conservation.**

is usually the outcome of a three-way relationship between the wildlife producer (i.e. private landholder or community), the outfitter, and the regulatory authority (e.g. Government), and that socioeconomic structures and histories differ around the world. So do strategies for renegotiating arrangements for power sharing, benefit sharing, and responsibility. However, none of this alters the fact that conservation is more likely where benefits accrue to landholders, and certification should promote this principle.

Metric 2: Monitoring harvest and ecosystem sustainability

It is easy to get entangled in the theoretical complexity of harvest sustainability, using expensive data and models for such things as population demography, catch per unit of effort and population surveys. In practice, however, two measures – trophy quality and age of animals harvested – have proven to be simple to collect and biologically robust indicators of sustainability. Hunters select carefully for trophy quality, a biological measurement that is also a vital market indicator of product quality. This makes it an ideal metric for a hunting certification system. After trophy quality, the second easiest measurement is the age of animals. Some species can be assigned an age by simply counting horn annual growth rings (e.g. mountain sheep, Festa-Bianchet & Lee, this volume, Chapter 6). Other species require teeth to be sectioned, which is easy and inexpensive. While trophy size and age are robust indicators on their own, in practice it usually adds value to triangulate this information with, say, catch-effort records, aerial surveys and survey results.

Unfortunately, even good metrics tend to break down at the extremes. Some private landowners manage species intensively for trophy quality to the exclusion of any broad-based ecological benefits. For example, a few outfitters in the US raise trophy quality white-tailed deer in feedlots while others even breed hybrids as 'new' types of trophies. Similarly, the small but injudicious market for 'canned hunting' in South Africa suggests that some hunters use trophy size (and maybe cheapness) to the exclusion of hunting skill or the quality of the wilderness experience, which is widely considered to be inherent to hunting. Canned hunting is certainly not part of conservation hunting, and despite being uncommon it severely damages the public reputation of hunting.

While retaining trophy quality as a key indicator, the inclusion of some measure of habitat quality, habitat management practices and ecological trends would be useful to avoid trophy quality being the sole criterion. However, defining habitat criteria quickly generates a complex ecological debate and an unmanageable and unaffordable set of indicators. Perhaps the best approach would be for the landholder to define the geographic area of the hunting concession and any manipulative habitat practices, and the customer report back on infractions to the integrity of this area in a semi-standardised way.

Metric 3: Ethics

Practices such as canned lion hunting, while ecologically inconsequential, are so emotive that they can have substantial consequences on markets and the formulation of regulations. The subject of ethics in the context of hunting is a cultural minefield (Dickson, this volume, Chapter 4), but we view concepts such as humane killing and fair chase (as well as the fair distribution of benefits) to be legitimate and helpful ethical variables to be included in certification criteria. However, some ethical arguments contain cultural bias and exacerbate inequitable power relations that already favour global elites over local people (Klooster, 2006), so we suggest that certification limits itself to ensuring that hunting is humane and, more contentiously, to the principles of the 'fair chase' – although the latter has more to do with arcane notions of 'sportsmanship' than any biodiversity conservation or social equity objectives.

Metric 4: The quality of the hunt, hunting area and outfitting services

The quality of the outfitting services, quarry (e.g. number and quality of species available) and hunting areas are already heavily selected for in the recreational hunting market. As noted earlier, these choices are supported by a remarkable information system consisting of trade shows, person-to-person information, references from past clients, websites and reports. Thus the market is already dealing effectively with relevant information, which is complicated, multi-dimensional and subject to personal taste. The market is also already providing information describing product quality at a level of complexity that would be difficult to incorporate into a certification programme. Most hunters use this information to evaluate the *product* they are buying, not the underlying production *process*; and are usually unfamiliar with the economic, ecological and political dimensions of sustainability. We suggest that the most appropriate approach would be to educate hunters to understand these issues, and incorporate indicators into the market information which already guides their purchasing decisions. Therefore a working system of certification would require two activities: the education of hunters (especially about the need for benefit sharing) and the provision of relevant information in a standardised way, and preferably in a way that subjects hunters to peer pressure.

Given the nature and size of the trophy hunting industry (Sharp & Wollscheid, this volume, Chapter 2) and the high impact of hunting expenditure on conservation incentives, a small investment incorporating 'certification' into hunting information systems is likely to be extremely effective – with the caveat that we will have to assume that hunters are socially and environmentally responsible away from home. Lindsey *et al.* (in press) suggest that they are, at least to some extent, and that social responsibility is increasing.

Discussion

Certification is a relatively new economic instrument for altering and influencing the governance of natural resources. It addresses otherwise intractable governance challenges by shifting power relations away from bureaucrats towards customers and civil society, and by increasing transparency in the governance of natural resources. We believe that certification could improve the way markets serve conservation goals, and that there is considerable merit in testing if certification can improve conservation hunting and the problems of governance that often affect its performance.

Our approach to certification differs from other initiatives to develop principles and criteria in two ways. First, we strongly emphasise using a few, powerful and simple (not simplistic) metrics. We recognise that other approaches may have great value in different contexts (see Box 20.1) but we eschew complex lists of criteria and indicators that arise, often in the absence of an explanatory model, and which are commonly difficult, if not impossible to implement. Second, we have reversed the order and importance of socio-economic and biological criteria. This follows the observation that institutions, power and incentives determine the process upon which sustainable use rests. Biological criteria can measure the outcome of these processes, but seldom drive them operationally.

Instead of certification *per se* we also suggest that certification-type information could be incorporated into information systems that already guide the purchase of hunting. Educating hunters about the principles of sustainable

Box 20.1   Principles, criteria and indicators for sustainable recreational hunting in Europe (Prepared by Friedrich Reimoser, Jim Casaer and Wolfgang Lexer)

Hunting influences the genetic diversity and composition of game populations as well as the behaviour, life history and well-being of individual animals. It also affects wildlife habitats and ecosystem processes (Zeiler, 1996). In European multiple-use landscapes these impacts can result in conflicts with other land users. Partly as a result, public acceptance of hunting is decreasing in many countries (Myrberget, 1990).

While tools for assessing the sustainability of hunting should be kept as simple as possible, hunting is a complex activity and any consideration of sustainability must consider its ecological, economic and social dimensions including its spatial and temporal scales, its ecosystem impacts and its effects on non-quarry species, habitats, neighbouring land and other countryside stakeholders. Wildlife resources, hunting and many other land uses are connected in an interwoven system of dynamic interactions, interdependencies and feedback mechanisms, and if we only assess the impact of hunting on the quarry species we deal with only one part of the complex system in which hunting is embedded.

To address this complexity, principles, criteria and indicators (PCI) are being developed as tools to evaluate the sustainability of recreational hunting activities in Europe. PCI differ in their purpose, geographical scope and status. Some are formulated as guidelines, others are developed to the level of measurable indicators allowing quantitative evaluations (see Table 20.A). In Austria a set of principles, criteria, sub-criteria and indicators have been created to enable hunters to assess the sustainability of their own practices (Lexer *et al.*, 2005). This set includes 13 principles, 24 criteria and 51 sub-criteria with indicators (Forstner *et al.*, 2006). For example, the third principle states that 'The natural genetic diversity of game species is conserved and fostered by means of an appropriate hunting practice'. This has two criteria: (a) 'There are no hunting-related limitations to the conservation and fostering of the natural genetic variability of game species', and (b) 'Native wildlife populations are not altered by the introduction of and blending with exotic wildlife'. Criterion (a) is amplified by two sub-criteria: (i) 'Existence of aims relating to the aesthetics of trophies in shooting guidelines' and (ii) 'Selective hunting of wildlife according to certain natural characteristics of individuals'. Each of the criteria and sub-criteria has related indicators (see Figure 20.A). Recently the European chapter of the IUCN/SSC Sustainable Use Specialist Group has built on the Austrian experience to create guidelines applicable throughout Europe.

These approaches are designed to encourage interactions and learning within a wide range of countryside stakeholders as well as allowing hunters to demonstrate sustainability more objectively. We believe that these European approaches to the assessment of hunting sustainability are broadly relevant worldwide at the level of principles and guidelines, but they will be less transferable where criteria and indicators are concerned. In the latter case, adjustments to differing framework conditions are required.

Table 20.A   European approaches to governance of sustainable hunting. Assessment type: G – guideline, P – Principles, C – criteria, I – indicators.

Project (institution, reference)	Assessment type	Year	Who controls application	Geo-graphical coverage	Development method	Scale (assessment unit)	Responsible for hunting sustainability	Current status
Austrian Environmental Agency; Research Institute of Wildlife Ecology, Vienna; Forstner et al. 2001, 2006 http://www.biodiv.at/chm/jagd (interactive self-evaluation)	P, C, I	2001 2006	Self-evaluation of hunting-management units and hunters (voluntary)	Austria, Central Europe	Participatory stakeholder process	Hunting ground, game management unit (aggregations to larger scales possible)	Persons responsible for management unit (hunters, landowners, leaseholders)	Amended edition published
IUCN ESUSG - Wisper http://www.iucn.org/themes/ssc/susg/	P, G	2006 ongoing	Not defined (non-binding)	Europe	Interdisciplinary experts	Entity of use (estate, management unit)	Not defined	1st version published; additionally a draft charter for Falconry on website

INBO, Flemish Government http://www.inbo.be	P, C, I	2004 ongoing	Flemish authorities	Flemish region, Belgium	Interdisciplinary experts, participation process (stakeholders, NGOs, government bodies)	Game management unit	Hunter, game manager	Under development (publishing scheduled for 2007)
Pilot Wildlife estates – ELO (Network of European Estates) http://www.elo.org	Charter, label	2005 ongoing	Not defined (self-binding through voluntary endorsement)	Europe	Interdisciplinary experts	Private or public estates	Estate manager	Under development
ISWI-MAB Biosphere Reserve 'Wienerwald' Lexer et al. 2006 http://www.oeaw.ac.at/english/forschung/programme/mab.html	P, C, I	2005 ongoing	Stakeholder self-control (incl. BR management) of their land use activities in relation to hunting & wildlife	Biosphere Reserve (BR) Wienerwald, Austria	Participatory process	Biosphere Reserve region; hunting grounds	Stakeholders (land users), hunters, landowners, leaseholders, BR manager	End of project 2008
CIC Hunting Tourism Programme http://www.cic-wildlife.org	P (G, C, I)	2006 ongoing	Still open	Global	Interdisciplinary experts	Global; (planned: region; hunting management unit)	Management unit; operators, outfitters, policy makers	1st draft, under development

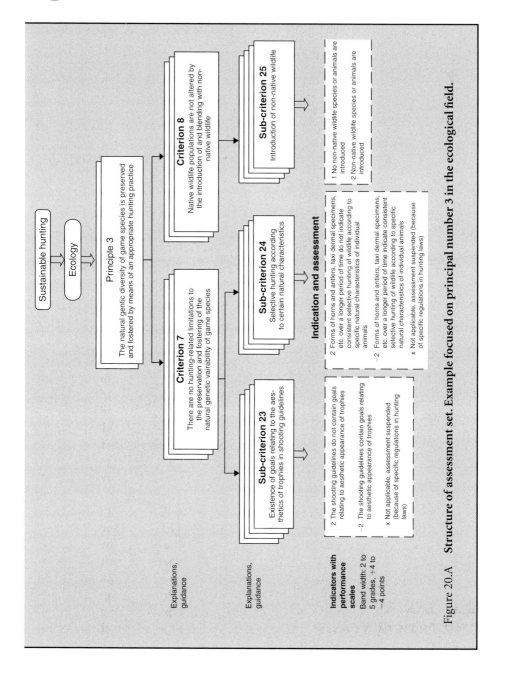

Figure 20.A   Structure of assessment set. Example focused on principal number 3 in the ecological field.

hunting, and the metrics that measure it, is an opportunity to apply adaptive management in practice: the performance criteria represent the underlying explanatory model, while openly available information drives the decision made by producers, consumers and regulators and is likely to promote debate about, and improvements in, the underlying explanatory models.

Given that recreational hunting happens around the world in many contexts (Leader-Williams, this volume, Chapter 1), we need to consider whether a common certification mechanism is possible, or if each circumstance needs a tailored approach. The set of indicators that we have suggested is largely targeted to improve hunting where landholders have few rights, regulation is weak or corrupt, and civic oversight is fragile. They apply particularly to developing countries where conservation hunting is one of the few effective ways of giving incentives for wildlife conservation yet is often misgoverned. However, there are also strong motivations for certification emerging in countries that have highly effective systems of control and very often well-managed systems of public hunting. Metrics 2, 3 and 4 are certainly relevant. The main question is the usefulness of metric 1 because in many of these countries there is no intent for the landholder to receive direct benefit. However, as the global conservation narrative is rapidly changing to incorporate the principles of sustainable use and social equity, the assumption that landholders in developed countries should provide this public service unremunerated is being questioned, including from the perspective of conservation effectiveness. Whatever one's take on this matter, wildlife conservation would benefit from increased transparency and understanding with respect to precisely who bears the costs and benefits.

We believe that a unified system of certification could be devised for conservation hunting, but recognise that the immediate need is for a system applied specifically for hunting in developing countries where hunting dollars can have a significant impact on conservation (and poverty and governance) if properly channeled.

We would like to change the way conservationists think about sustainability criteria, to emphasise that simplicity arises out of a cogent model of how conservation hunting works, and that issues of power and incentive need to be privileged over previously biocentric criteria. Nevertheless, we would be arrogant to believe that designing criteria outside of a participatory, experimental and adaptive testing process would be effective. Pronouncing criteria, no matter how technically merit-worthy, can be counterproductive: people resist solutions that are imposed on them, and this would jeopardise the commitment that is associated with genuine participation.

In this regard, perhaps the most important conclusion we can bring to this paper is that the collaborative and political process of developing criteria is likely to be at least as important as the technical merits of the final criteria. It is more important to design an inclusive and fair process for developing the principles and criteria underlying certification than to design the criteria themselves. Moreover, a sound stakeholder process is likely to create better criteria and a process designed by landholders and customers (rather than NGOs) will be most likely to succeed. Considerable care would need to be taken to ensure the participation of wildlife producers (be these private, community or state landholders) and businesses that add value by providing services (e.g. outfitters and agents), as well as customers (the hunting clients). While many of these participants have a strong intuitive understanding of sustainable processes, there is also an important role for technicians and scholars to explain the underlying (economic, political, organisational and ecological) processes of sustainable use in simple, workable terms.

Certification will not happen on its own. It needs to be championed, with sufficient investment to develop and agree performance criteria, to educate participants about these, and to maintain information systems, preferably in partnership with the hunting industry itself.

Finally, a word of caution is required: a very real danger of a certification approach is that the process could be captured by those who oppose hunting rather than those who wish to see it improved as a conservation tool. We would not wish to see this happen.

Notes

1 http://www.huntingreport.com/trophy_gallery.cfm
2 http://www.cbd.int/sustainable/addis-principles.shtml

References

Bond, I. (1999) *CAMPFIRE as a Vehicle for Sustainable Rural Development in the Semi-arid Communal Lands of Zimbabwe: Incentives for Institutional Change.* PhD Thesis, Faculty of Agriculture, University of Zimbabwe, Harare.

Chabal, P. & Daloz, J.P. (1999) *Africa Works: Disorder as Political Instrument.* James Currey, Oxford.

Child, B. (2002) Review of African wildlife and livelihoods: the promise and performance of community conservation. *Nature*, 415, 581–582.

Child, G. (1995) *Wildlife and People: The Zimbabwean Success.* WISDOM Foundation, Harare.

Forstner, M., Reimoser, F., Lexer, W., Heckl, F. & Hackl, J. (2006) *Kriterien und Indikatoren einer nachhaltigen Jagd.* Erweiterte Fassung, Wien.

Forstner, M., Reimoser, F., Hackl, J. & Heckl, F. (2001) *Kriterien und Indikatoren einer nachhaltigen Jagd.* Monograph M-158. Umweltbundesamt, Vienna.

Grindle, M.S. & Thomas, J.W. (1991) *Public Choices and Policy Change: The Political Economy of Reform in Developing Countries.* Johns Hopkins University Press, Baltimore.

Hulme, D. & Murphree, M. (2001) *African Wildlife and Livelihoods: The Promise and Performance of Community Conservation.* James Currey, Oxford.

Hutton, J.M. (1995) Developing international markets for wildlife products: facing public opinion and policy constraints in the Northern Hemisphere. In *The Commons without the Tragedy? Strategies for Community Based Natural Resources Management in Southern Africa.* Proceedings of the Regional Natural Resources Management Programme Annual Conference, 3–6 April 1995, Kasane, Botswana, ed. L. Rihoy. SADC Wildlife Technical Coordination Unit, Lilongwe.

Klooster, D. (2006) Environmental certification of forests in Mexico: the political ecology of a nongovernmental market intervention. *Annals of the Association of American Geographers*, 96, 541–565.

Lexer, W., Reimoser, F., Hackl, J., Heckl, F. & Forstner, M. (2005) Criteria and indicators of sustainable hunting – the Austrian assessment approach. *Wildlife Biology in Practice*, 1, 163–183.

Lindsey, P.A., Roulet, P.A. & Romañach, S.S. (2007) Economic and conservation significance of the trophy hunting industry in sub-Saharan Africa. *Biological Conservation*, 134, 455–469.

Lindsey, P.A., Alexander, R., Frank, L.G., Mathieson, A. & Romañach, S.S. (2006) Trophy hunter preferences can create market incentives for wildlife conservation. *Animal Conservation*, 9(3), 283–291.

Murphree, M. (2000) Constituting the commons: Crafting sustainable commons in the new millennium. In *Multiple Boundaries, Borders and Scale. Eighth Biennial Conference of the International Association for the Study of Common Property (IASCP)*, Bloomington, Indiana, USA.

Myrberget, S. (1990) Wildlife management in Europe outside the Soviet Union. *NINA Utredning*, 18, 1–47.

Ostrom, E. (1990) *Governing the Commons: The Evolution of Institutions for Collective Action.* Cambridge University Press, Cambridge.

Upton, C. & Bass, S. (1995) *The Forest Certification Handbook.* Earthscan, London.

Zeiler, H. (1996) *Jagd und Nachhaltigkeit.* Monograph M-73. Umweltbundesamt, Vienna.

Conclusion

Conservation, Livelihoods and Recreational Hunting: Issues and Strategies

William M. Adams[1], Barney Dickson[2],
Holly T. Dublin[3] and Jon Hutton[2]

[1]Department of Geography, University of Cambridge, Cambridge, UK
[2]UNEP-World Conservation Monitoring Centre, Cambridge, UK
[3]South African National Biodiversity Institute, Centre for Biodiversity Conservation, Cape Town, South Africa

Hunting and conservation

Historically, the ideological and practical links between conservation and recreational hunting run deep (Adams, this volume, Chapter 8). Through the 19th and 20th centuries, recreational hunters in Europe, North America and the territories of the former British Empire have repeatedly adopted conservationist positions, motivated in part, no doubt, by the desire to maintain a ready supply of animals to be hunted. Conservation, in turn, has drawn on the science of wildlife management practised by recreational hunters and has adopted administrative models such as protected areas that emerged, in some parts of the world, in the context of recreational hunting (Adams, this volume; Booth & Cumming, this volume, Chapter 17). These links have been premised

Recreational Hunting, Conservation and Rural Livelihoods: Science and Practice, 1st edition.
Edited by B. Dickson, J. Hutton and W.M. Adams. © 2009 Blackwell Publishing,
ISBN 978-1-4051-6785-7 (pb) and 978-1-4051-9142-5 (hb).

upon recreational hunters distancing themselves from subsistence and commercial hunting. One of their means for doing so has been the sporting code, with its notions of 'fair chase' and the 'true sportsman'. Conservationists have often been much less tolerant of 'baser' forms of killing. Here, the threat of over-hunting, combined with the taint of the poacher's illegality and increasing concerns about cruelty, remove the patina of acceptability that has, in some times and places, covered recreational hunting.

Nevertheless, the relations between recreational hunting and conservation have remained fraught. The central reason why this is so is that recreational hunting – while often claiming to deliver conservation benefits – rests on the *killing* of individual animals. For many this seems absolutely contrary to any conservationist ethos. Addressing this central paradox remains one of the chief challenges for those who would defend recreational hunting on the grounds of its contribution to conservation. Moreover, for much of the second half of the 20th century, one of the ways in which conservation grew was by harnessing public concerns in the West about the cruelty and lack of order associated with certain types of hunting (epitomised by the poaching of mammals in national parks). This was not fertile ground in which recreational hunters, claiming to hold to the virtues of fair chase, clean kill and sustainable harvest, could press their case.

It could be argued that hunters helped found the modern conservation movement (Adams, this volume) but, with a few exceptions, recreational hunters have not been a loud voice in recent times amidst the well-organised NGOs and growing popular movements concerned with conservation and animal welfare. Recreational hunters and their organisations, where they exist, have become accustomed to working with landowners and regulatory officials, but not to communicating with the general public. This relative invisibility has become impossible to maintain. A variety of factors have contributed to bringing recreational hunting to the fore. These include the growing – and very public – opposition from animal welfare and animal rights movements and, on the other side, the use of elite tourist hunting as a source of revenue in high-profile community-based natural resource management projects in southern Africa, Pakistan and elsewhere.

In discussing these issues, this book has tried to do two things. The first has been to provide a fuller, more informed picture of recreational hunting, with respect to its history, scale, social context and diversity. As understood here, recreational hunting includes not only well-known activities such as big game safari hunting and pheasant shooting in the UK, but also niche pursuits like falconry, as well as a very popular form of recreation, angling, that is not always included under the umbrella of recreational hunting.

The second task has been to consider the major challenges that recreational hunting faces. Foremost among these is its relationship with conservation, but also important are the contributions (or lack of them) that recreational hunting in developing countries makes to the livelihoods of (often poor) local populations, and the criticisms of hunting that come from those with an ethical concern about its effect on the hunted animals. In discussing these issues the book has attempted both to analyse the nature of the challenges and to consider the ways in which those supporting recreational hunting might respond.

There are inevitably many gaps. We have not attempted to provide a systematic survey of the different types of recreational hunting, nor have any of the authors attempted to arrive at a figure for its global economic significance. There are also geographical limitations. Our studies of hunting and its challenges focus on – or use examples from – Europe, North America, southern and East Africa and South Asia. There is nothing about recreational hunting in Latin America and East Asia, and very little about Australasia.

Nevertheless, we have tried to assemble a range of insights and positions, bringing different arguments and bodies of evidence into play in ways that have not been done previously. What have we learned?

Conservation science

One thing is quite clear from the various studies here. On the central question of the impact of recreational hunting on biodiversity conservation, there is no simple one-size-fits-all answer. Recreational hunting takes place in a wide variety of contexts: rural Germany and rural Namibia are very different places, with different patterns of social capital, different developmental needs, different pressures and different opportunities. In some contexts, recreational hunting has made a real contribution to conservation strategies. In others it is has either been neutral, irrelevant or, occasionally, unhelpful. In all cases the question 'does it work' gets the answer 'it depends'. The devil, as always, is in the detail. What, then, does the capacity of hunting to contribute to conservation depend on?

In outline, at least, the science of hunting is well understood (Milner-Gulland et al., this volume, Chapter 5). Issues of harvest levels and population dynamics are complex, but have been analysed for an increasing number of species. The impacts of selection on the genetics of populations are also complex (especially because they may not at first be obvious), but again, the basic parameters of impacts and their management are understood at increasingly sophisticated levels of resolution. There are also issues of wider impacts on non-target species, secondary system-wide impacts of killing quarry species, and parallel impacts of other human influences (including persistent pollutants, habitat loss and climate change). The fact that many ecosystems exhibit non-linear behaviour adds scientific excitement to the study of these impacts, but the kinds of questions that need to be asked are broadly clear, and being addressed by research. Hunting is just one among many anthropogenic pressures on animal populations, and their combined impact is a key issue for conservation and ecosystem science.

In addition to the effects of removing certain individuals from a population, there are also the impacts of the management practice associated with recreational hunting. Once certain species become identified as valuable within the ecosystem, there is an inevitable tendency to adjust ecosystem management to favour them. This is likely to happen whether hunting is organised as a business activity on economic criteria, or simply as recreation, with favoured quarry species. It might involve ecosystem management (for example changing burning regimes to favour particular species), or changing patterns of mortality (e.g. protecting the eggs or young of quarry species). These changes might be advantageous in biodiversity terms if the quarry species itself is threatened, or if, as in the case of the red partridge (Aebischer, this volume, Chapter 12), the ecological requirements of the quarry species (in this case, insects for chicks to eat) allow a wider range of other species to survive as well. More damaging is the so-called 'gamekeeper effect', whereby predators or competitors of target species are persecuted: the pressure of illegal killing of birds of prey is a critical conservation problem for many species in the UK, for example. There can also be significant impacts from artificial propagation of quarry species, for example the introduced pheasant in English woodlands, threatening biodiverse woodland ground flora, or the release of exotic species for the sole purpose of hunting (Mahoney, this volume, Chapter 16). Arguably there is a slippery slope between the management of wild species, the farming of wild species for hunting, and fully fledged domestication. Practices such as 'canned hunting' (the shooting of a captive-bred animal released into a confined area) are not just morally repugnant and perverse in sporting terms but can be taken as the first step towards domestication.

Even where problems are more clearly 'scientific', such as the appropriate quota of quarry species that can be sustainably hunted, there are, of course, questions about how much science you need. Do all decisions about quotas need to be based on regular population surveys and must this involve scientifically qualified staff? If so, who is to train these people and pay for them, and how are local communities to be given power to make decisions about their wildlife resources if they have to defer to some higher scientific authority at every turn? There is increasing experience of local quota-setting by lay (non-scientific) authorities.

Livelihoods

In recent years conservationists have been directly challenged to show that the maintenance of wild biodiversity, particularly in developing countries, contributes to reducing poverty at the local level. Recreational hunting, when viewed as a tool for delivering conservation, has inevitably faced the same challenge. Indeed, because elite tourist safari hunting, for example, may bring some of the world's wealthiest individuals into close proximity with some of its poorest, the challenge can present itself in a particularly stark way. But the impact of recreational hunting on livelihoods is less fully researched than its effects on biodiversity. Many have claimed that supporting conservation objectives through recreational hunting provides win–win situations, where hunting fees sustain both a safari industry and rural communities, generating employment and significant levels of income in regions lacking effective alternatives (e.g. African drylands), while still sustaining wildlife populations. However, as Jones (this volume, Chapter 10) indicates, outcomes may be more mixed. Some claims have been wildly optimistic, failing to report on the high levels of external subsidy (e.g. from aid donors) or unfair and unequal distribution of benefits or costs, or failing to compare hunting-based income with that available from other sources, or with the level of local need. Moreover, hunting and conservation projects (even those labelled 'community-based natural resource management') vary a great deal in terms of what they actually deliver on the ground. Countries such as Tanzania, with vast Game Management Areas set aside by the Government, and with revenues rather inefficiently captured by the State (Leader-Williams *et al.*, this volume, Chapter 18), offer very different mixes of costs and benefits to projects closely or collaboratively managed by local people. Yet even within programmes nominally run by 'communities' there can be fiercely unequal sharing of benefits, and the concentration of power in the hands of a few.

Other issues besides the distribution of benefits amongst local people may arise (Jones, this volume). There are questions about the rights of local people themselves to hunt, their access to legal hunting opportunities, and whether these may be compromised by programmes based on raising revenue from tourist hunters. For many local or indigenous people hunting may be as much a way of life – from which they derive pleasure – as it is a source of subsistence or economic return. In Nunavut, for example, debates about sport hunting of polar bears saw those for whom hunters were a potentially valuable source of revenue pitted against those who would prefer to hunt for themselves (Jones, this volume). This, in turn, raises the question of whether this type of 'traditional' hunting should be regarded as subsistence, or recreational, hunting, or whether, in social contexts where there is not a sharp division between work and leisure, these concepts may lose their purchase.

Questions of social and economic benefits also arise elsewhere. In countries like the United States or a number of European countries (the Scandinavian countries, for example), hunting appeals to a relatively wide range of socioeconomic groups. Even in the UK, where most hunting is confined to the ranks of the landowning elite and their wealthy business clients and friends, there are important (and complex) networks of social benefit (for example through employment in depressed rural regions), and participation in recreational fishing is popular across social classes.

Institutions and governance

Clearly, under the right circumstances, hunting can contribute to conservation and to local livelihoods, although its potential to do so varies considerably from place to place. Critical factors in determining this potential are the institutions in place in different contexts and their functional integrity.

Government regulation, whether national or international, is one option. The role of CITES in regulating the export and import of hunting trophies is important (Rosser, this volume, Chapter 19). However, the efficacy of CITES in this area, as in others, depends how well it is implemented at the national level. Where there are good regulatory institutions at the national level, CITES can work well, but where they are absent, it is difficult to remedy the situation through international action alone. The safari hunting industry generates large flows of foreign exchange, and in poor countries where governance is weak such income is highly attractive but also hard to manage for the public good (Leader-Williams *et al.*, this volume).

These governance problems have led some in the direction of market-based solutions such as certification, particularly in countries where state institutions are weak (Wall & Child, this volume, Chapter 15; Child & Wall, this volume, Chapter 20). If hunting operations were certified (or not) according to whether they satisfied certain criteria, then hunters would be able, so the argument goes, to choose those operations that are sustainable and equitable. There are some potential difficulties with introducing certification schemes for hunting, including the vexed question of who is to pay for them.

Certification schemes can shade into various less stringent forms of self-regulation, often built around codes of conduct or norms and standards for hunters to help encourage them to behave in ways that further conservation effectively. Such codes are not at all new, and have been proposed repeatedly in the past, to teach hunters 'proper' sporting behaviour and to proscribe abuse (Adams, this volume). There is currently intense interest among hunting organisations in standards and certification.

Ethics

The challenges to recreational hunting already discussed raise ethical questions about such issues as the value of biodiversity and our obligations to reduce poverty. However, the challenge that comes from those who think that hunting contravenes our obligations to individual animals is the most openly and explicitly ethical in nature. One of the difficulties for those who would defend recreational hunting against such challenges is that the objections take different forms (Dickson, this volume, Chapter 4). Another is that defenders of hunting will have to decide whether it is possible to meet the critic half-way or whether it is necessary to refute the objection altogether. In the case of objections based on concerns about animal welfare, defenders of recreational hunting might wish to accept that this is a legitimate concern and either argue that hunting does not harm the welfare of the hunted animals in the way alleged, or work to reform hunting methods to ensure that suffering is reduced.

This type of pragmatic response would not help at all to allay the arguments of those who think – perhaps on the grounds that animals have rights – that it is straightforwardly and absolutely wrong to kill individual wild animals for recreation. Any defender of recreational hunting has little choice but to confront this objection head on, to try to show that it is misplaced and to offer an alternative and persuasive account of human obligations to wild animals.

There are other ethical objections to recreational hunting, particularly those that focus on the motives of hunters, but these are harder to pin down. What the controversies over recreational hunting make clear is that there is little consensus either within individual societies, or at the global level, on the ethically appropriate attitudes towards wild animals.

Social licence to operate

Those who see a future for recreational hunting as a strategy that contributes to the conservation of biodiversity and the welfare of poor people in the developing world face a number of challenges. One way to understand the predicament of recreational hunters is to hold that their primary need – a social licence to operate – has not been attained. They need both to establish a substantive evidence-based case that hunting can make a positive contribution to conservation – and to livelihoods – and win the support of domestic and international constituencies that it should be allowed the opportunity to do so.

At the very least the acquisition of a social licence will require recreational hunters to provide much more information about all aspects of their practices. In summarising some of the information that is already available, Sharp & Wollscheid (this volume, Chapter 2) have also highlighted the many gaps that exist. But it is not just a matter of saying more about what they do. There is also a requirement to make the case for recreational hunting, and that, in turn, will require action to reform and improve recreational hunting practices.

The challenges facing recreational hunting are of different sorts and, depending on the political or social setting, defenders of the practice may need to demonstrate that recreational hunting is sustainable, or that it contributes to the livelihoods of the poor, or that it does not contravene our obligations to wild animals or, possibly, all of those things. Moreover, it is not just that the criteria for acquiring the social licence are various. Those criteria may themselves be under dispute. In many societies where recreational hunting is under attack, views about our ethical obligations to wild animals are in a state of considerable flux. The debate is evolving and, as has been noted, there is presently no consensus about what the ethical standards are that recreational hunting needs to meet. Finally, there is also uncertainty about who is responsible for issuing the social licence. While the concept of 'a social licence' itself implies that it is the court of public opinion that makes the decision, there may be doubt about the role of different groups within that court. What weight is to be given to the judgement of conservation scientists that recreational hunting is sustainable? Public opinion on this issue – and with it the opinions of politicians – may be moved by less determinate considerations than those offered by conservation science.

However, the speculations of the previous paragraph about the causes of changes in societal views of hunting have taken us outside the scope of this book. The rather more modest conclusion here is that where conditions are right, where ecological and biological impacts are small and human social benefits large, where hunters are closely regulated (or self-regulated) and where governance is transparent, open and effective, then recreational hunting can contribute to conservation and may be seen as another tool in the toolbox of approaches.

In order to increase the number of places that currently meet these conditions, then one option is clear: identify best practice and conditions for success and replicate them more widely. Hunting is by no means a universal panacea, and it brings with it real ethical dilemmas for many nature-loving people. But it can work as a conservation tool if it is used intelligently and with care.

Index

Page numbers in **bold** represent tables, those in *italics* represent figures.